Pressure-Induced Phase Transformations

Pressure-Induced Phase Transformations

Editor

Daniel Errandonea

MDPI • Basel • Beijing • Wuhan • Barcelona • Belgrade • Manchester • Tokyo • Cluj • Tianjin

Editor
Daniel Errandonea
Universidad de Valencia
Spain

Editorial Office
MDPI
St. Alban-Anlage 66
4052 Basel, Switzerland

This is a reprint of articles from the Special Issue published online in the open access journal *Crystals* (ISSN 2073-4352) (available at: https://www.mdpi.com/journal/crystals/special_issues/Pressure_Phase-Transitions).

For citation purposes, cite each article independently as indicated on the article page online and as indicated below:

LastName, A.A.; LastName, B.B.; LastName, C.C. Article Title. *Journal Name* **Year**, *Article Number*, Page Range.

ISBN 978-3-03936-816-7 (Hbk)
ISBN 978-3-03936-817-4 (PDF)

Contents

About the Editor

Daniel Errandonea, Professor Dr., is a full professor with the Department of Applied Physics of University of Valencia (Spain). He is an Argentinean-born physicist (married with two sons) who received a M.S. from the University of Buenos Aires (1992) and a Ph.D. from the University of Valencia (1998). He has authored/co-authored over 270 research articles in refereed scientific journals including *Nat. Commun.*, *Phys. Rev. Lett.*, and *Advanced Science*, which have attracted 7500+ citations. His work on materials under extreme conditions of pressure and temperature has implications for fundamental and applied research. Among other subjects, during the last decade, Prof. Errandonea has comprehensively explored phase transitions in ternary oxides, semiconductors, metals, and related materials. Some of his accomplishments include the determination of high-pressure phase transitions, high-pressure and high-temperature phase diagrams (including melting curves), and the study of their implications in the physical properties of materials. Prof. Errandonea is a fellow of the Alexander von Humboldt Foundation, winning the Van Valkenburg Award and IDEA Prize, among others. He is presently a member of the MALTA Consolider and the EFIMAT Teams, and on the executive committee of the International Association for the Advancement of High Pressure Science and Technology (AIRAPT).

Editorial

Pressure-Induced Phase Transformations

Daniel Errandonea

Departamento de Física Aplicada-ICMUV, Universitat de València, Calle Dr. Moliner 50, 46100 Burjassot, Spain; daniel.errandonea@uv.es

Received: 7 July 2020; Accepted: 9 July 2020; Published: 10 July 2020

The study of phase transitions in solids under high pressure conditions is a very active and vigorous research field. In recent decades, thanks to the development of experimental techniques and computer simulations, a plethora of important discoveries has been made under high-pressure conditions. Many of the achievements accomplished in recent years affect various research fields, from solid-state physics to chemistry, materials science, and geophysics. They not only contribute to a deeper understanding of solid-sold phase transitions but also to a better understanding of melting under compression.

This Issue collects thirteen contributions, starting with the paper of Bing Li et al. [1]. This paper presents a Raman and X-ray diffraction study of C_{60} fullerene. The authors showed that C_{60} underwent a phase transition from a face-centered cubic structure to a single cubic structure at around 0.3 GPa. They also report evidence of dimerization of C_{60} at about 3.2 GPa.

In the second article, Andreas Tröster et al. [2] introduce a new version of the Landau theory, which is based on symmetry-adapted finite strains, which results in a substantial simplification of the original formulation. These authors apply their theoretical development to the high-pressure phase transition of the perovskite $KMnF_3$, characterizing in detail the cubic-tetragonal transition that occurs around 3.4 GPa.

In recent years, hydrated sulfates and their high-pressure behavior have attracted a large amount of interest due to their great importance in exploring the interior structure of icy satellites, such as Europa, Ganymede, and Callisto. In their work, Linfei Yang et al. [3] studied epsomite ($MgSO_4 \cdot 7H_2O$), a representative of magnesium-bearing hydrated sulfates. The reported results could contribute to understanding the interior structure, composition, and physical properties of the icy satellites. The most interesting finding of this work is the observation that epsomite undergoes a three-step dehydration reaction under high pressure, and its dehydration temperature gradually increases with pressure.

Melting under compression is one of the most challenging subjects of high-pressure studies. In this Issue, Samuel Baty et al. [4] report density-functional studies of molybdenum (Mo) and tungsten (W), two metals with a high melting temperature. In particular, they demonstrate the topological equivalence of both Group 6B elements. Phase diagrams have been extended to 2000 (Mo) GPa and 2500 (W) GPa, respectively. For both metals, the authors propose the existence of two solid structures: the ambient body-centered cubic and a high-pressure double hexagonal close-packed. The solid–solid transition occurs at 660 GPa in Mo and 1060 GPa in W.

In the subsequent contribution, Raquel Chuliá-Jordán et al. [5] reported laser-heating diamond-anvil cell studies of CO_2 and carbonate. They present experimental data that evidences the chemical reactivity between rhenium (Re) and tungsten (W) with carbon dioxide (CO_2) and carbonates at temperatures above 1300 °C and pressures above 6 GPa. Metal oxides and diamond are identified as reaction products. Recommendations to minimize non-desired chemical reactions in high-pressure high-temperature experiments are given.

Another high-pressure and high-temperature study is the contribution by Ezenwa and Secco [6]. The study is focused on the electrical resistivity and thermal conductivity behavior of iron (Fe) at core conditions. The results of such studies are important for understanding planetary interior thermal evolution as well as characterizing the generation and sustainability of planetary dynamos. The study also discusses the behavior of Fe, cobalt (Co), and nickel (Ni), at the solid–liquid melting transition. In particular, the authors report the thermal conductivity difference on the solid and liquid sides of Mercury's inner core boundary and discuss the implications of their findings on the modeling of the adiabatic heat flow of on the inner core side.

An interesting problem is the behavior of organic compounds under high-pressure. In such compounds, a small pressure of the order of a few GPa could have drastic consequences. In the contribution to this Issue, Mahesta and Mochizuki [7] investigate, using molecular dynamics simulations, the spontaneous homogeneous melting of benzene (C_6H_6) under a pressure of 1.0 GPa. They propose the existence of an apparent stepwise transition via a metastable crystal phase, unlike the direct melting observed at ambient pressure. The transition to the metastable phase is achieved by rotational motions, without the diffusion of the center of mass of benzene. These results could open the door to interesting novel findings on the behavior hydrocarbons under compression.

In another original article, Enrico Bandiello et al. [8] studied the high-pressure behavior of zircon-type $NdVO_4$. In particular, these authors report on optical spectroscopic measurements in pure $NdVO_4$ crystals at pressures up to 12 GPa. They correlated the influence of pressure on the crystal structure and pressure-induced structural transition with the high-pressure behavior of the fundamental absorption band gap and the Nd^{3+} absorption bands. The experiments indicate that a phase transition takes place near 5 GPa. Bandiello et al. also have also determined the pressure dependence of the band-gap and discussed the behavior of the Nd^{3+} absorption lines under compression. Important changes in the optical properties of $NdVO_4$ occur at the phase transition, which, according to Raman measurements, corresponds to a zircon to monazite phase change. In particular, in these conditions a collapse of the band gap occurs, changing the color of the crystal. The results of the study are analyzed in comparison with those deriving from previous studies on $NdVO_4$ and related vanadates.

In addition to the eight original research articles described above, this Special Issue also includes five review articles. In the first one, Anzellini and Boccato [9] present an extensive review of the laser-heated diamond-anvil cell technique for university laboratories and synchrotron applications. In the last two decades, the laser-heated diamond anvil cell method, combined with different characterization techniques, has become an extensively used tool for studying pressure-temperature-induced evolution of various physical and chemical properties of materials. In their review, the authors present and discuss the general challenges associated with the use of laser-heated diamond-anvil cells. This discussion is combined with examples of the recent progress in the use of this tool, combined with synchrotron X-ray diffraction and absorption spectroscopy.

Another article focused on recent developments of experimental techniques is the one of Popov et al. [10]. This review is devoted to synchrotron X-ray radiation Laue diffraction, a widely used diagnostic technique for characterizing the microstructure of materials. The authors describe in detail the current status of this powerful technique, including experimental routines and data analysis. They also present results from some case studies and a description of the new experimental setup at the High-Pressure Collaborative Access Team (HPCAT) facility at the Advanced Photon Source, specifically dedicated for in situ and in operando microstructural studies by Laue diffraction under high pressure.

In another of the reviews, Denis Rychov [11] presents a description of the progress made by him and his collaborators on computational studies of high-pressure phase transitions in molecular crystals. The advantages and disadvantages of different approaches are discussed, and the interconnection of experimental and computational methods is highlighted. Finally, challenges and possible ways for progress in high-pressure phase transition research of organic compounds are briefly discussed.

Biesner and Uykur authored another review paper [12], which focused on quantum spin liquids, which are prime examples of strongly entangled phases of matter with unconventional exotic excitations. In particular, they discussed how strong quantum fluctuations prohibit the freezing of the spin system. They also discussed frustrated magnets, which are candidates to search for the quantum spin liquids.

The main topic of the review is the ability of pressure to influence the magnetic phases. The authors review experimental progress in the field of pressure-tuned magnetic interactions, showing that chemical or external pressure is a suitable parameter to create exotic states of matter.

In their recent article, Prof. Manjon et al. [13] conduct an extensive review of the progress they have made in recent years on pressure-induced phase transition in sexquioxides. These compounds constitute a large subfamily of ABO_3–type compounds, which have many different crystal structures due to their large diversity of chemical compositions. They are very important for Earth and Materials Sciences, thanks to their presence in our planet's crust and mantle, and their wide variety of technological applications. Recent discoveries, hot spots, controversial questions, and future directions of research are highlighted in the article.

In summary, the articles presented in this Special Issue are representative of some of the lines of a topic as broad as high-pressure research as well as of its importance in different scientific fields, and cover aspects including structural, electronic, and magnetic transitions. They touch on the latest advancements in several aspects related to the behavior of matter under high-pressure. The included papers show that impressive progress has been made recently on high-pressure research.

Conflicts of Interest: The authors declare no conflict of interest.

References

1. Li, B.; Zhang, J.; Yan, Z.; Feng, M.; Yu, Z.; Wang, L. Pressure-Induced Dimerization of C_{60} at Room Temperature as Revealed by an *In Situ* Spectroscopy Study Using an Infrared Laser. *Crystals* **2020**, *10*, 182. [CrossRef]
2. Tröster, A.; Schranz, W.; Ehsan, S.; Belbase, K.; Blaha, P. Symmetry-Adapted Finite Strain Landau Theory Applied to KMnF$_3$. *Crystals* **2020**, *10*, 124. [CrossRef]
3. Yang, L.; Dai, L.; Li, H.; Hu, H.; Hong, M.; Zhang, X. The Phase Transition and Dehydration in Epsomite under High Temperature and High Pressure. *Crystals* **2020**, *10*, 75. [CrossRef]
4. Baty, S.; Burakovsky, L.; Preston, D. Topological Equivalence of the Phase Diagrams of Molybdenum and Tungsten. *Crystals* **2020**, *10*, 20. [CrossRef]
5. Chuliá-Jordán, R.; Santamaría-Pérez, D.; Marqueño, T.; Ruiz-Fuertes, J.; Daisenberger, D. Oxidation of High Yield Strength Metals Tungsten and Rhenium in High-Pressure High-Temperature Experiments of Carbon Dioxide and Carbonates. *Crystals* **2019**, *9*, 676. [CrossRef]
6. Ezenwa, I.C.; Secco, R.A. Fe Melting Transition: Electrical Resistivity, Thermal Conductivity, and Heat Flow at the Inner Core Boundaries of Mercury and Ganymede. *Crystals* **2019**, *9*, 359. [CrossRef]
7. Mahesta, R.; Mochizuki, K. Stepwise Homogeneous Melting of Benzene Phase I at High Pressure. *Crystals* **2019**, *9*, 279. [CrossRef]
8. Bandiello, E.; Sánchez-Martín, J.; Errandonea, D.; Bettinelli, M. Pressure Effects on the Optical Properties of NdVO$_4$. *Crystals* **2019**, *9*, 237. [CrossRef]
9. Anzellini, S.; Boccato, S. A Practical Review of the Laser-Heated Diamond Anvil Cell for University Laboratories and Synchrotron Applications. *Crystals* **2020**, *10*, 459. [CrossRef]
10. Popov, D.; Velisavljevic, N.; Somayazulu, M. Mechanisms of Pressure-Induced Phase Transitions by Real-Time Laue Diffraction. *Crystals* **2019**, *9*, 672. [CrossRef]
11. Rychkov, D.A. A Short Review of Current Computational Concepts for High-Pressure Phase Transition Studies in Molecular Crystals. *Crystals* **2020**, *10*, 81. [CrossRef]
12. Biesner, T.; Uykur, E. Pressure-Tuned Interactions in Frustrated Magnets: Pathway to Quantum Spin Liquids? *Crystals* **2020**, *10*, 4. [CrossRef]
13. Manjón, F.J.; Sans, J.A.; Ibáñez, J.; Pereira, A.L.J. Pressure-Induced Phase Transitions in Sesquioxides. *Crystals* **2019**, *9*, 630. [CrossRef]

Article

Pressure-Induced Dimerization of C_{60} at Room Temperature as Revealed by an *In Situ* Spectroscopy Study Using an Infrared Laser

Bing Li [1], Jinbo Zhang [2,3], Zhipeng Yan [4], Meina Feng [5,6], Zhenhai Yu [7] and Lin Wang [8,*]

[1] College of Physics, Changchun Normal University, Changchun 130032, China; libing@ccsfu.edu.cn
[2] College of Physical Science and Technology, Yangzhou University, Yangzhou 225002, China; hatong024@163.com
[3] Department of Physics, HYU High Pressure Research Center, Hanyang University, Seoul 04763, Korea
[4] Center for High Pressure Science and Technology Advanced Research (HPSTAR), Shanghai 201203, China; zhipeng.yan@hpstar.ac.cn
[5] Institute of High Energy Physics, Chinese Academy of Sciences (CAS), Beijing 100049, China; fengmn@ihep.ac.cn
[6] Spallation Neutron Source Science Center, Dongguan 523803, China
[7] School of Physical Science and Technology, ShanghaiTech University, Shanghai 201210, China; yuzhh@shanghaitech.edu.cn
[8] Center for High Pressure Science (CHiPS), State Key Laboratory of Metastable Materials Science and Technology, Yanshan University, Qinhuangdao 066004, China
* Correspondence: linwang@ysu.edu.cn

Received: 19 January 2020; Accepted: 28 February 2020; Published: 7 March 2020

Abstract: Using in situ high-pressure Raman spectroscopy and X-ray diffraction, the polymerization and structure evaluation of C_{60} were studied up to 16 GPa at room temperature. The use of an 830 nm laser successfully eliminated the photo-polymerization of C_{60}, which has interfered with the pressure effect in previous studies when a laser with a shorter wavelength was used as excitation. It was found that face-centered cubic (fcc) structured C_{60} transformed into simple cubic (sc) C_{60} due to the hint of free rotation for the C_{60} at 0.3 GPa. The pressure-induced dimerization of C_{60} was found to occur at about 3.2 GPa at room temperature. Our results suggest the benefit and importance of the choice of the infrared laser as the excitation laser.

Keywords: fullerenes; polymerization; pressure-induced; Raman; infrared laser

1. Introduction

C_{120}, which is formed with two C_{60} cages via a 2 + 2 cyclo-addition reaction, has attracted considerable interest due to its unique geometry and interesting physical and chemical properties [1–4]. The formation of C_{120} can be induced by the mechanochemical reaction of C_{60} with KCN [1], the high-pressure reaction of C_{60} at a certain temperature range [2], and the light irradiation of C_{60} crystals if the energy of the photon is higher than the band gap (1.7 eV, ~730 nm) of the sample [5]. Compared to all the aforementioned methods, high pressure is much more effective and has been widely studied [4,6–9]. Because the formation of bonds between neighboring C_{60} molecules is a thermally activated process, the polymerization reaction is very slow at room temperature [4,9]. The majority of high-pressure studies have shown that the dimerization of C_{60} can only form under the conditions of high temperature (373–473K) and high pressure (1.5 GPa) [4,9,10]. However, few studies have provided evidence, suggesting that C_{60} polymerizes at room temperature [2]. Therefore, whether C_{60} can polymerize at room temperature is still an open question.

Raman spectroscopy is a very important diagnostic technique in the study of polymerization of C_{60} [4–13]. In-situ Raman studies on C_{60} at high pressure using a visible laser have been carried out for pressures over 15 GPa [12,13]. An abrupt change in the slope of the Raman modes of C_{60} was found at around 2.5 GPa. It was claimed that the change was due to the phase transition from the partially ordered simple cubic (sc) phase to the rotation free orientationally ordered sc phase of C_{60}, which occurred at 2.5 GPa at room temperature [12]. No signs of the pressure-induced polymerization of C_{60} were found, even though the pressure was up to 15 GPa [12]. However, it should be noted that the selection of an excitation laser for Raman studies on C_{60} is important because it is known that a laser with a wavelength shorter than 730 nm will introduce photo-induced polymerization of C_{60} [4,11]. Therefore, to successfully eliminate the effect of light irradiation, an excitation laser with a much lower energy (or a longer wavelength) should be used. Unfortunately, almost all in-situ Raman studies of C_{60} under high pressure, which were carried out at room temperature, employed visible lasers (400–700 nm) as the excitation. The use of these high-frequency lasers will induce photo-polymerization and interfere with pressure effect, as also found in the polymerization process of butadiene and 2-(hydroxyethyl)methacrylate under high pressure [14,15]. Therefore, it was necessary to carry out an in-situ study to investigate the pressure-induced dimerization using a laser with much lower energy as the excitation.

It is also noted that the band gap of C_{60} decreases as the pressure increases. It has been well determined that the gap closing rate under high pressure is about -0.05 eV/GPa [16,17]. In this study, an infrared laser with an 830 nm (~1.49 eV) wavelength, which would not induce polymerization as a pressure lower than about 4 GPa, was selected as a probe in order to study the behavior of C_{60} under cool compression (at room temperature) by carrying out in-situ Raman experiments. Our obtained results showed unambiguous evidence, indicating the dimerization of C_{60} at a pressure of about 3.2 GPa at room temperature. This result was different to that obtained using visible lasers as excitations, showing the pressure effect.

2. Experiments

C_{60} with purity higher than 99.9% was purchased from the Wuda Sanwei Carbon Cluster Corporation, China. It was used without any further treatment. A Mao-type diamond-anvil cell was used to generate high pressure for the samples [18,19]. The C_{60} crystals were loaded into a hole with a diameter of 120 μm that was drilled in a T301 stainless steel gasket, preindented to a thickness of 50 μm. A methanol-ethanol-water mixture in the volume ratio of 16:3:1 was used as pressure-transmitting medium [20]. The pressure at the sample chamber was determined using the shift of the ruby fluorescence [21]. The Raman spectra were recorded by a Raman spectrometer (Renishaw inVia, UK) with an 830 nm laser as the excitation. The system was well calibrated using a strain-free Si wafer. Two different locations of the sample were studied at each pressure point, and the obtained results were consistent. To avoid the heating effect of the laser, a power of <1 mW was used for the Raman measurements. The collection time of each spectrum was 60 s.

In-situ high-pressure angle-dispersive x-ray diffraction (ADXD) experiments were carried out with the high-pressure collaborative access team (HPCAT), 16ID-B station at the Advanced Photon Source facility, in Argonne National Laboratory. The focused monochromatic beam with dimensions of approximately 5 μm × 5 μm in full width at half maximum (FWHM) was utilized for the ADXD measurements. The wavelength of the X-ray was 0.34531 Å. The diffraction patterns were collected with a MAR345 image plate with a pixel size of 100 μm, and they were processed using standard techniques. The pressures were obtained from the equation of state of Au, which was loaded along with the sample.

3. Results and Discussions

The Raman spectra of different pressures are shown in Figure 1A and B. Based on the spectrum of the ambient conditions, 10 peaks with positions at 270.0 cm^{-1}, 430 cm^{-1}, 493.0 cm^{-1}, 708.0 cm^{-1},

772.0 cm^{-1}, 1099.0 cm^{-1}, 1248.0 cm^{-1}, 1426.0 cm^{-1}, 1469.0 cm^{-1}, and 1573.0 cm^{-1} were found. They were indexed as eight Hg and two Ag modes of C_{60} [7,11]. It is necessary to point out that the Raman spectrum measured at ambient pressure showed that the $A_g(2)$ pentagonal pinch mode was at 1469.0 cm^{-1}, which is the characteristic frequency for pristine C_{60}. It is known that this vibration mode will shift to a lower frequency in polymerized C_{60} [4,7]. Therefore, this indicated that the starting sample was monomeric C_{60}. As the pressure increased at a relatively low-pressure range, the peaks shifted gradually, as shown in the pressure dependences of the peak positions. It was also found that the peaks became broad and the intensities decreased as the pressure increased further. As marked in Figure 1A, some peaks split and some new peaks appeared, indicating that the symmetry of the C_{60} was reduced at high pressure. As the pressure reached up to about 3.75 GPa, the peaks with frequencies of less than 600 cm^{-1} became too broad and weak to be recognized. The peaks of $H_g(3)$, $H_g(4)$, $A_g(2)$, and $H_g(8)$ persisted up to 15 GPa.

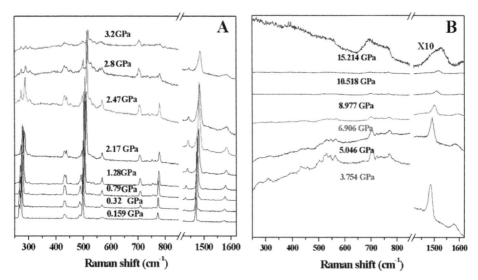

Figure 1. Raman spectra of C_{60} under different pressures: low pressure range (**A**) and higher pressures (**B**).

The pressure dependencies of the Raman shifts were analyzed, as shown in Figure 2 and Table 1. Because $A_g(2)$ is the characteristic pinch mode of the pentagon rings in C_{60}, its pressure dependence will be discussed first. As seen from the figure, two obvious changes in the entire range of pressure were found. One took place at around 0.3 Gpa, and another occurred at about 3.2 GPa. As shown in Table 1, the pressure dependence of $A_g(2)$ was 8.2 cm^{-1}/GPa below 0.3 GPa. It was reduced to 5.4 cm^{-1}/GPa as the pressure increased. As the pressure rose up to about 3.2 GPa, an obvious softening was observed for the $A_g(2)$. Based on the pressure dependences of all the other peaks shown by the dash lines in Figure 2 and the table, it could be found that all the changes of the peaks took place at the same pressure. Taking $H_g(3)$ as an example, it shifted to a higher frequency at < 0.3 GPa, and it turned to red shifts as the pressure increased further. These changes suggested that some transitions must have taken place in the sample at 0.3 GPa and 3.2 GPa. Based on the previous studies, the C_{60} cages rotated freely at each site in the fcc structure at ambient pressure. The free rotation was hinted at as the pressure reached about 0.4 GPa, causing the structural transition from fcc to sc [4,22]. This happened at a similar pressure to that of our result. Therefore, we suggest that the change at 0.3 GPa was induced by the phase transition of C_{60} from the fcc structure to the sc structure. This was verified by our in-situ X-ray diffraction and the refinements, as shown in Figure 3. The crystal structure at 0.2 GPa could be indexed as an fcc structure. As the pressure increased, several new peaks appeared, indicating a phase transition. As shown in Figure 3C, the new phase could be indexed as sc, consistent with the previous report and our findings for the Raman study.

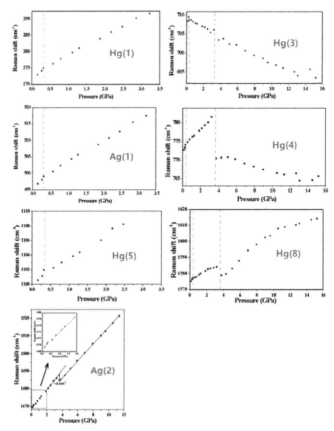

Figure 2. The pressure dependence of the Raman shifts: $H_g(1)$, $H_g(3)$, $A_g(1)$, $H_g(4)$, $H_g(5)$, $H_g(8)$, and $A_g(2)$.

Table 1. The slopes of Raman shifts of C_{60} versus pressure at room temperature.

Modes	Pressure Dependences of Raman Shift (cm^{-1}/GPa)				
	0–0.3 GPa	0.3–3.2 GPa		>3.2 GPa	
$H_g(1)$	11.9 (2.2)	5.8 (0.1)		—	
$H_g(2)$	—	0.34 (0.05)		—	
		2.3 (0.05)		—	
$A_g(1)$	13.2 (2.7)	6.3 (0.1)		—	
$H_g(3)$	9.6		−1.09 (0.04)		
$H_g(4)$	7.7 (2.4)	2.4 (0.1)		0.29 (0.06)	−0.59 (0.07)
$H_g(5)$	8.1 (2.9)	3.2 (0.1)		—	
$H_g(6)$	6.2 (4.1)	2.9 (0.2)		—	
$H_g(7)$	12.4 (5.5)	6.3 (0.1)		—	
$A_g(2)$	8.2 (0.5)	5.4 (0.1)		4.73 (0.04)	
$H_g(8)$	9.7 (0.2)	3.3 (0.1)	1.4 (0.04)	1.2	4.6 (0.2)

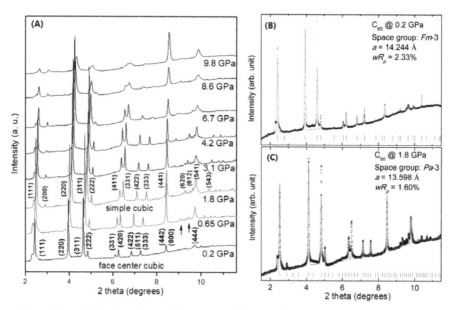

Figure 3. X-ray diffraction of C_{60} at high pressures (**A**), and refinements of the diffraction spectra at 0.2 GPa (**B**) and 1.8 GPa (**C**), respectively.

As mentioned previously, similar significant softening at 3.2 GPa could also be found in the $H_g(4)$ and $H_g(8)$. These results confirmed that the transition occurred in the sample at high pressure. Through careful analysis, a softening of about 4 cm^{-1} was found with $A_g(2)$. Based on our knowledge, the $A_g(2)$ of C_{60} shifts to a lower frequency as polymerization occurs [4,6,7,9]. The shift is about 4 cm^{-1} for the dimerization of C_{60}. This suggests that the pristine C_{60} underwent a pressure-induced dimerization as the pressure increased to 3.2 GPa. To confirm the dimerization of C_{60} at about 3.2 GPa, we further studied the samples decompressed from different maximum pressures by Raman spectroscopy using 830 nm laser that could not induce photo-dimerization of the C_{60} at room temperature. The Raman spectra measured after the decompression from different pressures of 2.94 GPa, 3.48 GPa, and 16.6 GPa for the samples are shown in Figure 4(a)–(c), respectively. Based on the comparison with the Raman spectrum of pristine C_{60}, no obvious change was found in the Raman spectrum of the C_{60} recovered from 2.94 GPa. It was noted that the position of the $A_g(2)$ mode was still at 1469 cm^{-1}, which indicated that the C_{60} quenched from 2.94 GPa was monomeric C_{60}. However, it was remarkable that the $A_g(2)$ peak split into two peaks as the pressure went over about 3.2 GPa. The split was able to be fitted into two peaks, with their positions at 1464 cm^{-1} and 1469 cm^{-1}. Additionally, the intensity of the peak at 1464 cm^{-1} grew as the pressure increased over this pressure. As mentioned above, the peak at 1464 cm^{-1} was from C_{60} dimers and the peak at 1469 cm^{-1} was of monomeric C_{60} [4,6,7,9]. Furthermore, as marked in Figure 4, some new peaks that were observed in the spectra of the sample recovered from 3.48 GPa and 16.6 GPa also proved the existence of the dimerization of C_{60} in the samples. Since the use of the 830 nm laser could not induce the photo-polymerization of C_{60} even at the pressure of 4 GPa and room temperature, these results indicated that the pressure did induce the dimerization of C_{60} as the pressure went up to around 3.2 GPa at ambient temperature.

Compared with the results that were obtained using a 514.5 nm laser as the excitation, the results obtained in this study were very different. As shown, both the Raman spectra of the sample under high pressure and the quenched samples with different pressures proved that the pressure-induced dimerization of C_{60} occurred at a pressure of about 3.2 GPa. However, this phenomenon was not found in the previous investigations, even for the pressures of up to 15 GPa. Based on the comparison of the two experiments, the major difference was the wavelength of the excitation. Therefore, the selection of the wavelength of a laser is crucial for spectroscopic studies of the behavior of C_{60} under high pressure.

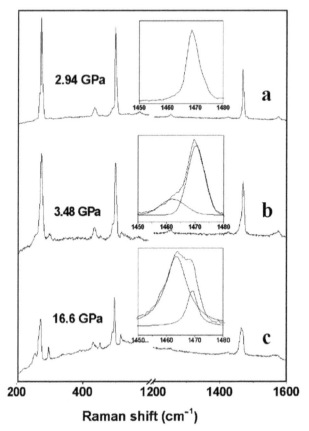

Figure 4. The Raman spectra of the samples recovered from different pressures of 2.94 GPa (**a**), 3.48 GPa (**b**), and 16.6 GPa (**c**).

In summary, we carried out in situ Raman studies of C_{60} under high pressure using an infrared laser as the excitation, which could not induce photo-polymerization even for a pressure of up to 5 GPa at room temperature. Our result showed that the C_{60} underwent a phase transition from an fcc structure to an sc structure at around 0.3 GPa. This result agreed with the previous neutron diffraction data very well. The high-pressure-induced dimerization of C_{60} at room temperature was first observed by an in-situ experiment using the infrared laser as the excitation. The pressure of the dimerization of C_{60} at room temperature was about 3.2 GPa. Our results were very different from those of the previous ones obtained using a visible laser as the excitation, indicating the benefit and importance of the choice of the infrared laser as the excitation laser.

Author Contributions: Conceptualization, L.W.; software, B.L. and Z.Y. (Zhenhai Yu); validation, L.W. and Z.Y. (Zhenhai Yu); formal analysis, B.L.; investigation, B.L., J.Z., Z.Y. (Zhipeng Yan), and M.F.; resources, L.W.; data curation, B.L. and J.Z.; writing—original draft preparation, L.W., B.L.; writing—review and editing, L.W., B.L.; supervision, L.W.; project administration, L.W.; funding acquisition, L.W. All authors have read and agreed to the published version of the manuscript.

Funding: This work research wais mainly supported by Natural Science Foundation of China (Grant No. 11404036, 11874076), National Science Associated Funding (NSAF, Grant No. U1530402), "the 13th Five-year" Planning Project of Jilin Provincial Education Department Foundation (No.20190504), and Science Challenging Program

(Grant No. TZ2016001). Portions of this work were performed at HPCAT (Sector 16), which is supported by DOE-BES under Award No. DE-FG02-99ER45775. This research used resources of the Advanced Photon Source, a U.S. DOE Office of Science User Facility operated for the DOE Office of Science by Argonne National Laboratory under Contract No. DE-AC02-06CH11357.

Conflicts of Interest: The authors declare no conflict of interest. The funders had no role in the design of the study; in the collection, analyses, or interpretation of data; in the writing of the manuscript, or in the decision to publish the results.

References

1. Wang, G.U.; Komatsu, K.; Murata, Y.; Shiro, M. Synthesis and X-ray structure of dumb-bell-shaped C120. *Nature* **1997**, *387*, 583–586. [CrossRef]
2. Markin, A.V.; Smirnova, N.N.; Lebedev, B.V.; Lyapin, A.G.; Kondrin, M.V.; Brazhkin, V.V. Thermodynamic and dilatometric properties of the dimerized phase of a C_{60} fullerene. *Phys. Solid State* **2003**, *45*, 802–808. [CrossRef]
3. Fagerstrom, J.; Stafstrom, S. Fromation of C_{60} dimers: A theoretical study of electronic structure and optical absorption. *Phys. Rev. B* **1996**, *53*, 13150–13158. [CrossRef]
4. Sundqvist, B. Fullerenes under high pressure. *Adv. Phys.* **1999**, *48*, 1–134. [CrossRef]
5. Suzuki, M.; Iida, T.; Nasu, K. Relaxation of exciton and photoinduced dimerization in crystalline C_{60}. *Phys. Rev. B* **2000**, *61*, 2188–2198. [CrossRef]
6. Pei, C.Y.; Wang, L. Recent progresses on high-pressure and high temperature studies on fullerenes and related materials. *Matter Radiat. Extrem.* **2019**, *4*, 028201. [CrossRef]
7. Pei, C.Y.; Feng, M.N.; Yang, Z.X.; Yao, M.G.; Yuan, Y.; Li, X.; Hu, B.W.; Shen, M.; Chen, B.; Sundqvist, B.; et al. Quasi 3D polymerization in C_{60} bilayers in a fullerene solvate. *Carbon* **2017**, *124*, 490–505. [CrossRef]
8. Wang, L. Solvated fullerenes, a new class of carbon materials suitable for high-pressure studies: A review. *J. Phys. Chem. Solids* **2015**, *84*, 85–95. [CrossRef]
9. Blank, V.D.; Buga, S.G.; Dubitsky, G.A.; Serebryanaya, N.R.; Popov, M.Y.; Sundqvist, B. High-pressure polymerized phases of C_{60}. *Carbon* **1998**, *36*, 319–343. [CrossRef]
10. Davydov, V.A.; Kashevarova, L.S.; Rakhmanina, A.V.; Senyavin, V.M.; Ceolin, R.; Szwarc, H.; Allouchi, H.; Agafonov, V. Spectroscopic study of pressure-polymerized phases of C_{60}. *Phys. Rev. B* **2000**, *61*, 11936–11945. [CrossRef]
11. Wang, L.; Liu, B.B.; Liu, D.D.; Yao, M.G.; Hou, Y.Y.; Yu, S.D.; Cui, T.; Li, D.M.; Zou, G.T.; Iwasiewicz, A.; et al. Synthesis of Thin, Rectangular C60 Nanorods Using m-Xylene as a Shape Controller. *Adv. Mater.* **2006**, *18*, 1883–1888. [CrossRef]
12. Meletov, K.P.; Christofilos, D.; Ves, S.; Kourouklis, G.A. Pressure-induced orientational ordering in C60 single crystals studied by Raman spectroscopy. *Phys. Rev. B* **1995**, *52*, 10090–10096. [CrossRef] [PubMed]
13. Li, Y.; Rhee, J.H.; Singh, D.; Sharma, S.C. Raman spectroscopy and x-ray diffraction measurements of C60 compressed in a diamond anvil cell. *Phys. Rev. B* **2003**, *68*, 024106. [CrossRef]
14. Citroni, M.; Ceppatelli, M.; Bini, R.; Schettino, V. Laser-induced selectivity for dimerization versus polymerization of butadiene under high pressure. *Science* **2002**, *295*, 2058–2060. [CrossRef] [PubMed]
15. Evlyukhin, E.; Museur, L.; Traore, M.; Nikitin, S.M.; Zerr, A.; Kanaev, A. Laser-assisted high-pressure-induced polymerization of 2-(hydroxyethyl)methacrylate. *J. Phys. Chem. B* **2015**, *119*, 3577–3582. [CrossRef]
16. Snoke, D.W.; Syassen, K.; Mittelbach, A. Optical absorption spectrum of C60 at high pressure. *Phys. Rev. B* **1993**, *47*, 4146–4148. [CrossRef]
17. Moshary, F.; Chen, N.H.; Silvera, I.F.; Brown, C.A.; Dorn, H.C.; Vries, M.S.; Bethune, D.S. Gap reduction and the collapse of solid C60 to a new phase of carbon under pressure. *Phys. Rev. Lett.* **1992**, *69*, 466–469. [CrossRef]
18. Mao, H.K.; Hemley, R.J. Experimental studies of Earth's deep interior: Accuracy and versatility of diamond-anvil cells. *Philos. Trans. A* **1996**, *354*, 1315–1332.
19. Mao, H.K.; Chen, X.J.; Ding, Y.; Li, B.; Wang, L. Solids, liquids, and gases under high pressure. *Rev. Mod. Phys.* **2018**, *90*, 15007. [CrossRef]
20. Klotz, S.; Chervin, J.-C.; Munsch, P.; Le Marchand, G. Hydrostatic limits of 11 pressure transmitting media. *J. Phys. D Appl. Phys.* **2009**, *42*, 075413. [CrossRef]

21. Mao, H.K.; Xu, J.; Bell, P.M. Calibration of the ruby pressure gauge to 800 kbar under quasi-hydrostatic conditions. *J. Geophys. Res.* **1986**, *91*, 4673–4676. [CrossRef]
22. Snoke, D.W.; Raptis, Y.S.; Syassen, K. Vibrational modes, optical excitations, and phase transition of solid C60 at high pressure. *Phys. Rev. B* **1992**, *45*, 14419–14422. [CrossRef] [PubMed]

Article

Symmetry-Adapted Finite Strain Landau Theory Applied to KMnF$_3$

Andreas Tröster [1,*], **Wilfried Schranz** [1], **Sohaib Ehsan** [2], **Kamal Belbase** [2] and **Peter Blaha** [2]

[1] Faculty of Physics, University of Vienna, Boltzmanngasse 5, A-1090 Vienna, Austria; wilfried.schranz@univie.ac.at
[2] Institute of Materials Chemistry, Vienna University of Technology, Getreidemarkt 9, A-1060 Wien, Austria; sohaib.ehsan@tuwien.ac.at (S.E.); kamal.belbase@tuwien.ac.at (K.B.); peter.blaha@tuwien.ac.at (P.B.)
* Correspondence: andreas.troester@univie.ac.at

Received: 27 January 2020; Accepted: 11 February 2020; Published: 17 February 2020

Abstract: In recent years, finite strain Landau theory has been gradually developed as both a conceptual as well as a quantitative framework to study high pressure phase transitions of the group-subgroup type. In the current paper, we introduce a new version of this approach which is based on symmetry-adapted finite strains. This results in a substantial simplification of the original formulation. Moreover, it allows for replacing the clumsy use of truncated Taylor expansions by a convenient functional parametrization. Both the weaknesses of the traditional Landau approach based on infinitesimal strains as well as the major improvements made possible by our new parametrization are illustrated in great detail in an application to the ambient temperature high pressure transition of the perovskite KMnF$_3$.

Keywords: high pressure; phase transitions; Landau theory; nonlinear elasticity theory; perovskites

---

1. Introduction

Through many decades, the Landau theory (LT) of phase transitions [1,2] (PTs) has proven to be one of the most valuable conceptual tools for understanding PTs of the group-subgroup type. In particular, the field of structural PTs abounds even with successful quantitative applications, and, for many classes of materials, complete collections of the corresponding coupling coefficients have been gathered in the literature (for ferroelectrics see, e.g., Appendix A of Ref. [3]). Effects of spontaneous strain that generally accompany temperature-driven structural PTs are sufficiently parameterized in terms of infinitesimal strain tensor components defined with respect to a baseline, which is obtained by extrapolating the generally small thermal expansion changes of the high symmetry reference phase. The corresponding Landau potential then involves only terms up to harmonic order, and any temperature dependence of the relevant parameters (high symmetry phase elastic constants and other coupling constants) can usually be completely neglected [4].

The situation changes drastically for high pressure phase transitions (HPPTs). Spontaneous strain components may still be numerically small, but they now must be defined with respect to a P-dependent base line. The total strain measured with respect to the ambient pressure reference state must then be calculated from a nonlinear superposition of finite background and spontaneous strain (see Equation (11) below). Furthermore, the Landau potential may be truncated beyond harmonic order only if calculated with respect to this P-dependent elastic background reference system. Therefore, neither the elastic constants nor the other elastic couplings can be assumed to be P-independent. In a high pressure context, clinging to the familiar infinitesimal strain Landau toolbox may result not only in a quantitatively, but also qualitatively completely erroneous description.

Crystals **2020**, *10*, 124; doi:10.3390/cryst10020124   www.mdpi.com/journal/crystals

It is not easy to construct a mathematically consistent and yet practically useful version of Landau theory taking into account the inherent nonlinearities and anharmonic effects that accompany HPPTs. In recent years, however, such a theory, for which we have coined the name finite strain Landau theory (FSLT), has been successfully developed. FSLT constitutes a careful extension of Landau theory beyond coupling to infinitesimal strain, fully taking into account the nonlinear elastic effects at finite strain. Its capabilities have been demonstrated in a number of applications to HPPTs [5–10]. However, as it stands, the numerical scheme underlying FSLT is still quite involved, and many practical workers in the field of HPPTs may be hesitant to go through the mathematical hardships it seems to pose. It is the purpose of this paper to show that FSLT is drastically simplified by switching from a formulation in terms of Cartesian Lagrangian strains to one in terms of symmetry-adapted finite strains. The enormous reduction of overall complexity of the approach as well as the vastly reduced numerical requirements of our new version of FSLT are demonstrated on the example of the HPPT in the perovskite $KMnF_3$ (KMF).

2. Review of Experimental Results on the Cubic-to-Tetragonal Transition of KMF

In what follows, we focus on the antiferrodistortive high pressure phase transition of KMF from the cubic perovskite aristophase $Pm\bar{3}m$ to a tetragonal $I4/mcm$ phase at room temperature which was experimentally investigated in Ref. [11] by X-ray diffraction up to 30 GPa. Since the ambient pressure Landau theory also provides a limiting reference frame for the description of the high pressure transition, we start our discussion with a detailed survey of the corresponding Landau theory.

According to Ref. [12], a similar transition observed at ambient pressure and temperature $T_c = 186.5$ K is weakly first order but close to critical, and the mechanism underlying these transitions is the same as in the well-studied $T = 105$ K cubic-to-tetragonal transition in strontium titanate, i.e., octahedral tilting with a critical wavevector at the R-point of the cubic Brillouin zone. Furthermore, in Ref. [12], it is argued that, even though a further structural transition to an orthorhombic phase related to further octahedral tilting at the M-point of the Brillouin zone at $T_N = 87$ K is accompanied by antiferromagnetism, this coincidence between structural and magnetic transition temperatures may just be accidental, and actually there seems to be essentially no coupling between structural and magnetic order parameters (OPs) for these transitions. In passing we note that at $T = 82$ K there is a further transition to an orthorhombic canted ferromagnet [12].

As discussed in Ref. [12], the $Pm\bar{3}m \to I4/mcm$ symmetry reduction corresponds to the isotropy subgroup of the three-dimensional irreducible representation R_4^+ of $Pm\bar{3}m$ with respect to the order parameter direction $(Q, 0, 0)$, the corresponding Landau expansion to sixth order being

$$
\begin{aligned}
F = \ & \frac{A}{2}\left(Q_1^2 + Q_2^2 + Q_3^2\right) + \frac{B}{4}\left(Q_1^4 + Q_2^4 + Q_3^4\right) + \frac{B_2}{4}\left(Q_1^2 Q_2^2 + Q_2^2 Q_3^2 + Q_1^2 Q_3^2\right) \\
& + \frac{C}{6}\left(Q_1^6 + Q_2^6 + Q_3^6\right) + \frac{C_2}{6}\left(Q_1^2 Q_2^4 + Q_2^2 Q_3^4 + Q_1^2 Q_3^4 + Q_2^2 Q_1^4 + Q_3^2 Q_2^4 + Q_3^2 Q_1^4\right) + \frac{C_3}{6} Q_1^2 Q_2^2 Q_3^2 \\
& + \lambda_a^{(0)}(Q_1^2 + Q_2^2 + Q_3^2)\epsilon_a + \lambda_t^{(0)}(2Q_1^2 - Q_2^2 - Q_3^2)\epsilon_t + \frac{K^{(0)}}{2}\epsilon_a^2 + \frac{\mu^{(0)}}{2}\epsilon_t^2
\end{aligned}
\tag{1}
$$

with volume and tetragonal symmetry-adapted strains $e_a = e_{11} + e_{22} + e_{33}$, $e_t = \frac{2e_{33} - e_{11} - e_{22}}{\sqrt{3}}$ and bulk and longitudinal shear modulus related to the bare cubic elastic constants by $K^{(0)} = (C_{11}^{(0)} + C_{12}^{(0)})/3$, $\mu^{(0)} = (C_{11}^{(0)} - C_{12}^{(0)})/2$. Targeting a transition into a single tetragonal domain where $(Q_1, Q_2, Q_3) \equiv (Q, 0, 0)$ and $\epsilon_{22} = \epsilon_{33}$, we have

$$
\begin{aligned}
F = \ & \frac{A}{2}Q^2 + \frac{B}{4}Q^4 + \frac{C}{6}Q^6 + \lambda_a^{(0)}Q^2\epsilon_a + 2\lambda_t^{(0)}Q^2\epsilon_t \\
& + \frac{K^{(0)}}{2}\epsilon_a^2 + \frac{\mu^{(0)}}{2}\epsilon_t^2
\end{aligned}
\tag{2}
$$

In a standard Landau approach, the coefficients $B, C, \lambda_a^{(0)}, \lambda_t^{(0)}$ are assumed to be independent of temperature (and pressure), while for A the ansatz

$$A = A_0(T - T_0) \tag{3}$$

with T-independent coefficients A_0, T_0 is made and quantum saturation has been neglected [13]. The elastic equilibrium conditions $-P = \frac{1}{V_0} \frac{\partial F}{\partial \bar{\varepsilon}_a}\big|_{\bar{\varepsilon}_a, \bar{\varepsilon}_t}$ and $0 = \frac{1}{V_0} \frac{\partial F}{\partial \bar{\varepsilon}_t}\big|_{\bar{\varepsilon}_a, \bar{\varepsilon}_t}$ amount to

$$\bar{\varepsilon}_a = e_a(P) - \frac{\lambda_a^{(0)}}{K^{(0)}} Q^2, \qquad \bar{\varepsilon}_t = -\frac{2\lambda_t^{(0)}}{\mu^{(0)}} Q^2 \tag{4}$$

with a background volume strain

$$e_a(P) = 3e(P), \quad e(P) = -\frac{P}{3K^{(0)}} \tag{5}$$

Performing a Legendre transform yields the Gibbs potential

$$G = \frac{A_R}{2} Q^2 + \frac{\tilde{B}}{4} Q^4 + \frac{C}{6} Q^6 - \frac{P^2}{2K^{(0)}} \tag{6}$$

where

$$A_R = A - 6\lambda_a^{(0)} e(P) = A - \frac{2\lambda_a^{(0)} P}{K^{(0)}} \tag{7a}$$

$$\tilde{B} = B - \frac{2(\lambda_a^{(0)})^2}{K^{(0)}} - \frac{8(\lambda_t^{(0)})^2}{\mu^{(0)}} \tag{7b}$$

The values of coefficients $A_0 = 63.118\,\text{kPa/K}$, $\tilde{B} = -1.308\,\text{MPa}$ and $C = 13.032\,\text{MPa}$ at $P = 0$, which imply a first order phase transition at $T_c = 186.15\,\text{K}$, have been determined from caloric measurements by Salje and coworkers [13]. In principle, numerical values for the OP-strain coupling coefficients $\lambda_a^{(0)}, \lambda_t^{(0)}$ may be extracted from experimental data on the temperature evolution of spontaneous strains $\bar{\varepsilon}_a, \bar{\varepsilon}_t$. Unfortunately, however, for KMF, this is easier said than done. At room temperature, the thermal expansion data $a_{\text{cubic}}(T)$ of Ref. [14] are observed to perfectly reproduce the value of the cubic lattice constant $a_{\text{cubic}}(T_R)$ at ambient pressure as determined in Ref. [11]. However, the measurement data of thermal lattice parameter irregularities $a(T), c(T)$ around $T_c \approx 186\,\text{K}$ (Refs. [15–18]) available in the literature appear to be in rather poor mutual agreement. As Figure 1 illustrates, while a discontinuous behavior of the lattice parameters is clearly visible in all three data sets, the absolute values of the measured unit cell parameters differ considerably, yet none of the data sets seem to be compatible with extrapolating the thermal expansion data of Ref. [14].

Not unexpectedly, the agreement in relative splitting between the a-and c-axis in the tetragonal phase appears to be better, albeit far from perfect. Nevertheless, the cubic parts of the data of Refs. [15,18] exhibit slopes similar to the low temperature extrapolation of the thermal expansion data of Ref. [14]. In order to be able to collapse the data onto a common "master set", we thus shifted the data of Refs. [15,18] by constant absolute offsets to match the extrapolated baseline of Ref. [14] in an optimal way, treating the seemingly more precise measurements of Sakashita et al. (Refs. [16,17]) as an outlier. Figure 2 illustrates our results for a corresponding effort.

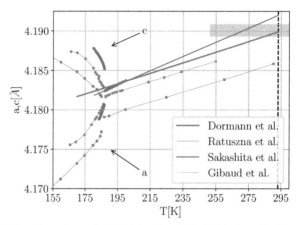

Figure 1. Compilation of experimental unit cell data from Refs. [15] (red) and [18] (green) (data range restricted to $T > 155$ K) [16,17] (blue) and [14] (gray). For comparison, the value (including error bars) of the room temperature (indicated by the vertical dashed line) lattice constant at ambient pressure as measured in Ref. [11] is illustrated by the gray horizontal area.

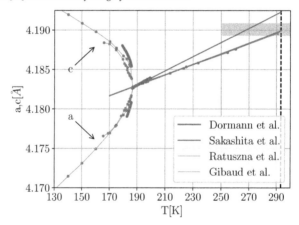

Figure 2. Collapse of data from Refs. [15] (red) and [18] (green), onto the low temperature expansion of the thermal expansion data of Ref. [14] (gray) by constant vertical shifts. The positive and negative branches of the data correspond to values ϵ_t and ϵ_a from the various references, respectively. As in Figure 1, the room temperature ambient pressure lattice constant of Ref. [11] is indicated by a horizontal gray bar for comparison. The shifted data from Ref. [16,17] (blue) clearly appear to be at odds with the other measurements.

With a meaningful baseline $a_{\text{cubic}}(T)$ for unit cell parameters $a(T), c(T)$ in place, we calculate the spontaneous strain components $\epsilon_1 = a/a_{\text{cubic}} - 1$ and $\epsilon_3 = c/a_{\text{cubic}} - 1$ and thus the (infinitesimal) spontaneous volume and tetragonal strains

$$\epsilon_a = 2\epsilon_1 + \epsilon_3 = \frac{2a + c}{a_{\text{cubic}}} - 3, \tag{8}$$

$$\epsilon_t = \frac{2(\epsilon_3 - \epsilon_1)}{\sqrt{3}} = \frac{2}{\sqrt{3}} \frac{c - a}{a_{\text{cubic}}}, \tag{9}$$

respectively. According to Equation (4), when plotted against $Q^2(t)$ at $P = 0$, ϵ_a and ϵ_t should resemble straight lines with slopes $-\lambda_a^{(0)}/K^{(0)}$ and $-2\lambda_t^{(0)}/\mu^{(0)}$, respectively. Figure 3 illustrates corresponding fits with results

$$\lambda_a^{(0)}/K^{(0)} = 0.002, \qquad \lambda_t^{(0)}/\mu^{(0)} = -0.005 \tag{10}$$

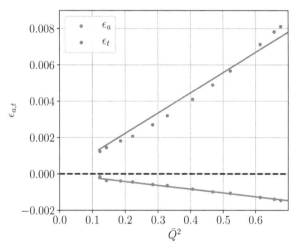

Figure 3. Fits of $\epsilon_a(T)$ and $\epsilon_t(T)$ against $Q^2(T)$ according to Equations (4).

With the Landau theory of the temperature-driven transition at ambient pressure available, we are tempted to analyze the ambient temperature HPPT based on the same framework. In Ref. [11], the variation of the pseudo-/cubic lattice constants of KMF under pressure at room temperature was measured with X-ray scattering (Figure 4) and the cubic part of the data was fitted to a simple Murnaghan equation of state (EOS) with $K_0 = 64$ GPa and $V_0 = 73.608^3$. This provides a baseline to determine (Lagrangian) spontaneous strains $\widehat{\epsilon}_a \approx \epsilon_a, \widehat{\epsilon}_t \approx \epsilon_t$ (Figure 5).

Comparing the thermal and pressure-induced spontaneous strains shown in Figures 2 and 5, respectively, we note that there is some spontaneous thermal volume strain ϵ_a while practically $\widehat{\epsilon}_a \approx 0$ for the pressure-induced case. Furthermore, even though it may be difficult to directly relate temperature and pressure scales, the pressure-induced tetragonal spontaneous strain $\widehat{\epsilon}_t$ is observed to be roughly one order of magnitude larger than its thermal counterpart ϵ_t. From the perspective of traditional Landau theory based on infinitesimal strain coupling, these findings are difficult to explain. Based on Equation (4), there are in principle two ways to alter the magnitude of spontaneous strains. One may either change the value of the couplings $\lambda_a^{(0)}/K^{(0)}$ and $\lambda_t^{(0)}/\mu^{(0)}$ or find a mechanism to increase the magnitude of $\bar{Q}^2$ which, of course, implicitly depends on all Landau coefficients. Since these are actually only known for $T = 186.5$ K, one could assume a thermal drift in $\lambda^{(0)}$ towards zero to be responsible for the vanishing of $\hat{\epsilon}_a$ at room temperature (more than counteracting against the thermal drift of K_0 which is generally expected to decrease with increasing T). For the tetragonal strain, a similar mechanism seems to be difficult to conceive, however. On the one hand, we would need to increase $\lambda_t^{(0)}$ dramatically to explain the large values of $\hat{\epsilon}_t$. On the other hand, Equation (7b) indicates that such an increase would send parameter $\widetilde{B}$ to negative values much larger than those found for the thermal transition, resulting in a pronounced first order character of the HTTP. This, however, is not observed. The only remedy therefore seems to find a way to increase $\bar{Q}^2$. Calculated from a standard 2–4 Landau potential, $\bar{Q}^2$ would be inversely proportional to $1/\widetilde{B}$. This may explain why advocates of an orthodox Landau description frequently resort to assuming HPPT's to be near a tricritical point, explicitly postulating some ad-hoc pressure dependence $\widetilde{B} = \widetilde{B}(P)$ induced somehow by unspecified higher order coupling effects.

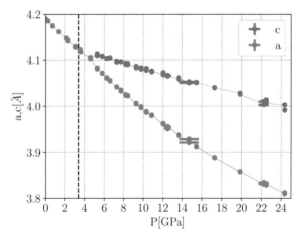

Figure 4. Pseudo-/cubic lattice constants of KMF under pressure at room temperature as measured in Ref. [11] with X-ray scattering. The transition pressure estimate $P_s \approx 3.4$ GPa put forward in Ref. [11], which is indicated by the dashed vertical line, is clearly too low.

Further difficulties arise if we try to reconcile the observed value of P_c with the prediction of standard Landau theory. In Ref. [11], a brute force fit based on the above assumptions of a second order transition close to a tricitical point produced an estimate of $P_c \approx 3.4$ GPa. In principle, for a second order or slightly first order phase transition, this pressure should coincide or be somewhat lower than the pressure value $P_0(T_R)$ at which $A_R(T_R, P_0(T_R))$ vanishes at room temperature T_R. Unfortunately, however, this is incompatible with extrapolating the Landau parametrization of Hayward et al. [19] to room temperature. In fact, inserting the parameter value (10) into Equation (7a) yields $P_0(T_R) \approx 1.7$ GPa, which is completely at odds with $P_c \approx 3.4$ GPa as reported in Ref. [11]. To reach this transition pressure at room temperature would require reducing our previous result $\lambda_a^{(0)}/K^{(0)} = 0.002$ obtained at $T = 186$ K by a full factor of 2 (cf. Figure 6). However, even then, any unbiased reader should have a hard time believing that the bare data of Figure 5 should indicate a continuous transition at $P_c = 3.4$ GPa. In summary, we hope to have demonstrated that standard Landau theory is completely inadequate to describe the HPPT of KMF unless one is willing to sacrifice any numerical meaning to Landau theory, leaving us with all coupling parameters as essentially unknown and with ad hoc pressure dependencies at room temperature.

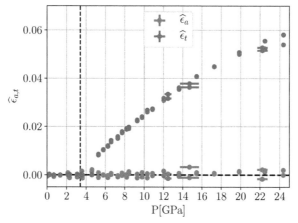

Figure 5. Lagrangian strains $\hat{\epsilon}_a, \hat{\epsilon}_t$ calculated from unit cell data and Murnaghan fit of cubic part according to Ref. [11].

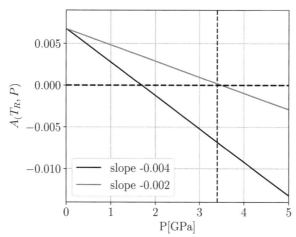

Figure 6. $A_R(T = T_R, P)$ as defined in Equation (10) for correct slope $-2\lambda_a^{(0)}/K^{(0)} = -0.004$ in comparison to a slope of $-2\lambda_a^{(0)}/K^{(0)} \approx -0.002$ assumed in Ref. [11].

3. A Quick Review of FSLT

In a nutshell, in a generic high pressure experiment, a given crystal is observed to change under application of high hydrostatic pressure P from an ambient pressure "laboratory" state X to a deformed state to be denoted as $\hat{X} = \hat{X}(P)$ with an accompanying total (Lagrangian) strain η_{ij}. Frequently, a high pressure phase transition manifests itself in such an experiment through the observation of relatively small strain anomalies on top of a much larger "background strain" that in itself is unrelated to the actual transition. Recognizing that the concept of strain is always defined with respect to a chosen elastic reference state, in Ref. [5], a corresponding background system $\widehat{X} = \widehat{X}(P)$, defined as the (hypothetical) equilibrium state of the system with the primary OP clamped to remain zero, was introduced. Let α_{ik} and $e_{ik} = \frac{1}{2} \left(\sum_n \alpha_{ni} \alpha_{nk} - \delta_{ik} \right)$ denote the deformation and Lagrangian strain tensors from X to $\widehat{X}$, respectively. The total strain η_{ij} may then be disentangled as a nonlinear superposition

$$\eta_{ij} = e_{ij} + \alpha_{ki} \widehat{\epsilon}_{kl} \alpha_{lj} \tag{11}$$

of the—generally large—Lagrangian background strain and a relatively small spontaneous strain $\widehat{\epsilon}_{ij}$ *measured relative to the floating "background" reference state* $\widehat{X}$. Determining the proper background strain e_{ij} in the resulting reference scheme

$$X \xrightarrow[\underset{a,\eta}{\underbrace{}}]{a,e} \widehat{X} \xrightarrow{\widehat{a},\widehat{\epsilon}} \widehat{\widehat{X}} \tag{12}$$

is thus a crucial step in correctly identifying the actual spontaneous strain, which in turn is mandatory in a successful application of the concepts of Landau theory. Effectively "subtracting" the elastic baseline, the strategy of FSLT therefore consists of setting up Landau theory *within the background reference system* $\widehat{X}$. Based on the reasonable assumption that a harmonic expansion with pressure-dependent elastic constants $C_{ijkl}[\widehat{X}]$ suffices to capture the elastic energy originating exclusively from the relatively small spontaneous strain $\widehat{\epsilon}_{kl}$, one arrives at the Landau free energy

$$\frac{F(Q, \widehat{\epsilon}; \widehat{X})}{V[\widehat{X}]} = \Phi(Q; \widehat{X}) + \sum_{\mu \geq 1} Q^{2n} d_{ij}^{(2\mu)}[\widehat{X}] \widehat{\epsilon}_{ij} + \frac{F_0(\widehat{\epsilon}; \widehat{X})}{V[\widehat{X}]} \tag{13}$$

where we have assumed a scalar OP Q for simplicity, and

$$\frac{F_0(\widehat{\epsilon}; \widehat{X})}{V[\widehat{X}]} \approx \sum_{ij} \tau_{ij}[\widehat{X}] \widehat{\epsilon}_{ij} + \frac{1}{2} \sum_{ijkl} C_{ijkl}[\widehat{X}] \widehat{\epsilon}_{ij} \widehat{\epsilon}_{kl} \tag{14}$$

denotes the pure spontaneous strain-dependent elastic free energy at hydrostatic external stress $\tau_{ij}[\widehat{X}] = -\delta_{ij} P$. For the pure OP potential part, we assume the traditional Landau expansion

$$\Phi(Q; \widehat{X}) = \frac{A[\widehat{X}]}{2} Q^2 + \frac{B[\widehat{X}]}{4} Q^4 + \frac{C[\widehat{X}]}{6} Q^6 + \dots \tag{15}$$

with P-dependent coefficients yet to be determined. In Ref. [9], it was explicitly shown that the harmonic structure of $\frac{F_0(\widehat{\epsilon}; \widehat{X})}{V[\widehat{X}]}$ yields the equilibrium spontaneous strain

$$\widetilde{\widehat{\epsilon}}_{mn} = - \sum_{\nu=1}^{\infty} \bar{Q}^{2\nu} \sum_{ij} d_{ij}^{(2\nu)}[\widehat{X}] S_{mnij}[\widehat{X}] \tag{16}$$

where the compliance tensor $S_{mnij}[\widehat{X}]$ is defined as the tensorial inverse of the so-called *Birch coefficients* [20,21]

$$B_{ijkl}[\widehat{X}] = C_{ijkl}[\widehat{X}] + \frac{1}{2} \left(\tau_{jk}[\widehat{X}] \delta_{il} + \tau_{ik}[\widehat{X}] \delta_{jl} + \tau_{jl}[\widehat{X}] \delta_{ik} + \tau_{il}[\widehat{X}] \delta_{jk} - 2\tau_{ij}[\widehat{X}] \delta_{kl} \right) \tag{17}$$

of the background system $\widehat{X}$ which effectively take over the role of the elastic constants at finite strain. Furthermore, if we eliminate $\bar{\hat{\epsilon}}_{mn}$ from $\frac{F(Q,\hat{\epsilon};\widehat{X})}{V[\widehat{X}]}$ by this formula, we obtain the renormalized pure OP potential

$$
\begin{aligned}
\Phi_R(Q;\widehat{X}) \quad:=\quad & \Phi(Q;\widehat{X}) - \sum_{\mu,\nu=1}^{\infty} \frac{2\mu}{2\mu+2\nu} \left(\sum_{ijkl} d_{ij}^{(2\mu)}[\widehat{X}] S_{ijkl}[\widehat{X}] d_{kl}^{(2\nu)}[\widehat{X}] \right) Q^{2(\mu+\nu)} \\
\equiv\quad & \frac{A_R[\widehat{X}]}{2} Q^2 + \frac{B_R[\widehat{X}]}{4} Q^4 + \frac{C_R[\widehat{X}]}{6} Q^6 + \cdots
\end{aligned}
\tag{18}
$$

from which the equilibrium OP $\bar{Q}$ can be determined by minimization.

Information on the P-dependence of elastic constants $C_{ijkl}[\widehat{X}]$ is usually available from density functional theory (DFT) or may be extracted from experimental measurements. At this stage, it therefore remains to relate the potential coefficients $A[\widehat{X}]$, $B[\widehat{X}],\dots$ and $d_{ij}^{(2\mu)}[\widehat{X}]$, which are still defined with respect to the reference system $\widehat{X}$ i.e., P-dependent. In a generic application of FSLT, however, one starts from knowledge of an ambient pressure Landau potential, i.e., the lowest order coefficients of the free energy

$$
\begin{aligned}
\frac{F(Q,\eta;X)}{V(X)} \quad=\quad & \Phi(Q;X) + \sum_{\mu=1}^{\infty} Q^{2\mu} \left(\sum_{ij} d_{ij}^{(2\mu,1)} \eta_{ij} + \frac{1}{2!} \sum_{ijkl} d_{ijkl}^{(2\mu,2)} \eta_{ij}\eta_{kl} + \frac{1}{3!} \sum_{ijklmn} d_{ijklmn}^{(2\mu,3)} \eta_{ij}\eta_{kl}\eta_{mn} + \cdots \right) \\
& + \frac{1}{2!} \sum_{ijkl} C_{ijkl}^{(2)} \eta_{ij}\eta_{kl} + \frac{1}{3!} \sum_{ijklmn} C_{ijklmn}^{(3)} \eta_{ij}\eta_{kl}\eta_{mn} + \cdots
\end{aligned}
\tag{19}
$$

(defined at $\tau_{ij}[X] \equiv 0$) with P-independent coefficients are assumed to be known, which obviously places constraints on the possible P-dependence of the above coefficients $A[\widehat{X}]$, $B[\widehat{X}],\dots$ and $d_{ij}^{(2\mu)}[\widehat{X}]$. To explore the relations between the two set of coefficients defined in the P-dependent background system $\widehat{X}$ and the laboratory system X, we insert the nonlinear superposition relation (11) into (19) and compare coefficients. Following Ref. [9], we content ourselves to just include OP-strain couplings of type $Q^2\hat{\epsilon}_{ij}$ and obtain

$$
A[\widehat{X}] = \frac{1}{J(\alpha)} \left[A[X] + 2 \left(\sum_{ij} d_{ij}^{(2,0)} e_{ij} + \frac{1}{2!} \sum_{ijkl} d_{ijkl}^{(2,1)} e_{ij}e_{kl} + \frac{1}{3!} \sum_{ijklmn} d_{ijklmn}^{(2,2)} e_{ij}e_{kl}e_{mn} + \cdots \right) \right]
\tag{20a}
$$

$$
d_{st}^{(2)}[\widehat{X}] = \frac{1}{J(\alpha)} \left(\sum_{ij} \alpha_{si} d_{ij}^{(2,0)} \alpha_{tj} + \sum_{ijkl} \alpha_{si} d_{ijkl}^{(2,1)} \alpha_{tj}e_{kl} + \sum_{ijklmn} \alpha_{si} d_{ijklmn}^{(2,2)} \alpha_{tj}e_{kl}e_{mn} + \cdots \right)
\tag{20b}
$$

in addition to the trivial relations $B[\widehat{X}] = B[X]/J(\alpha)$, $C[\widehat{X}] = C[X]/J(\alpha)$, where $J(\alpha) = \det(\alpha_{il})$.

The above parametrization scheme, although mathematically correct, is certainly not very convenient for applications in which the background reference state $\widehat{X}$ is of high symmetry. For the most important example of a cubic high-symmetry phase, the above equations simplify considerable, since the deformation tensor $\alpha_{ij} \equiv \alpha\delta_{ij}$ is diagonal, and so is the Lagrangian background strain $e_{ij} \equiv e\delta_{ij}$ with $e = \frac{1}{2}(\alpha^2 - 1)$, which yields

$$
A[\widehat{X}] = \frac{1}{\alpha^3} \left[A[X] + 2 \left(e \sum_i d_{ii}^{(2,0)} + \frac{e^2}{2!} \sum_{ik} d_{iikk}^{(2,1)} + \frac{e^3}{3!} \sum_{ikm} d_{iikkmm}^{(2,2)} + \cdots \right) \right]
\tag{21a}
$$

$$
d_{st}^{(2)}[\widehat{X}] = \frac{1}{\alpha} \left(d_{st}^{(2,0)} + e \sum_k d_{stkk}^{(2,1)} + e^2 \sum_{km} d_{stkkmm}^{(2,2)} + \cdots \right)
\tag{21b}
$$

These formulas are as far as we can get without committing to a specific set of irreducible representations that determine the symmetry-allowed couplings in the Landau potential. Since the background strains $e(P) \sim P + O(P^2)$, they effectively represent a set of highly interrelated power series in powers of P. Unfortunately, this also comes with all the inherent drawbacks. On the one hand, going beyond the lowest order terms, which may be taken from an previously determined ambient pressure Landau potential, we are forced to truncate the above series at rather low order to limit the number of additional unknown parameters. Of course, such truncated series inevitable diverge for increasing values of P. In addition, if we want to employ the theory for the purpose of fitting experimental data, we would like to fix certain experimental observables, in particular the pressure P_0 at which $A_R[\hat{X}]$ vanishes. Given the above parametrization, this is obviously difficult to do.

In Section 2, we have demonstrated the considerable structural simplification of a standard ambient pressure Landau approach upon replacing Cartesian strain tensors by symmetry-adapted strains. Remarkably, it turns out that, despite the additional complicated nonlinearities contained in formulas (21a) and (21b), similar manipulations may also be carried out in FSLT and are found to yield equally substantial structural simplifications. In the rest of the paper, the resulting scheme will be derived and illustrated by describing the HPPT of KMF.

4. Symmetry-Adapted FSLT: The Cubic-to-Tetragonal HP Phase Transition of KMF

For $\nu = 0, 1, 2 \ldots$, we set

$$\lambda_a^{(\nu)} \equiv \frac{1}{3} \sum_{k_1 \ldots k_\nu} \sum_i d_{iik_1 k_1 \ldots k_\nu k_\nu}^{(2,\nu)} \tag{22a}$$

$$\lambda_t^{(\nu)} \equiv \sum_{k_1 \ldots k_\nu} \frac{d_{33k_1 k_1 \ldots k_\nu k_\nu}^{(2,\nu+1)} - d_{11k_1 k_1 \ldots k_\nu k_\nu}^{(2,\nu)}}{2\sqrt{3}} \tag{22b}$$

in accordance with

$$\lambda_a[\hat{X}] \equiv \lambda_a(e) \equiv \frac{1}{3} \sum_s d_{ss}^{(2)}[\hat{X}] \tag{23a}$$

$$\lambda_t[\hat{X}] \equiv \lambda_t(e) \equiv \frac{d_{33}^{(2)}[\hat{X}] - d_{11}^{(2)}[\hat{X}]}{2\sqrt{3}} \tag{23b}$$

Equations (21a) and (21b) are then replaced by

$$\alpha^3 A[\hat{X}] = A[X] + 6 \sum_{\nu=0}^{\infty} \frac{\lambda_a^{(\nu)}}{(\nu + 1)!} e^{\nu+1} \tag{24a}$$

$$\alpha \lambda_{a,t}[\hat{X}] = \sum_{\nu=0}^{\infty} \lambda_{a,t}^{(\nu)} e^{\nu} \tag{24b}$$

Furthermore, we recall from nonlinear elasticity theory [20] that for a cubic system the compliance tensor $S_{ijkl}[\hat{X}]$ and the bulk modulus $K[\hat{X}] = (B_{1111}[\hat{X}] + 2B_{1122}[\hat{X}])/3$ at finite pressure are related by $\sum_i S_{iijj}[\hat{X}] = S_{1111}[\hat{X}] + 2S_{1122}[\hat{X}] = 1/3K[\hat{X}]$, while, for the longitudinal shear modulus, $\mu[\hat{X}] = (B_{1111}[\hat{X}] - B_{1122}[\hat{X}])/2$ the relation $S_{1111}[\hat{X}] - S_{1122}[\hat{X}] = 1/2\mu[\hat{X}]$ holds. If we replace the Cartesian equilibrium spontaneous strain components (16) by their symmetry-adapted volume and tetragonal counterparts using these relations, they acquire the representations

$$\tilde{\bar{\epsilon}}_a = -\frac{\lambda_a[\hat{X}]}{K[\hat{X}]}\bar{Q}^2, \qquad \tilde{\bar{\epsilon}}_t = -\frac{2\lambda_t[\hat{X}]}{\mu[\hat{X}]}\bar{Q}^2 \tag{25}$$

Furthermore, it is easy to check the identity

$$\sum_{ijkl} d_{ij}^{(2)}[\hat{X}]S_{ijkl}[\hat{X}]d_{kl}^{(2)}[\hat{X}] = \frac{\lambda_a^2[\hat{X}]}{K[\hat{X}]} \tag{26}$$

which allows for similarly rewriting the renormalized potential $\Phi_R(Q;\hat{X})$ as

$$\Phi_R(Q;\hat{X}) = \Phi(Q;\hat{X}) - \frac{Q^4}{4}\left(\frac{2\lambda_a^2[\hat{X}]}{K[\hat{X}]} + \frac{8\lambda_t^2[\hat{X}]}{\mu[\hat{X}]}\right) \tag{27}$$

Combining these equations with Equations (24a) and (24b), we can summarize the symmetry-adapted parametrization of the coefficients of the renormalized Landau potential (18), whose minimization determines the equilibrium OP $\bar{Q}$ with

$$A_R[\hat{X}] = \frac{A[X]}{\alpha^3} + \frac{6\Delta_a[\hat{X}]}{\alpha^3} \tag{28a}$$

$$B_R[\hat{X}] = \frac{B[X]}{\alpha^3} - \frac{2(\lambda_a^{(2)}[\hat{X}])^2}{K[\hat{X}]} - \frac{8\lambda_t^2[\hat{X}]}{\mu[\hat{X}]} \tag{28b}$$

in addition to $C_R[\hat{X}] = \tilde{C}[\hat{X}]$, where we introduced the function

$$\Delta_a[\hat{X}] \equiv \Delta_a(e) \equiv \sum_{\nu=0}^{\infty} \frac{\lambda_a^{(\nu)}}{(\nu+1)!}e^{\nu+1} \tag{29}$$

Note the close formal similarity of Equations (28a), (28b) and (29) with their infinitesimal counterparts (7a) and (7b). In a generic application of this theory with cubic high symmetry, the lowest order Landau coefficients $A[X], B[X], C[X]$ and the lowest order strain-OP coupling coefficients $\lambda_{a,t}^{(0)}$ can be taken from an existing ambient pressure Landau theory. Furthermore, the (diagonal) background deformation components $\alpha = \alpha(P)$ and the resulting Lagrangian strain $e = e(P)$ may be determined by fitting a suitable EOS to the cubic unit cell volume data. Such a fit also immediately yields the pressure-dependent bulk modulus $K[\hat{X}] \equiv K(P)$. It is only for the pressure-dependence of the longitudinal shear modules $\mu[\hat{X}] = \mu(P)$ that we are truly forced to resort to additional input from DFT. To determine the EOS and the elastic constants of KMnF$_3$, we have performed a series of fairly standard DFT calculations. We refer to Appendix A for further details of these simulations.

Since we do not need to maintain the highest possible precision in determining $\mu(P)$ at room temperature but can content ourselves with a reasonable approximation, we use the following heuristic strategy to promote the DFT result $\mu_{\text{DFT}}(P)$ from $T = 0$ to ambient temperature. Figure 7 shows a comparison of the bulk moduli $K_{\text{DFT}}(P)$ calculated from DFT to the result for $K(P)$ obtained from the Murnaghan fit of the data published in Ref. [11]. Numerically, one finds the fraction of these moduli stays entirely within the narrow bounds $0.918 \leq K(P)/K_{\text{DFT}}(P) \leq 0.92$ over the whole interval $0 \leq P \leq 20\,\text{GPa}$. We therefore postulate a similar behavior for $\mu(P)$, setting

$$\mu(P) \equiv \frac{K(P)}{K_{\text{DFT}}(P)} \cdot \mu_{\text{DFT}}(P) \tag{30}$$

Figure 8 illustrates the resulting behavior of $\mu(P)$.

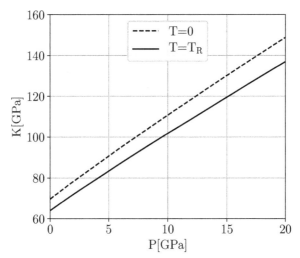

Figure 7. Comparison of pressure-dependent bulk modulus $K(P)$ at room temperature $T = T_R$ extracted from the Murnaghan fit of the cubic part of the unit cell data of Ref. [11] to the $T = 0$ result obtained from DFT.

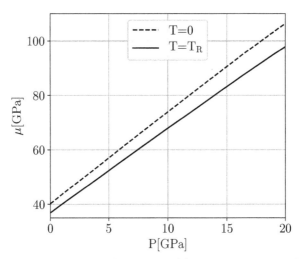

Figure 8. Pressure-dependent longitudinal shear modulus $\mu(P)$ at room temperature $T = T_R$ as determined by extrapolating the corresponding DFT result from $T = 0$ to $T = T_R$ based on Formula (30).

With all other ingredients in place, this leaves only the higher order coefficients $\lambda_{a,t}^{(\nu)}, \nu > 0$ as the remaining undetermined parameters of FSLT. It therefore seems that these parameters must be treated as unknown fit parameters in a practical application. In the next section, however, we will introduce a much more convenient and powerful approach.

5. Efficient Parametrization

Comparing the above symmetry-adapted system of equations to what we had before, a considerable structural simplification is obvious. However, the following drawbacks seem to persist:

- Truncating the functions $\lambda_{a,t}[\widehat{X}]$ defined by Equation (24b) at finite order $e^{\nu} = e^{\nu}(P)$ still results in divergent behavior with increasing P.
- In addition to a given set of strain measurements which we would want to feed to least squares fits based on Landau theory, rather accurate experimental information on the critical temperature T_c or the critical pressure P_c is frequently available from complementary experimental measurements. Unfortunately, least squares fitting procedures of strain measurements with an unconstrained value of T_c or P_c frequently tend to displace T_c or P_c and thus degrade the accuracy of the fit in the transition region. At least for a second order transition, the critical point is, of course, determined by the zero of the quadratic coefficient A_R of the Landau expansion. Therefore, we would like to be able to explicitly constrain the behavior of A_R, possibly fixing both its zero and/or slope at zero as a function of T or P. Unfortunately, based on a set of interrelated truncated power series, this is still hard to do.

In what follows, we propose a new scheme which finally allows for practically overcoming all of these problems in a single push. Let us start by taking a second look at Figure 6. Of course, the correct functional form of $A_R[\widehat{X}] \equiv A_R(T_R, P)$ that we are looking for should be that of a well-behaved function passing through zero around $P_c \approx 4$ GPa. At $P = 0$, however, it should start out with roughly twice the initial slope of the purely linear yellow curve if we are to retain the ambient pressure Landau parameters. The pressure dependence of the correct function $A_R(T_R, P)$ must therefore be far from linear. Experimentally, there is no indication of a re-entrant behavior below 25 GPa, so we do not expect any second crossing point in this pressure range. Unfortunately, our numerical tests quickly revealed that, using a truncated version of $\Delta_a(e(P))$ as defined in (29), it seems virtually impossible to meet these requirements unless one is willing to go to prohibitively high truncation order, thus introducing a plethora of unknown fit parameters and the accompanying horrific numerical problems.

A way out is to propose a reasonable function $\Delta_a(e(P))$ in closed form that meets all of the above requirements while still containing some adjustable parameters that offer a certain amount of variational flexibility to allow improvement by least squares fitting. Of course, there are many ways to do this, and the choices are only limited by the reader's ingenuity. In fact, since the summand $\nu = 0$ of $\Delta_a(x)$ contributes the lowest order coefficient $x\lambda_a^{(0)}$, which is usually fixed from knowledge of an existing ambient pressure Landau theory, all candidate functions $\Delta_a(x)$ that start out like

$$\Delta_a(x) = \lambda_a^{(0)} x + O(x^2) \qquad (31)$$

qualify as candidates for a meaningful function $\Delta_a(x)$. In our current application to KMF, we parametrize

$$A_R \equiv A_R(T_R, e(P)) = \frac{A_R(T_R) + 6\Delta(e(P); b, c, d)}{\alpha^3(P)} \qquad (32)$$

by introducing the function

$$\Delta(e; b, c, d) := ce + 2bf_d(e) \qquad (33)$$

with parameters b, c, d combining a linear part with slope c and a nonlinear contribution $f_d(e)$, which remains yet to be specified. Our general parametrization strategy is then as follows:

- As a function of e, $A_R(T_R, e)$ should start out with slope $s = -6\lambda_a^{(0)} \equiv c + 2bf_d'(0)$, so parameter b can be traded for the slope s of $\Delta(e; b, c, d)$ at $e = 0$.
- Suppose further that $A_R(T_R, e)$ should vanish at a critical value $e_0 = e(P_0)$ of the background strain e. For $\Delta(e; b, c, d)$, this implies the constraint equation $\alpha + \Delta(e_0; b, c, d) \equiv 0$ where $\alpha \equiv A_R(T_R, 0)/6$, which may be solved for parameter c.

These steps eliminate parameters b, c in favor of the constants α and e_0, leaving d as the single remaining free variational parameter. We still need to reconcile this parametrization of A_R with that of the function $\lambda_a[\widehat{X}]$ as defined in Equation (24b). We focus on the power series part

$$\Lambda(x) \equiv \sum_{\nu=0}^{\infty} \lambda_a^{(\nu)} x^\nu \tag{34}$$

which is based on the same set of coefficients $\lambda_a^{(\nu)}$ as $\Delta_a(x)$ but lacks the accompanying factorials $1/\nu!$. These factorials can, however, easily be taken care of. Observe that

$$\int_0^\infty dt\, e^{-t} \frac{(tx)^{\nu+1}}{(\nu+1)!} = x^{\nu+1} \tag{35}$$

Therefore, using the *Borel transform*

$$(\mathcal{B}\Delta)(x) \equiv \int_0^\infty dt\, e^{-tx} \Delta(tx) \tag{36}$$

we may relate

$$\Lambda(x) = \frac{(\mathcal{B}\Delta)(x)}{x} \tag{37}$$

It remains to specify a suitable function $f_d(x)$. Beyond producing a reasonable function $A_R(T_R, e)$, the job profile for recruiting such a function includes at least two basic requirements:

- It would be highly desirable to be able to compute the corresponding Borel transform $(\mathcal{B}f_d)(x)$ in closed form.
- $f_d(x)$ should also allow for solving the equation $\alpha + \Delta(e_0; b, c, d) \equiv 0$ explicitly.

For the present goal of understanding the HPPT in KMF, we came up with the choice

$$f_d(e) := -1 + \sqrt{1 - d^2 e} \tag{38}$$

(note that $e(P) < 0$ for $P > 0$) which meets both of these minimal requirements, since, in this case, $b = (c - s)/d^2$ and

$$c = \frac{s\left(\sqrt{1 - d^2 e_0} - 1\right) - \alpha d^2/2}{d^2 e_0/2 + \sqrt{1 - d^2 e_0} - 1} \tag{39}$$

In this way, we have completely bypassed Taylor series expansions and their various accompanying drawbacks. Figure 9 illustrates the remaining variational freedom in our chosen parametrization.

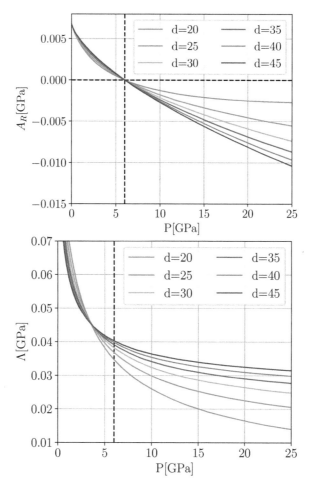

Figure 9. Illustration of the remaining d-dependence of the parametrization of $A_R(T_R, e(P))$ by Equation (32) (upper panel) and the resulting function $\Lambda(e(P))$ as given by the Borel transform Equation (37) (lower panel). Parameters b and c were eliminated in favor of parameter $\lambda_a^{(0)} = 0.128$ GPa as prescribed from ambient pressure LT and a chosen pressure parameter $P_0 = 6.0$ GPa.

In summary, at this stage, the function $\Delta_a[\hat{X}]$ which governs the behavior of $A_R[\hat{X}]$ and—through a Borel transform—also provides the coupling function $\lambda_a[\hat{X}]$ between the spontaneous volume strain $\tilde{\epsilon}_a$ and $\bar{Q}^2$ according to Equation (25) has been specified up to a single free parameter d. The remaining coupling function $\lambda_a[\hat{X}] \equiv \lambda_a(P)$ explicitly determines the proportionality between the spontaneous tetragonal strain $\tilde{\epsilon}_a$ and $\bar{Q}^2$, However, both $\lambda_a[\hat{X}]$ and $\lambda_t[\hat{X}]$ also enter implicitly into the spontaneous strain via its implicit dependence on the quartic coefficient $B_R[\hat{X}]$ of the renormalized Landau potential as given by Equation (28b), and, apart from the reasonable requirement $\lim_{P \to 0} \lambda_t[\hat{X}] = \lambda_t^{(0)}$, nothing is known in advance about its pressure dependence, such that introducing a truncated Taylor series and least squares fitting seems unavoidable. However, we can actually do a lot better than this. Let us consider experimental high pressure spontaneous strain data in the form of n data points $(P_i, \widehat{\epsilon}_{a,\exp}(P_i), \widehat{\epsilon}_{a,\exp}(P_i))_{i=1}^n$. In fact, at a given prescribed pressure value P_i and with all other parameters in place, we can regard the room temperature values $\widehat{\epsilon}_a = \widehat{\epsilon}_a(\lambda_t(P_i))$ and $\widehat{\epsilon}_t = \widehat{\epsilon}_t(\lambda_t(P_i))$ as functions of the-unknown-function values $\lambda_t(P_i)$. The "best" value $\lambda_t(P_i)$ matching the data point $(P_i, \widehat{\epsilon}_{a,\exp}(P_i), \widehat{\epsilon}_{a,\exp}(P_i))$ may then be determined by numerically minimizing the function

$$
\begin{aligned}
s_i(\lambda_t(P_i)) \quad := \quad & w_a \left[\widehat{\epsilon}_{a,\exp}(P_i) - \widehat{\epsilon}_a(\lambda_t(P_i)) \right]^2 \\
& + w_t \left[\widehat{\epsilon}_{t,\exp}(P_i) - \widehat{\epsilon}_t(\lambda_t(P_i)) \right]^2
\end{aligned}
\tag{40}
$$

with weights w_a, w_t suitable adjusted to counterbalance size differences between $\widehat{\epsilon}_a$ and $\widehat{\epsilon}_t$. Carried out for all $i = 1, \ldots, n$, this prescription results in a collection of n "optimal" values $\lambda_t(P_i)$ from which one may hope to recover the full function $\lambda_t(P)$ by interpolation. Figure 10 shows the result of our corresponding effort for KMF. Amazingly, we observe that all values $\lambda_t(P_i)$ seem to accumulate on a straight line whose extrapolation $P \to 0$ perfectly passes through the point $(0, \lambda_t^{(0)})$, which is just the limiting value imposed by ambient pressure Landau theory. We believe that this behavior is not coincidental but a strong indication that the present parametrization is internally consistent and correct.

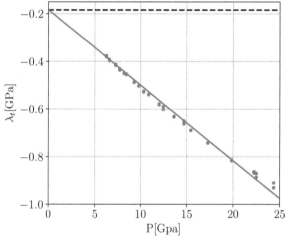

Figure 10. Results of minimizing the sum (40) using the relative weights $(w_a, w_t) = (10, 1)$ and pressure parameters $P_0 = 6\,$GPa, $d = 40$. The dashed horizontal line indicates the limiting value $\lambda_t(P = 0) \equiv \lambda_t^{(0)}$.

A simple linear fit of $\lambda_t(P)$ therefore completes our Landau parametrization. Our results for the pressure dependence of the couplings $A_R(P), \widetilde{B}(P), \lambda_{a,t}(P)$ and the resulting pressure dependence of the equilibrium OP $\bar{Q}(P)$ are illustrated in Figure 11. Note that in the present description the transition appears to be of first rather than second order, with $P_0 = 6\,$GPa yielding a transition pressure $P_c \approx 4.5\,$GPa. Finally, the resulting parametrization of the spontaneous strain is compared to experiment in Figure 12.

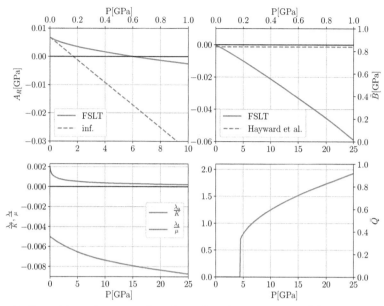

Figure 11. Upper left panel: pressure dependence of Landau parameters $A_R(P)$ as compared to the simple linear behavior of infinitesimal strain LT assumed in Equation (7a). Upper right panel: pressure dependence of $\bar{B}(P)$. The dashed horizontal line indicates the corresponding value from the parametrization of Hayward et al. [19] taken at $T = 186.5$ K, which is slightly displaced from our limiting value at $P = 0$ since we are taking into account thermal softening of $K(P)$ and $\mu(P)$ for room temperature. Lower left panel: pressure dependence of coupling parameters $\lambda_a(P)/K(P)$ and $\lambda_t(P)/\mu(P)$. Lower right panel: resulting behavior of the equilibrium order parameter $\bar{Q}(P)$ (right).

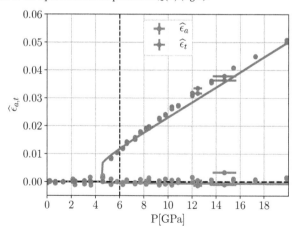

Figure 12. Parametrization of spontaneous strains $\hat{\epsilon}_a, \hat{\epsilon}_t$ in comparison to experimental data from Ref. [11]. The thin vertical line indicates the pressure parameter $P_0 = 6$ GPa.

6. Discussion

For the high pressure community, our present paper contains some good news and some bad news—first, the bad news. We hope to have demonstrated convincingly that classical Landau theory is usually completely inadequate for "explaining" experimental findings in the field of high pressure phase transitions, and the conclusions drawn from it will often be misleading at best. Taking the example of KMF, both the first order character of the transition and the correct value of the transition pressure are completely obscured by sticking to Landau theory with infinitesimal strain coupling, even if one is willing to distort a pre-existing ambient pressure Landau parametrization beyond recognition. The good news, however, is that with the development of FSLT a mathematically consistent alternative incorporating nonlinear elasticity has recently become available. Up to now, however, FSLT has been rather complicated in structure, which quite likely scared off many potential users and thus did not lead to the widespread use that its developers were initially hoping for. The present paper, which exploits the enormous simplifications that arise by passing (i) from Cartesian to symmetry-adapted finite strains and (ii) by virtue of (i), from truncated Taylor expansions to a functional parametrization. These improvements should pave the way for routine use of our theory in successfully describing HPPTs. In particular, the present paper illustrates that, while the former version of FSLT involved delicate least-squares fitting procedures with a large number of unknown fit parameters and dealing with all the inadequacies implicit in the use of truncated Taylor expansions, in the present scheme, the unknown pressure dependencies can be systematically determined one-by-one in a step-wise, almost "deterministic" manner.

Admittedly, even the present parametrization of the HPPT in KMF is still less than perfect. For instance, the coupling function $\lambda_a(P)/K(P)$ shown in the lower left panel of Figure 11 exhibits a steep initial decrease with increasing pressure. Since no spontaneous strain data are available for this pressure region, this does not change the physical values produced by the theory. However, it hints at a sub-optimal choice Equation (38) for the auxiliary function $f_d(x)$. The reader is invited to come up with an improved candidate function.

More importantly, in a full-blown application of FSLT, we should be able to predict e.g., the (P, T) cubic-to-tetragonal phase boundary of KMF and compute pressure—and temperature-dependent elastic constants. In principle, the ability to do so depends mainly on the successful construction of a pressure—and temperature-dependent baseline, i.e., the cubic EOS $V_{\text{cubic}} = V_{\text{cubic}}(P, T)$. In a previous paper [10], this task has been successfully carried out for the perovskite $PbTiO_3$ by combining zero temperature DFT calculations (see Appendix A for details) with the Debye approximation as implemented in the GIBBS2 package [22,23] to incorporate effects of thermal expansion (recently, we learned [24] that a similar approach also seems to work for $MgSiO_3$). Unfortunately, our corresponding efforts to derive $V_{\text{cubic}}(P, T)$ for KMF along the same lines have failed so far, however. This failure manifests itself e.g., in the inability to simultaneously reproduce the thermal baseline at $P = 0$ measured experimentally and the ambient temperature EOS, even if we allowed for the introduction of a constant compensating background pressure which is frequently introduced in DFT calculations to compensate inadequacies of an employed exchange-correlation functional. Ref. [25] states that perovskites with octahedral tilting generally do not show an appreciable coupling between structural and magnetic order parameters. Nevertheless, this statement obviously does not exclude effects due to a coupling between magnetic degrees of freedom and the background volume strain. Since the Debye approximation is based exclusively on phonons, we may speculate that in KMF residual magnetic effects may be responsible for additional thermal energy consumption. An investigation of this problem is currently underway.

Finally, it may be argued that our current theory also does not "explain" the origins of the involved nonlinear P-dependencies on a fundamental level. In approaches based on infinitesimal strain couplings, similarly looking P-dependent couplings are also introduced, but in a more or less completely ad hoc manner, blaming their existence on rather unspecific "higher order strain couplings". As a rule, such an approach results in a mathematically inconsistent theory. In contrast, the nonlinearities that arise in FSLT result for different but well-defined reasons. On the one hand, there are couplings between powers of the background strain $e(P)$ and the Landau potential $\Phi_R(Q; \widehat{X})$ "floating" on this background strain, and there is a pressure-dependence of the elastic moduli $K(P)$ and $\mu(P)$, which is in principle accessible e.g., to DFT calculations. This leaves the task of "explaining" the residual nonlinearities in the functions $\lambda_{a,t}(P)$ describing the couplings between order parameter and spontaneous strains. In contrast to blaming their existence on the effects of some unspecified higher order couplings, our present theory provides a practical way to numerically determine these functions. In addition, we see no reason in principle as to why such P-dependent coupling constants could not eventually be extracted from DFT calculations along the general philosophy laid out in Refs. [26,27] and the subsequent follow-up literature.

Author Contributions: A.T.: concept, main theory, and writing; W.S.: additional theory, comparison to previous experimental data; S.E., K.B., and P.B.: DFT calculations. All authors have read and agreed to the published version of the manuscript.

Funding: A.T., S.E., and K.B. acknowledge support by the Austrian Science Fund (FWF) Project P27738-N28. W.S. acknowledges support by the Austrian Science Fund (FWF) Project P28672-N36. S.E. acknowledges support from H.E.C., Pakistan.

Acknowledgments: A.T. would like to thank Ronald Miletich-Pawliczek for useful discussions.

Conflicts of Interest: The authors declare no conflict of interest.

Abbreviations

The following abbreviations are used in this manuscript:

AFN	antiferromagnetic
DFT	density functional theory
EOS	equation of state
FSLT	finite strain Landau theory
HPPT	high pressure phase transition
KMF	$KMnF_3$
LDA	local density approximation
LT	Landau theory
OP	order parameter
PT	phase transition
NM	nonmagnetic

Appendix A. DFT Calculation Details

The EOS and the elastic constants in the cubic phase of KMF were calculated using the WIEN2K DFT package, which is an all-electron code based on the (linearized) augmented plane-wave and local orbitals [(L)APW+lo] basis representation of the Kohn–Sham equations [28] of DFT. Here, we only content ourselves with a quick outline of the basic ideas and refer to Refs. [29,30] and the monograph Ref. [31] for more details.

In the (L)APW+lo method, the crystal's unit cell is partitioned into a set of atomic spheres surrounding the nuclei and a remaining interstitial region. Inside these atomic spheres, the wave functions are expanded into atomic-like basis functions, i.e., numerical radial functions times spherical harmonics while they are represented by plane waves throughout the interstitial region. These two regions are glued together by requiring continuity of the basis functions in value (and, depending on the flavor of the (L)APW+lo method, in radial slope) across the sphere boundaries. These LAPW calculations require making a couple of choices regarding cell size, standard parameter values, etc. In detail, the calculations of the present work were done with $R_{MT}^{min} K_{max} = 9$ and atomic sphere radii of 2.1, 2.0, and 1.6 bohr for K, Mn and F, respectively. The energy separation between core and valance states used was −6.0 Ry. Inside the sphere, the maximum angular momentum used in the spherical expansions was $L_{max} = 10$, while the charge density in the interstitial was Fourier expanded up to a cutoff of $G_{max} = 14 (a.u.)^{-1}$.

As to the use of exchange-correlation functionals, we have performed calculations using the standard local density approximation (LDA) [32] and three functionals of the generalized gradient approximation, namely, PBE from Perdew et al. [33,34], its solid-state optimized version PBEsol [35], and WC from Wu and Cohen [36]. For LDA and PBE, we observed the well-known tendency to underestimate and overestimate the lattice constants of solids, respectively, while PBEsol and WC produced more accurate results in between LDA and PBE [37,38].

To study the effect of magnetism on our results, we undertook calculations for a standard non-magnetic (NM) and ferromagnetic (FM) cubic perovskite unit cell with five atoms as well as an for a cubic supercell with 10 atoms in antiferromagnetic (AFM) structures of A-type, C-type and G-type [39]. After sufficient testing, we settled for a k-mesh sampling of $10 \times 10 \times 10$ k-points for all types of structures. In FM and AFM structure, we observed a linearization error which is inherent to the basis functions inside the spheres. To overcome this problem, we use the second energy derivative of the radial part (HDLO) for d electrons for Mn atom (for more detail see Ref. [40]). Using this setup, we identified that G-type AFM structure to have lowest energy.

Cubic elastic elastic constants were calculated with the help of the WIEN2k add-on package by Charpin [41]. Results at $T = 0$ are compiled in Table A1. In particular, we conclude that FM and AFM structures give similar value of lattice parameters and bulk modulus, which suggests that the specific magnetic ordering is not overly important for these quantities. In passing, we note that non-magnetic KMF is found to be a metal (with all functionals), but all magnetic structures lead to insulators (with all functionals). Simulations with PBEsol+U with U = 4eV in the AFM G-type phase would even lead to slightly better agreement with experiment for lattice constants and bulk modulus, but overall the effect is small and we therefore used the PBEsol results in Figures 7 and 8.

Table A1. Lattice constant (Å) and bulk modulus (GPa) of cubic $KMnF_3$ for different methods used in this work.

	Present Work			Other Works	
	NM	FM	AFM (G-Type)	PBE	Expt.
Lattice constant					4.185 [42]
LDA	3.89	4.11	4.09		
PBEsol	3.96	4.19	4.17		
PBE	4.04	4.26	4.24	4.19 [43]	
PBEsol + U (U = 4 ev)		4.20	4.19		
Bulk modulus					
LDA	117.2	83.7	88.4		
PBEsol	97.6	70.4	69.6		
PBE	83.3	62.3	63.1		
PBEsol + U (U=4 ev)		68.2	67.7		

Estimating an additional softening due to finite temperature with various flavors of the Debye approximation which are implemented in the GIBBS2 software [22,23], the best agreement with the cubic part of the experimental data of Ref. [11] was reached for the combination of G-type antiferromagnetic structure and PBEsol functional.

References

1. Landau, L.; Lifshitz, E.; Pitaevskii, L. *Statistical Physics Part I*; Butterworth and Heinemann: Oxford, UK, 2001.
2. Tolédano, J.; Tolédano, P. *The Landau Theory of Phase Transitions*; World Scientific: Singapore, 1987.
3. Rabe, K.; Ahn, C.; Triscone, J.M. (Eds.) Physics of Ferroelectrics. In *Topics in Applied Physics*; Springer: Berlin/Heidelberg, Germany, 2007; Volume 105.
4. Salje, E. *Phase Transitions in Ferroelastic and Coelastic Crystals*; Cambridge University Press: Cambridge, UK, 1990.
5. Tröster, A.; Schranz, W.; Miletich, R. How to Couple Landau Theory to an Equation of State. *Phys. Rev. Lett.* **2002**, *88*, 055503. [CrossRef]
6. Koppensteiner, J.; Tröster, A.; Schranz, W. Efficient parametrization of high-pressure elasticity. *Phys. Rev. B* **2006**, *74*, 014111. [CrossRef]
7. Schranz, W.; Tröster, A.; Koppensteiner, J.; Miletich, R. Finite strain Landau theory of high pressure phase transformations. *J. Phys. Condens. Matter* **2007**, *19*, 275202. [CrossRef]
8. Tröster, A.; Schranz, W. Landau theory at extreme pressures (invited paper for contribution to the special edition of "FERROELECTRICS") (birthday edition in honour of V. Ginzburg's 90th birthday). *Ferroelectrics* **2007**, *354*, 208. [CrossRef]
9. Tröster, A.; Schranz, W.; Karsai, F.; Blaha, P. Fully Consistent Finite-Strain Landau Theory for High-Pressure Phase Transitions. *Phys. Rev. X* **2014**, *4*, 031010. [CrossRef]
10. Tröster, A.; Ehsan, S.; Belbase, K.; Blaha, P.; Kreisel, J.; Schranz, W. Finite-strain Landau theory applied to the high-pressure phase transition of lead titanate. *Phys. Rev. B* **2017**, *95*, 064111. [CrossRef]
11. Guennou, M.; Bouvier, P.; Garbarino, G.; Kreisel, J.; Salje, E. Pressure-induced phase transition(s) in $KMnF_3$ and the importance of the excess volume for phase transitions in perovskite structures. *J. Phys. Condens. Matter* **2011**, *23*, 485901. [CrossRef]
12. Carpenter, M.A.; Becerro, A.I.; Seifert, F. Strain analysis of phase transitions in (Ca, Sr) TiO_3 perovskites. *Am. Mineral.* **2001**, *86*, 348–363. [CrossRef]
13. Salje, E.K.H.; Zhang, M.; Zhang, H. Cubic–tetragonal transition in $KMnF_3$: IR hard-mode spectroscopy and the temperature evolution of the (precursor) order parameter. *J. Physics: Condens. Matter* **2009**, *21*, 335402.
14. Dormann, E.; Copley, J.R.D.; Jaccarino, V. Temperature dependence of the MnF_2 and $KMnF_3$ lattice parameters from room temperature to the melting point. *J. Phys. C Solid State Phys.* **1977**, *10*, 2767–2771. [CrossRef]
15. Ratuszna, A.; Pietraszko, A.; Chelkowski, A.; Lukaszewicz, K. The Temperature Dependence of Lattice Parameters of KMeF3 and KMn0.9Me0.1F3 Compounds (Me = Mn2+, Co2+, and Ni2+). *Phys. Status Solidi (a)* **1979**, *54*, 739–743. [CrossRef]
16. Sakashita, H.; Ohama, N. A precursor effect in the lattice constant at the 186 K-structural phase transition in KMnF3. *Phase Transit.* **1982**, *2*, 263–276. [CrossRef]
17. Sakashita, H.; Ohama, N.; Okazaki, A. Thermal expansion and spontaneous strain of KMnF3 near the 186 K-structural phase transition. *Phase Transit.* **1990**, *28*, 99–106. [CrossRef]
18. Gibaud, A.; Shapiro, S.M.; Nouet, J.; You, H. Phase diagram of $KMn_{1-x}Ca_xF_3$ (x < 0.05) determined by high-resolution X-ray scattering. *Phys. Rev. B* **1991**, *44*, 2437–2443. [CrossRef]
19. Hayward, S.A.; Romero, F.J.; Gallardo, M.C.; del Cerro, J.; Gibaud, A.; Salje, E.K.H. Cubic-tetragonal phase transition in KMnF3: Excess entropy and spontaneous strain. *J. Phys. Condens. Matter* **2000**, *12*, 1133–1142. [CrossRef]
20. Wallace, D. *Thermodynamics of Crystals*; Dover: New York, NY, USA, 1998.
21. Morris, J.W.; Krenn, C.R. The internal stability of an elastic solid. *Philos. Mag. A* **2000**, *80*, 2827–2840. [CrossRef]
22. Otero-de-la-Roza, A.; Luaña, V. Gibbs2: A new version of the quasi-harmonic model code. I. Robust treatment of the static data. *Comput. Phys. Commun.* **2011**, *182*, 1708–1720. [CrossRef]
23. Otero-de-la-Roza, A.; Abbasi-Pérez, D.; Luaña, V. Gibbs2: A new version of the quasiharmonic model code. II. Models for solid-state thermodynamics, features and implementation. *Comput. Phys. Commun.* **2011**, *182*, 2232–2248. [CrossRef]

24. Liu, Z.; Sun, X.; Zhang, C.; Hu, J.; Song, T.; Qi, J. Elastic Tensor and Thermodynamic Property of Magnesium Silicate Perovskite from First-principles Calculations. *Chin. J. Chem. Phys.* **2011**, *24*, 703–710. [CrossRef]
25. Carpenter, M.A.; Salje, E.K.H.; Howard, C.J. Magnetoelastic coupling and multiferroic ferroelastic/magnetic phase transitions in the perovskite KMnF₃. *Phys. Rev. B* **2012**, *85*, 224430. [CrossRef]
26. Zhong, W.; Vanderbilt, D.; Rabe, K.M. Phase Transitions in BaTiO₃ from First, Principles. *Phys. Rev. Lett.* **1994**, *73*, 1861–1864. [CrossRef]
27. King-Smith, R.D.; Vanderbilt, D. First-principles investigation of ferroelectricity in perovskite compounds. *Phys. Rev. B* **1994**, *49*, 5828–5844. [CrossRef]
28. Kohn, W.; Sham, L.J. Self-Consistent Equations Including Exchange and Correlation Effects. *Phys. Rev.* **1965**, *140*, A1133–A1138. [CrossRef]
29. Blaha, P.; Schwarz, K.; Madsen, G.K.H.; Kvasnicka, D.; Luitz, J. *WIEN2k: An Augmented Plane Wave and Local Orbitals Program for Calculating Crystal Properties*; Vienna University of Technology: Vienna, Austria, 2001.
30. Schwarz, K.; Blaha, P.; Trickey, S. Electronic structure of solids with WIEN2k. *Mol. Phys.* **2010**, *108*, 3147–3166. [CrossRef]
31. Singh, D.; Nordström, L. *Planewaves, Pseudopotentials and the LAPW Method*, 2nd ed.; Springer: New York, NY, USA, 2006.
32. Perdew, J.P.; Wang, Y. Accurate and simple analytic representation of the electron-gas correlation energy. *Phys. Rev. B* **1992**, *45*, 13244–13249. [CrossRef]
33. Perdew, J.P.; Burke, K.; Ernzerhof, M. Generalized Gradient Approximation Made Simple. *Phys. Rev. Lett.* **1996**, *77*, 3865. [CrossRef]
34. Perdew, J.P.; Burke, K.; Ernzerhof, M. Generalized Gradient Approximation Made Simple [Phys. Rev. Lett. 77, 3865 (1996)]. *Phys. Rev. Lett.* **1997**, *78*, 1396. [CrossRef]
35. Perdew, J.P.; Ruzsinszky, A.; Csonka, G.I.; Vydrov, O.A.; Scuseria, G.E.; Constantin, L.A.; Zhou, X.; Burke, K. Restoring the Density-Gradient Expansion for Exchange in Solids and Surfaces. *Phys. Rev. Lett.* **2008**, *100*, 136406. [CrossRef]
36. Wu, Z.; Cohen, R.E. More Accurate Generalized Gradient Approximation for Solids. *Phys. Rev. B* **2006**, *73*, 235116. [CrossRef]
37. Tran, F.; Laskowski, R.; Blaha, P.; Schwarz, K. Performance on molecules, surfaces, and solids of the Wu-Cohen GGA exchange-correlation energy functional. *Phys. Rev. B* **2007**, *75*, 115131. [CrossRef]
38. Tran, F.; Stelzl, J.; Blaha, P. Rungs 1 to 4 of DFT Jacob's ladder: Extensive test on the lattice constant, bulk modulus, and cohesive energy of solids. *J. Chem. Phys.* **2016**, *144*, 204120. [CrossRef] [PubMed]
39. Hidaka, M.; Ohama, N.; Okazaki, A.; Sakashita, H.; Yamakawa, S. A comment on the phase transitions in KMnF3. *Solid State Commun.* **1975**, *16*, 1121–1124. [CrossRef]
40. Karsai, F.; Tran, F.; Blaha, P. On the importance of local orbitals using second energy derivatives for d and f electrons. *Comput. Phys. Commun.* **2017**, *220*, 230–238. [CrossRef]
41. Charpin, T. *A Package for Calculating Elastic Tensors of Cubic Phase Using WIEN*; Laboratory of Geometrix: Paris, France, 2001.
42. Ivanov, Y.; Nimura, T.; Tanaka, K. Electron density and electrostatic potential of KMnF₃: A phase-transition study. *Acta Crystallogr. Sect. B* **2004**, *60*, 359–368. [CrossRef]
43. Hayatullah; Murtaza, G.; Khenata, R.; Muhammad, S.; Reshak, A.; Wong, K.M.; Omran, S.B.; Alahmed, Z. Structural, chemical bonding, electronic and magnetic properties of KMF3 (M = Mn, Fe, Co, Ni) compounds. *Comput. Mater. Sci.* **2014**, *85*, 402–408. [CrossRef]

Article

The Phase Transition and Dehydration in Epsomite under High Temperature and High Pressure

Linfei Yang [1,2], Lidong Dai [1,*], Heping Li [1], Haiying Hu [1], Meiling Hong [1,2] and Xinyu Zhang [1,2]

[1] Key Laboratory of High-Temperature and High-Pressure Study of the Earth's Interior,
    Institute of Geochemistry, Chinese Academy of Sciences, Guizhou 550081, Guiyang, China;
    yanglinfei@mail.gyig.ac.cn (L.Y.); hepingli_2007@hotmail.com (H.L.); huhaiying@mail.gyig.ac.cn (H.H.);
    hongmeiling@mail.gyig.ac.cn (M.H.); zhangxinyu@mail.gyig.ac.cn (X.Z.)
[2] University of Chinese Academy of Sciences, Beijing 100049, China
*   Correspondence: dailidong@vip.gyig.ac.cn

Received: 15 January 2020; Accepted: 28 January 2020; Published: 30 January 2020

Abstract: The phase stability of epsomite under a high temperature and high pressure were explored through Raman spectroscopy and electrical conductivity measurements in a diamond anvil cell up to ~623 K and ~12.8 GPa. Our results verified that the epsomite underwent a pressure-induced phase transition at ~5.1 GPa and room temperature, which was well characterized by the change in the pressure dependence of Raman vibrational modes and electrical conductivity. The dehydration process of the epsomite under high pressure was monitored by the variation in the sulfate tetrahedra and hydroxyl modes. At a representative pressure point of ~1.3 GPa, it was found the epsomite ($MgSO_4 \cdot 7H_2O$) started to dehydrate at ~343 K, by forming hexahydrite ($MgSO_4 \cdot 6H_2O$), and then further transformed into magnesium sulfate trihydrate ($MgSO_4 \cdot 3H_2O$) and anhydrous magnesium sulfate ($MgSO_4$) at higher temperatures of 373 and 473 K, respectively. Furthermore, the established *P-T* phase diagram revealed a positive relationship between the dehydration temperature and the pressure for epsomite.

Keywords: epsomite; phase transition; dehydration reaction; Raman spectra; electrical conductivity; high pressure

---

1. Introduction

In recent decades, hydrated sulfates have attracted a large amount of interest due to their great importance in exploring the interior structure of icy satellites, such as Europa, Ganymede and Callisto. It was widely reported that hydrated sulfates might be dominant minerals in the interior of these icy satellites, which was proved by the infrared spectral result collected from the Galileo spacecraft [1,2]. In addition, the discovery that anhydrous sulfates occur in carbonaceous chondritic meteorites also provides evidence for the existence of hydrated sulfates in the icy mantle of satellites, and these hydrated sulfates can be formed during the accretion of icy satellites [3]. In consideration of the high-pressure and high-temperature environment in the interior of these icy satellites, it is possible that these hydrated sulfates undergo a series of pressure-induced phase transitions and dehydration reactions. More importantly, as illustrated by Fortes and Choukroun [4], these transformations would have great impacts on the internal structure and heat transport in the icy mantle. As a representative magnesium (Mg)-bearing hydrated sulfate for epsomite ($MgSO_4 \cdot 7H_2O$), the investigation into its optical and electrical properties under a high pressure and temperature could help us to deeply understand the interior structure, composition and physical property the icy satellites.

Many previous works have been devoted to studying the phase stability of epsomite under a high pressure and room temperature, but their results showed great inconsistencies. Livshits et al. investigated the high-pressure behavior of epsomite and found four phase transitions, occurring at ~0.4,

~1.2, ~1.6 and ~2.5 GPa, respectively [5]. Gromnitskaya et al. employed ultrasonic and neutron scattering measurements on epsomite to reveal its elastic property under a high pressure of up to ~3.0 GPa, and three phase transitions were identified at pressures of ~1.4, ~1.6 and ~2.5 GPa, respectively [6]. Grasset et al. established the phase relation of $MgSO_4$–H_2O under a pressure of up to ~2.0 GPa in a diamond anvil cell (DAC), but they only determined one phase transition for epsomite at ~0.6 GPa [7]. The results from Nakamura et al. indicated that there were no any phase transitions in epsomite up to ~4.5 GPa [8]. Therefore, it is still unclear how epsomite behaves under a high pressure and room temperature, which requires more detailed experimental investigations. Raman spectroscopy and electrical conductivity measurements have been demonstrated to be efficient methods to characterize the pressure-induced structural variations in various hydrated sulfates. In view of this, systematic investigations on the optical and electrical properties of epsomite under a high pressure and room temperature will provide a good deal of insight into its high-pressure behaviors.

It is a common phenomenon for hydrated sulfates to undergo dehydration reactions to form lower hydrates under a high temperature and pressure. A great number of previous high-pressure studies were undertaken, mainly concerned with the dehydration process of gypsum, chalcanthite and blödite, and all of them were found to have a positive relationship between the dehydration temperature and pressure [9–13]. For example, Yang et al. found that the dehydration temperature of gypsum gradually increased with the rise in pressure, and the dehydration boundary between gypsum and bassanite was determined to be P (GPa) = −23.708 + 0.050T (K) [13]. As a similar hydrated sulfate, it is possible that the dehydration temperature of epsomite would also be significantly affected by pressure. However, to the best of our knowledge, there are no relative experimental reports on the issue of epsomite dehydration under a high pressure and temperature.

The present study clarified the phase transition and dehydration reaction for epsomite under a high pressure and temperature, using Raman scattering and electrical conductivity measurements in a diamond anvil cell (DAC). Furthermore, we determined a P-T phase diagram for epsomite, which is crucial to understanding the pressure effect on the dehydration reaction of hydrated sulfates, and also has great implications for modeling the internal structure of icy satellites.

2. Materials and Methods

Natural epsomite was used as the starting sample for all of the high-pressure experiments, which was obtained from a phosphorus-bearing rock series in Songlin town, Zunyi county, Guizhou province. The high pressure and room temperature experiments were implemented in a symmetric diamond anvil cell (DAC) with a 300 μm anvil culet. In the high-temperature and high-pressure experiments, we placed an external resistive heater around the diamond to achieve the high temperature conditions. A K-type thermocouple wire was directly pasted onto the diamond surface in order to measure the temperature in the pressure chamber. A classic ruby pressure calibration formula from Mao et al. was adopted to accurately calculate the pressure in the sample chamber, based on the shifts of R_1 fluorescence lines [14]. Taking into account the inevitable influence of the high temperature on the pressure conditions in the sample chamber, the high temperature-corrected equation of pressure calibration was selected in the process of the high-temperature and high-pressure measurements [15]. The Raman spectra were collected by a micro-confocal Raman spectrometer combined with a 514.5 nm argon-ion laser as the excitation source. Helium was selected as the pressure medium to provide the hydrostatic environment in the sample chamber. The resolution of the Raman spectrometer and the repetition rate were 1 and 0.2 cm^{-1}, respectively. Each Raman spectra was obtained in the wave number range of 900–1200 cm^{-1} for the internal vibration modes of the sulfate and 3000–3800 cm^{-1} for the water molecular stretching modes. The acquisition time for each Raman spectra was set to 120 s. As for electrical conductivity experiments, a Solartron-1260 impedance/gain phase analyzer was used in this study to acquire the AC impedance spectrum. No pressure medium was used in order to avoid the introduction of impurity substances. The laser drilling technique was used to drill a 200 μm hole in the pre-indented T-301 stainless steel gasket, with a thickness of 50 μm. A mixture of boron nitride (BN) and epoxy was then compressed into the hole as the insulator, and another hole of 100 μm was drilled to provide an insulating sample chamber. More specific descriptions of the measurement procedure have been reported in previous works [16–18].

3. Results and Discussion

Under ambient conditions, three Raman peaks are observed to be located at positions of 983.3, 3322.6 and 3458.4 cm^{-1}, respectively. The sharp peak at 983.3 cm^{-1} is assigned to the v_1 symmetric stretching mode of the sulfate tetrahedra, and the other two bands at 3322.6 and 3458.4 cm^{-1} are ascribed to the v_1 and v_3 vibrational modes of water molecules, respectively. All of these acquired Raman modes for the epsomite are generally consistent with the previous data by Wang et al. [19].

Figure 1a displays the high-pressure Raman spectra of the epsomite in the SO$_4$ vibrational mode range at pressures of 0–12.8 GPa and room temperature. The corresponding pressure shifts for the v_1 (SO$_4$) mode are plotted in Figure 1b, from which one obvious inflection point can be obtained, at a pressure of ~5.1 GPa. From ~0 to ~5.1 GPa, the vibrational mode of v_1 (SO$_4$) slightly shifts towards the higher wave numbers, with a slope of 0.16 cm^{-1} GPa^{-1}. However, upon further compression from ~5.1 to ~12.8 GPa, the v_1 (SO$_4$) mode exhibits a strong pressure dependence with a relatively high slope value of 4.46 cm^{-1} GPa^{-1}. In the meantime, it can be seen that the full width at half maximum (FWHM) of the v_1 (SO$_4$) peak also shows a discontinuous change at the same pressure point of ~5.1 GPa. As shown in Figure 1c, the FWHM of the v_1 (SO$_4$) band remains at an almost constant value of about 8.3 cm^{-1} below ~5.1 GPa, but it gruadally increases with pressure above ~5.1 GPa. Figure 2 shows the Raman spectra of the epsomite in the OH-stretching mode range under a high pressure and room temperature. Obviously, both of the two OH-stretching peaks at 3322.6 and 3458.4 cm^{-1} suddenly shift to lower frequencies above ~5.1 GPa. Upon decompression, the Raman spectra for the recovered epsomite exhibit a similar characterization to those of the starting sample.

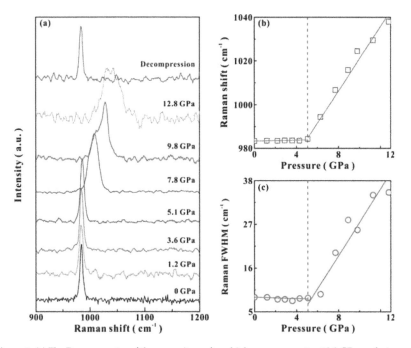

Figure 1. (a) The Raman spectra of the epsomite under a high pressure, up to ~12.8 GPa, and at room temperature, collected in the frequency range of 900–1200 cm^{-1}; (b) the pressure dependence of the

Raman shift for the ν_1 (SO$_4$) mode; (c) the full width at half maximum (FWHM) of the ν_1 (SO$_4$) peak as a function of the pressure.

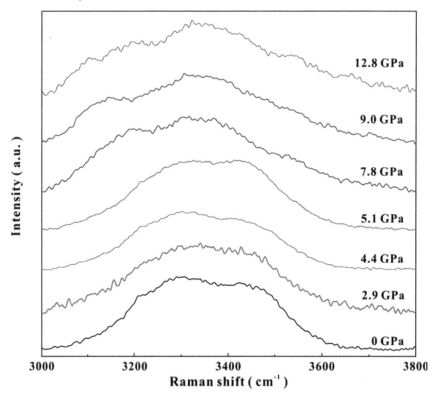

Figure 2. The high-pressure Raman spectra of the epsomite at room temperature in the range of 3000–3800 cm^{-1}.

In this study, we determined a pressure-induced structural phase transition in epsomite at ~5.1 GPa and at room temperature, based on the variation in Raman modes of SO$_4$ tetrahedra and water molecules. The recovery of the Raman spectra after decompression reflects that this pressure-induced phase transformation is reversible. As for the nature of this structural phase transition, no evidence shows that it is associated with the dehydration process of epsomite, because the Raman spectra obtained in the whole pressure range are not in agreement with the dehydration product of the epsomite. We think that this is possibly related to the distortion of the sulfate tetrahedra and the strengthening of the hydrogen bonding. A recent work by Gonzalez–Platas et al. revealed that the chalcomenite underwent an isostructural phase transition at around ~4.0 GPa, which is very close to the transition point of epsomite that was revealed in this study (~5.1 GPa) [20]. The similarities between the transition pressures are closely associated with their crystal structure, because both of them have an orthorhombic structure with the same space group of $P2_12_12_1$. In addition, other hydrated sulfates have also been revealed to undergo phase transitions under a high pressure and room temperature, such as gypsum, chalcanthite and mirabilite [12,21–23]. This implies that the pressure can play a very important role in tuning the structural and vibrational properties of most hydrated sulfates.

In order to further check the structural phase transition in epsomite, high-pressure electrical conductivity experiments were carried out in the pressure range of 0.5–12.4 GPa and under room temperature. Figure 3a presents the collected impedance spectra of epsomite in the frequency range from 10^{-1} to 10^7 Hz and at a series of pressure points, and the horizontal and vertical axes represent the real and imaginary parts of the complex impedance, respectively. Each spectra datum is composed of two semicircular arcs, which stand for grain interior and boundary contributions, respectively. All of these displayed impedance spectra were fitted using the equivalent circuit method in the Z-View software to obtain the grain interior resistance of the sample. The equivalent circuit was composed of a resistor (R) and a constant phase element (CPE) in parallel. The electrical conductivity for the grain interior contribution of sample was then calculated by the formula as follows:

$$\sigma = L/SR$$

where σ stands for the electrical conductivity, L is the distance between two electrodes, S represents the electrode and the R denotes the fitting resistance. As shown in Figure 3b, according to the obvious change in the slope of electrical conductivity, this diagram can be divided into two distinct regions: (i) in the pressure range of 0.5–4.6 GPa, the electrical conductivity of the epsomite monotonously increases with pressure, and the corresponding pressure coefficient is determined to be 0.06 S cm^{-1} GPa^{-1}; (ii) from ~4.6 to ~12.4 GPa, its electrical conductivity still shows an increasing tendency but with a smaller slope value of ~0.04 S cm^{-1} GPa^{-1}. The abrupt change in resistivity indicates the occurrence of a phase transition [24]. This transition point at ~4.6 GPa obtained from electrical conductivity measurements is slightly lower than the data from our high-pressure Raman spectroscopic experiments, which is possibly caused by the different pressure environment in the sample chamber. In comparison with the hydrostatic environment for the Raman measurements, it is non-hydrostatic for electrical conductivity experiments. There exists relatively high deviatoric stress in the non-hydrostatic compression, which has been reported to significantly influence the high-pressure behavior of materials [25]. Our electrical conductivity results provide another critical clue to support the occurrence of a pressure-induced phase transition in epsomite.

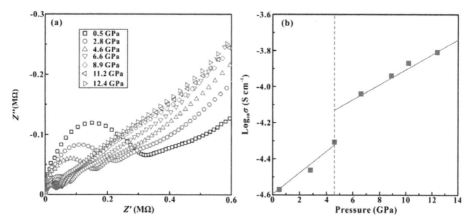

Figure 3. (**a**) The typical impedance spectra for the epsomite, measured at 0.5–12.4 GPa and room temperature; (**b**) the calculated electrical conductivity of the epsomite with increasing pressure.

To reveal the effect of the high pressure on the dehydration process of epsomite, we in situ measured a series of Raman spectra of the epsomite under high temperatures of up to ~623 K and at constant pressure points of ~0.8, ~1.3, ~3.7 and ~6.4 GPa. Figure 4 presents the Raman spectra for the epsomite in the temperature range of 293–573 K and at a pressure of ~1.3 GPa. Characterized by the variation in the temperature dependence of the sulfate group and hydroxyl modes, three important temperature points were well-determined at ~343, ~373 and ~473 K, respectively. At ~343 K, it can be clearly observed from Figure 4a that the v_1 (SO$_4$) peak at 983.3 cm^{-1} abruptly shifted to a lower frequency of 982.4 cm^{-1}. Moreover, we also detected some obvious variations in the two OH-stretching modes, which were changed into one mode at a position of ~3441.5 cm^{-1} (Figure 4b). These observed new peaks at ~982.4 and ~3441.5 cm^{-1} agreed well with the position of the characteristic Raman vibrational mode of hexahydrite (MgSO$_4$·6H$_2$O) and, hence, we think that the epsomite initially dehydrated to a new hydrous phase hexahydrite at ~343 K. At a higher temperature of ~373 K, the v_1 (SO$_4$) peaks suddenly broadened and moved to a higher wave number of 1023.2 cm^{-1}, meanwhile, two new OH-stretching modes also started to appear at positions of 3153.03 and 3341.8 cm^{-1}, respectively. The Raman spectra obtained at ~373 K exhibited similar peak characteristics to those of magnesium sulfate trihydrate (MgSO$_4$·3H$_2$O) [19]. Therefore, this provided robust evidence for the occurrence of another dehydration reaction, from hexahydrite to a new trihydrate phase. When the temperature was further enhanced, up to ~473 K, the complete dehydration of epsomite to form anhydrous magnesium sulfate (MgSO$_4$) was finally observed by the appearance of a new peak at 1022.9 cm^{-1} and the vanishment of OH-stretching bands [19].

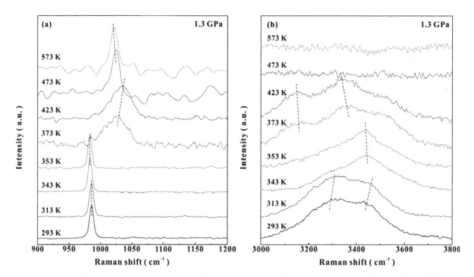

Figure 4. (**a**) The temperature-dependent Raman spectra of the epsomite at ~1.3 GPa for the sulfate internal vibration; (**b**) the high-temperature Raman spectra of the epsomite at ~1.3 GPa for the hydroxyl-stretching vibration.

Figure 5 shows the temperature-dependent Raman spectra of epsomite collected at conditions of 293–623 K and at a higher pressure of ~6.3 GPa. At temperatures below ~553 K, the mode of v_1 (SO$_4$) remained at an almost constant wave number, with an increasing temperature (Figure 5a). However, when the temperature was above ~553 K, this mode suddenly split into two new Raman peaks at 997.4 and 1020.0 cm^{-1}. In addition, the Raman spectra for the OH-stretching modes were also greatly changed after ~553 K, as shown in Figure 5b. These discontinuous variations in the Raman modes of the sulfate tetrahedra and water molecules are associated with the dehydration reaction from the epsomite to magnesium sulfate trihydrate (MgSO$_4$·3H$_2$O). The high-temperature Raman spectra of the epsomite at another two pressure points of ~0.8 and ~3.7 GPa are displayed in Figure 6. At ~0.8 GPa, the epsomite was found to undergo thermal dehydration reactions at temperatures of ~313, ~353 and ~423 K, respectively. For the dehydration process at ~3.7 GPa, the epsomite initially dehydrated to hexahydrite at ~373 K, and then transformed to magnesium sulfate trihydrate at ~453 K.

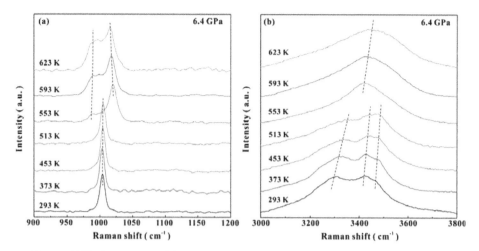

Figure 5. (a) The Raman spectra of the epsomite at 293–623 K and ~6.4 GPa in the sulfate vibrational range; (b) the collected high-temperature Raman spectra of the epsomite at ~6.4 GPa in the hydroxyl-stretching vibrational range.

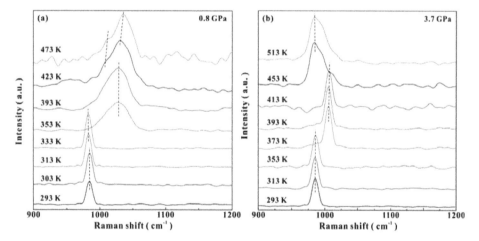

Figure 6. (a) The high-temperature Raman spectra of the epsomite at a pressure of ~0.8 GPa; (b) the Raman spectra of the epsomite with an increasing temperature at ~3.7 GPa.

On the basis of the data of dehydration temperatures obtained from the high-temperature and high-pressure Raman scattering experiments, a *P-T* phase diagram for epsomite was well established at a wide temperature range of 293–623 K and at pressures of 0.8–6.3 GPa. As shown in Figure 7, it is verified that the dehydration temperature of epsomite gradually increased with the rise in pressure. This positive relation between dehydration temperature and pressure was coincident with the result of other similar hydrated sulfates (gypsum, chalcanthite and blödite) [11–13]. In addition, we also found that the epsomite undergoes three-step dehydration reactions in the heating process: from epsomite ($MgSO_4 \cdot 7H_2O$) to hexahydrite ($MgSO_4 \cdot 6H_2O$) to magnesium sulfate trihydrate ($MgSO_4 \cdot 3H_2O$) to anhydrous magnesium sulfate ($MgSO_4$). The anhydrous $MgSO_4$ has been studied and it is stable up to 17.5 GPa [26]. The dehydration sequence of the epsomite revealed in this study is different from the result reported by Brotton et al. [27]. They observed the direct formation of magnesium sulfate trihydrate without the occurrence of hexahydrite in the heated epsomite at room pressure.

This discrepancy was possibly caused by some crucial factors, such as the heating rate, relative humidity and pressure condition, all of which have been demonstrated to significantly influence the dehydration process of hydrous minerals [9,28–30]. Furthermore, three dehydration boundaries were determined by the linear fitting of these data. The epsomite-hexahydrite dehydration boundary was determined to be P (GPa) = −14.645 + 0.048 T (K), and the dehydration boundary for the transformation from hexahydrite to magnesium sulfate trihydrate corresponded to P (GPa) = −9.172 + 0.028 T (K). As for the last dehydration reaction, from magnesium sulfate trihydrate to anhydrous magnesium sulfate, the corresponding boundary can be described by a linear equation: P (GPa) = −3.430 + 0.010 T (K).

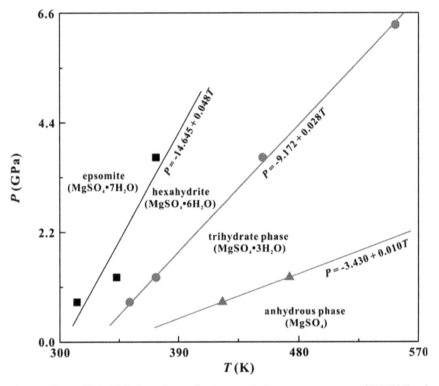

Figure 7. The established *P-T* phase diagram for the epsomite in temperature ranges of 293-623 K and pressure ranges of 0.8-6.4 GPa.

4. Conclusions

In summary, we performed in situ Raman spectroscopic and electrical conductivity measurements on epsomite to investigate its phase transition and dehydration process under high pressure. The results from room-temperature and high-pressure experiments demonstrated a structural phase transition for the epsomite, occurring at ~5.1 GPa, and the hydrogen bonding in the structure of the epsomite is strengthened after this transformation. The high-temperature and high-pressure Raman measurements were conducted at four pressure points (~0.8, ~1.3, ~3.7 and ~6.4 GPa) for the purpose of revealing the influence of pressure on the dehydration process of the epsomite. In the heating process of the epsomite at ~1.3 GPa, we successively observed the appearance of hexahydrite, magnesium sulfate trihydrate and anhydrous magnesium sulfate at ~343, ~373 and ~473 K. At ~6.4 GPa, the epsomite dehydrated to magnesium sulfate trihydrate at ~553 K. Therefore, it can be concluded that the epsomite undergoes a three-step dehydration reaction under a high pressure, and its dehydration temperature gradually increases with the pressure.

Author Contributions: Conceptualization, L.D. and L.Y.; methodology, L.D., Y.H. and H.L.; software, L.D.; validation, L.Y., M.H. and X.Z.; formal analysis, L.Y. and L.D.; investigation, L.D.; data curation, L.Y.; writing—original draft preparation, L.Y. and L.D.; writing—review and editing, L.D.; visualization, L.Y.; supervision, L.D., Y.H. and H.L.; project administration, L.D.; funding acquisition, L.D. Methodology, supervison, H.H. All authors have read and agreed to the published version of the manuscript.

Funding: This research was funded by the strategic priority Research Program (B) of the Chinese Academy of Sciences (18010401), Key Research Program of Frontier Sciences of CAS (QYZDB-SSW-DQC009), Hundred Talents Program of CAS, NSF of China (41774099 and 41772042), Youth Innovation Promotion Association of CAS (2019390), Special Fund of the West Light Foundation of CAS, and Postdoctoral Science Foundation of China (2018M643532).

Conflicts of Interest: The authors declare no conflict of interest.

References

1. McCord, T.B.; Hansen, G.B.; Matson, D.L.; Johnson, T.V.; Crowley, J.K.; Fanale, F.P.; Carlson, R.W.; Smythe, W.D.; Martin, P.D.; Hibbitts, C.A.; et al. Hydrated salt minerals on Europa's surface from the Galileo near-infrared mapping spectrometer (NIMS) investigation. *J. Geophys. Res.: Planet* **1999**, *104*, 11827–11851. [CrossRef]

2. McCord, T.B.; Hansen, G.B.; Hibbitts, C.A. Hydrated salt minerals on Ganymede's surface: Evidence of an ocean below. *Nature* **2001**, *5521*, 1523–1525. [CrossRef] [PubMed]

3. Fredriksson, K.; Kerridge, J.F. Carbonates and sulfates in CI chondrites: Formation by aqueous activity on the parent body. *Meteoritics* **1988**, *23*, 35–44. [CrossRef] [PubMed]

4. Fortes, A.D.; Choukroun, M. Phase behaviour of ices and hydrates. *Space Sci. Rew.* **2010**, *153*, 185–218. [CrossRef]

5. Livshits, L.D.; Genshaft, Y.S.; Ryabin, Y.N. Equilibrium diagram of the crystal hydrates of $MgSO_4$ at high pressure. *Russ. J. Inorg. Chem.* **1963**, *8*, 676–678.

6. Gromnitskaya, E.L.; Yagafarov, O.F.; Lyapin, A.G.; Brazhkin, V.V.; Wood, I.G.; Tucker, M.G.; Fortes, A.D. The high-pressure phase diagram of synthetic epsomite ($MgSO_4 \cdot 7H_2O$ and $MgSO_4 \cdot 7D_2O$) from ultrasonic and neutron powder diffraction measurements. *Phys. Chem. Miner.* **2013**, *40*, 271–285. [CrossRef]

7. Grasset, O.; Sotin, C.; Mousis, O.; Mevel, L. High pressure experiments in the system $MgSO_4$–H_2O: Implications for Europa. *Lunar Planet Sci. Conf.* **2000**, *31*, 1386.

8. Nakamura, R.; Ohtani, E. The high-pressure phase relation of the $MgSO_4$–H_2O system and its implication for the internal structure of Ganymede. *Icarus* **2011**, *211*, 648–654. [CrossRef]

9. Mirwald, P.W. Experimental study of the dehydration reactions gypsum-bassanite and bassanite-anhydrite at high pressure: Indication of anomalous behavior of H_2O at high pressure in the temperature range of 50–300 °C. *J. Chem. Phys.* **2008**, *128*, 074502. [CrossRef]

10. Comodi, P.; Kurnosov, A.; Nazzareni, S.; Dubrovinsky, L. The dehydration process of gypsum under high pressure. *Phys. Chem. Miner.* **2012**, *39*, 65–71. [CrossRef]

11. Comodi, P.; Stagno, V.; Zucchini, A.; Fei, Y.; Prakapenka, V. The compression behavior of blödite at low and high temperature up to ~10 GPa: Implications for the stability of hydrous sulfates on icy planetary bodies. *Icarus* **2017**, *285*, 137–144. [CrossRef]

12. Pu, C.; Dai, L.; Li, H.; Hu, H.; Zhuang, Y.; Liu, K.; Yang, L.; Hong, M. High–pressure electrical conductivity and Raman spectroscopy of chalcanthite. *Spectrosc. Lett.* **2018**, *51*, 531–539. [CrossRef]

13. Yang, L.; Dai, L.; Li, H.; Hu, H.; Zhuang, Y.; Liu, K.; Pu, C.; Hong, M. Pressure-induced structural phase transition and dehydration for gypsum investigated by Raman spectroscopy and electrical conductivity. *Chem. Phys. Lett.* **2018**, *706*, 151–157. [CrossRef]

14. Mao, H.K.; Xu, J.; Bell, P.M. Calibration of the ruby pressure gauge to 800 kbar under quasi-hydrostatic conditions. *J. Geophys. Res.* **1986**, *91*, 4673. [CrossRef]

15. Rekhi, S.; Dubrovinsky, L.S.; Saxena, S.K. Temperature-induced ruby fluorescence shifts up to a pressure of 15 GPa in an externally heated diamond anvil cell. *High Temp.-High Press.* **1999**, *31*, 299–305. [CrossRef]

16. Dai, L.; Zhuang, Y.; Li, H.; Wu, L.; Hu, H.; Liu, K.; Yang, L.; Pu, C. Pressure-induced irreversible amorphization and metallization with a structural phase transition in arsenic telluride. *J. Matter. Chem. C* **2017**, *5*, 12157–12162. [CrossRef]

17. Dai, L.; Liu, K.; Li, H.; Wu, L.; Hu, H.; Zhuang, Y.; Yang, L.; Pu, C.; Liu, P. Pressure-induced irreversible metallization accompanying the phase transitions in Sb_2S_3. *Phys. Rev. B* **2018**, *97*, 024103. [CrossRef]

18. Dai, L.; Pu, C.; Li, H.; Hu, H.; Liu, K.; Yang, L.; Hong, M. Characterization of metallization and amorphization for GaP under different hydrostatic environments in diamond anvil cell up to 40.0 GPa. *Rev. Sci. Instrum.* **2019**, *90*, 066103. [CrossRef]

19. Wang, A.; Freeman, J.J.; Jolliff, B.L.; Chou, I.M. Sulfates on Mars: A systematic Raman spectroscopic study of hydration states of magnesium sulfates. *Geochim. Cosmochim. Ac.* **2006**, *70*, 6118–6135. [CrossRef]

20. Gonzalez-Platas, J.; Rodriguez-Hernandez, P.; Muñoz, A.; Rodríguez-Mendoza, U.R.; Nénert, G.; Errandonea, D. A high-pressure investigation of the synthetic analogue of chalcomenite, $CuSeO_3 \cdot 2H_2O$. *Crystals* **2019**, *9*, 643. [CrossRef]

21. Huang, E.; Xu, J.A.; Lin, J.F.; Hu, J.Z. Pressure-induced phase transitions in gypsum. *High Pressure Res.* **2000**, *17*, 57–75. [CrossRef]

22. Knittle, E.; Phillips, W.; Williams, Q. An infrared and Raman spectroscopic study of gypsum at high pressures. *Phys. Chem. Miner.* **2001**, *28*, 630–640. [CrossRef]

23. Brand, H.E.A.; Fortes, A.D.; Wood, I.G.; Vočadlo, L. Equation of state and pressure-induced structural changes in mirabilite ($Na_2SO4 \cdot 10H_2O$) determined from ab initio density functional theory calculations. *Phys. Chem. Miner.* **2009**, *37*, 265–282. [CrossRef]

24. Errandonea, D.; Segura, A.; Martínez-García, D.; Muñoz-San Jose, V. Hall-effect and resistivity measurements in CdTe and ZnTe at high pressure: Electronic structure of impurities in the zinc-blende phase and the semimetallic or metallic character of the high-pressure phases. *Phys. Rev. B* **2009**, *79*, 125203. [CrossRef]

25. Errandonea, D.; Muñoz, A.; Gonzalez-Platas, J. Comment on "High-pressure x-ray diffraction study of YBO3/Eu3+, GdBO3, and EuBO3: Pressure-induced amorphization in GdBO3" [J. Appl. Phys. 115, 043507 (2014)]. *J. Appl. Phys.* **2014**, *115*, 216101. [CrossRef]

26. Benmakhlouf, A.; Errandonea, D.; Bouchenafa, M.; Maabed, S.; Bouhemadou, A.; Bentabet, A. New pressure-induced polymorphic transitions of anhydrous magnesium sulfate. *Dalton Trans.* **2017**, *46*, 5058–5068. [CrossRef]

27. Brotton, S.J.; Kaiser, R.I. In situ Raman spectroscopic study of gypsum ($CaSO_4 \cdot 2H_2O$) and epsomite ($MgSO_4 \cdot 7H_2O$) dehydration utilizing an ultrasonic levitator. *J. Phys. Chem. Lett.* **2013**, *4*, 669–673. [CrossRef]

28. Cardell, C.; Sánchez-Navas, A.; Olmo-Reyes, F.J.; Martín-Ramos, J.D. Powder X-ray thermodiffraction study of mirabilite and epsomite dehydration. Effects of direct IR-irradiation on samples. *Anan. Chem.* **2007**, *79*, 4455–4462. [CrossRef]

29. Donkers, P.A.J.; Linnow, K.; Pel, L.; Steiger, M.; Adan, O.C.G. $NaSO_4 \cdot 10H_2O$ dehydration in view of thermal storage. *Chem. Eng. Sci.* **2015**, *134*, 360–366. [CrossRef]

30. Nagase, K.; Yokobayashi, H.; Kikuchi, M.; Sone, K. Effects of heating rate (1–300°h^{-1}) on the non-isothermal thermogravimetry of $CuSO_4 \cdot 5H_2O$. *Thermochim. Acta* **1980**, *35*, 99–104. [CrossRef]

Article

Topological Equivalence of the Phase Diagrams of Molybdenum and Tungsten

Samuel Baty [1,†], **Leonid Burakovsky [2,*,†]** and **Dean Preston [3,†]**

1   Analytics Intelligence and Technology, Los Alamos National Laboratory, Los Alamos, NM 87545, USA;
    srbaty@lanl.gov
2   Theoretical Division, Los Alamos National Laboratory, Los Alamos, NM 87545, USA
3   Computational Physics Division, Los Alamos National Laboratory, Los Alamos, NM 87545, USA;
    dean@lanl.gov
*   Correspondence: burakov@lanl.gov; Tel.: +1-505-667-5222
†   These authors contributed equally to this work.

Received: 15 November 2019; Accepted: 26 December 2019; Published: 2 January 2020

Abstract: We demonstrate the topological equivalence of the phase diagrams of molybdenum (Mo) and tungsten (W), Group 6B partners in the periodic table. The phase digram of Mo to 800 GPa from our earlier work is now extended to 2000 GPa. The phase diagram of W to 2500 GPa is obtained using a comprehensive ab initio approach that includes (i) the calculation of the $T = 0$ free energies (enthalpies) of different solid structures, (ii) the quantum molecular dynamics simulation of the melting curves of different solid structures, (iii) the derivation of the analytic form for the solid–solid phase transition boundary, and (iv) the simulations of the solidification of liquid W into the final solid states on both sides of the solid–solid phase transition boundary in order to confirm the corresponding analytic form. For both Mo and W, there are two solid structures confirmed to be present on their phase diagrams, the ambient body-centered cubic (bcc) and the high-pressure double hexagonal close-packed (dhcp), such that at $T = 0$ the bcc–dhcp transition occurs at 660 GPa in Mo and 1060 GPa in W. In either case, the transition boundary has a positive slope dT/dP.

Keywords: phase diagram; quantum molecular dynamics; melting curve; Z methodology; multi-phase materials

---

1. Introduction

The high-pressure (HP) and high-temperature (HT) behavior of materials is of great importance for condensed matter physics and geophysics as well as for technological applications. In particular, the study of melting under extreme conditions is a subject of special interest. Among other topics, the HP melting of transition metals has been one of the main areas of research. Metals like tantalum (Ta), molybdenum (Mo), and nickel (Ni) have been studied both theoretically [1–10] and experimentally [11–25]. Research interest in HP-HT polymorphism in the body-centered cubic (bcc) transition metals has reemerged in connection with laser-heated diamond anvil cell (DAC) melting experiments in which melting curves with a small slope (dT/dP, where T stands for temperature and P stands for pressure) in the megabar pressure range have been determined [11–18]. These flat melting curves do not agree with the results of more recent experiments [19–25] and calculations [2,4,6,8–10]. We note that, in all the aforementioned experiments, P and T did not exceed 200 GPa and 5000 K, respectively. A pressure of $\sim$200 GPa was reached in the experimental study of Reference [21] and a temperature of $\sim$5000 K in those of both Reference [21] and Reference [24].

Several hypotheses have been proposed to explain apparent discrepancies between experiment and theory with regard to the flat melting curves. One of them suggests that the flat melting curves could in fact correspond to solid–solid (s–s) phase boundaries that occur under HP-HT below melting.

In particular, HP-HT solid–solid transitions have been predicted to take place in Ta [3,5,9] and Mo [1]. In contrast to these metals, there are only a few studies on tungsten (W). The experimental melting curve of W is one of the flat ones [13]; it has the initial slope of ~7 K/GPa and reaches ~4000 K at ~100 GPa, which contradicts both the initial slope of 44 K/GPa from isobaric expansion measurements [26] and the shock Hugoniot melting points of ~12,000 K at ~400 GPa [27]. The experimental melting curves of References [28–30] have even higher slopes of ~90, 75, and 60 K/GPa, respectively. As noted in Reference [30], large experimental errors of Reference [31] lead to ~50% error in the value of the corresponding slope, 70 ± 35 K/GPa. The other theoretical value of the slope is 28.7 K/GPa [32,33].

It is well known that some physical properties related to the phase diagrams of elements can be systematized across the corresponding groups or rows (or both) of the periodic table. Among those are the ambient melting point $T_m(0)$; the initial slopes of melting curve $dT_m/dP(0)$; and the ambient volume change at melt, $\Delta V_m(0)$, and latent heat of melting, $\Delta H_m(0)$. For example, elements of group 7B, Fe, Ru, and Os, exhibit a very clear systematics with regard to $T_m(0)$ and $dT_m/dP(0)$: $T_m(0)$ is, respectively, (in K) 1811, 2607, and ~3400 [34], and $dT_m/dP(0)$ is (in K/GPa) 30.0 [21], 40.7 ± 2.2 [35], and 49.5 [36]. That is, $T_m(0)$ is separated by ~800 K and $dT_m/dP(0)$ is separated by ~10 K/GPa. The analytic forms of the corresponding melting curves can also be systematized with regard to the numerical parameters involved: $1811 (1 + P/30.2)^{0.50}$ [21], $2607 (1 + P/(32.75 \pm 1.75))^{0.51}$ [35], and $3370 (1 + P/36.1)^{0.53}$ [36]. The proper knowledge of such systematics and similarities in material properties allows one to make predictions and to validate new results. For instance, the melting curve systematics for the elements of the third row of the periodic table discussed in Reference [10] allowed to predict the yet unknown melting curve of Hf, which the preliminary theoretical results seem to confirm [10].

The phase diagrams themselves can be topologically similar (look-alike) or, in some cases, even topologically equivalent. The topological similarity of the phase diagrams of Ti, Zr, and Hf is well known (more detail will follow). The phase diagrams of Si and Ge are topologically equivalent at low P : both contain semiconducting diamond and metallic tetragonal β-Sn solid structures, and the slopes of the corresponding solid–solid and solid–liquid phase boundaries are almost identical [37]; in either case, $dT_m/dP(0) < 0$. The proper knowledge of similarities of the phase diagrams can be useful in making predictions with regard to the phase diagram content and in offering suggestions as to what solid structures to look for in high-PT experiments.

The study of Frohberg [38] reveals that, for the elements of the three groups 4B, 5B, and 6B, namely, Ti, Zr, and Hf; V, Nb, and Ta; and Cr, Mo, and W, respectively, the corresponding $T_m(0)$ and $\Delta H_m(0)$ obey very clear systematics. In addition, the phase diagrams of 4B's Ti, Zr, and Hf are topologically similar; their only difference is in the slopes of the solid–solid transition boundaries between the three solid structures found on each of the three phase diagrams: hexagonal close-packed (hcp), hexagonal omega (hex-omega), and bcc. Since the phase diagrams of 5B's Nb and Ta may also be topologically similar, as we discuss below, it is natural to assume the topological similarity of the phase diagrams of Mo and W since they both belong to 6B, the last of these three groups.

Here, we present an extensive theoretical study of the phase diagram of W to a pressure of 2500 GPa (25 Mbar). We demonstrate that the phase diagram of W is topologically equivalent to that of Mo; the latter was considered in our previous publication [39] and is now extended to 2000 GPa (20 Mbar). As it turns out, the critical features of the two phase diagrams that make them equivalent, that is, the location of the solid–solid–liquid triple point and the analytic form of the solid–solid phase boundary, are revealed at much higher pressure that goes far beyond the P and T ranges of ~0–200 GPa and ~0–5000 K, respectively, that all the experimental studies on transition metals mentioned above have covered. Specifically, the solid–solid phase boundary between the two solid phases that are identical in both cases lies in a P range of ~650–1050 GPa for Mo and of ~1050–1700 GPa for W. The corresponding T ranges are, respectively, ~0–12,300 and ~0–23,700 K. We therefore expect that, in the near future, our findings may not be verified in experiment; they will remain theoretical predictions until pressures of ~1000 GPa (10 Mbar) and temperatures of ~10,000 K become experimentally attainable.

2. The Phase Diagram of Mo

The construction of the phase diagram of Mo using state-of-the-art theoretical techniques was discussed in detail in Reference [39], where we presented the phase diagram of Mo to 800 GPa. Here, we extend this phase diagram to 2000 GPa by calculating the melting curve of the double hexagonal close-packed (dhcp) structure of Mo, which is the stable solid phase of Mo at $P \geq 660$ GPa [39], and by determining the solid–solid phase transition boundary between the body-centered cubic (bcc) structure, which is the stable solid phase of Mo at $P \leq 660$ GPa, and the dhcp structure. Our results can be summarized as follows:

(i) the melting curve of bcc-Mo in the Simon–Glatzel form [40] (T_m in K, P in GPa):

$$T_m(P) = 2896 \left(1 + \frac{P}{36.6} \right)^{0.43},$$ (1)

(ii) the melting curve of dhcp-Mo in the Simon–Glatzel form (T_m in K, P in GPa):

$$T_m(P) = 1880 \left(1 + \frac{P}{24.5} \right)^{0.50},$$ (2)

Our melting curve of bcc-Mo, Equation (1), is in excellent agreement with another theoretical calculation, $T_m(P) = 2894(1 + P/37.2)^{0.433}$ [41]. For Equation (1), $dT_m/dP|_{P=0} = 34.0$ K/GPa, which is in excellent agreement with 32 K/GPa [26] or 34 ± 6 K/GPa [42], each from isobaric expansion measurements.

Figures 1 and 2 demonstrate the time evolution of T and P, respectively, in the Z method runs of the bcc-Mo melting point (P in GPa, T in K) $(P, T) = (1185, 12{,}770)$, which is one of the melting points that we calculated in the course of the present study. Figures 3 and 4 demonstrate the same for the dhcp-Mo melting point (1186, 13,430). These two points are chosen as examples and are shown in Figure 5 as open blue and green circles, respectively.

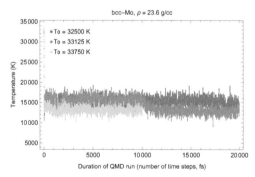

Figure 1. Time evolution of temperature for body-centred cubic (bcc) Mo in three QMD runs with initial temperatures (T_0s) separated by 625 K: The middle run is the melting run, during which T decreases from ~17,000 K for the superheated state to ~12,800 K for the liquid at the corresponding melting point.

The melting curves of bcc-Mo and dhcp-Mo cross each other at (P in GPa, T in K) $(P, T) = (1045, 12{,}320)$ which is the bcc–dhcp-liquid triple point. The choice of $T(P) = a(660 - P)^b$, $0 < b < 1$ as a functional form for the bcc–dhcp phase boundary leads to

(iii) the bcc–dhcp solid–solid phase transition boundary:

$$T(P) = 58.5 \, (P - 660)^{0.90},$$ (3)

which crosses the triple point as well as the point $(P, T) = (725, 2500)$ that comes from the solidification simulations using the inverse Z method; for more detail see Reference [39].

The phase diagram of Mo to 2000 GPa is shown in Figure 5. It includes the two melting curves, the solid–solid phase boundary, and the results of the solidification of liquid Mo into the final states of either solid bcc or solid dhcp using the inverse Z method. Figure 5 actually represents the extended version of Figure 11 of Reference [39] which covered a P range of 0–800 GPa.

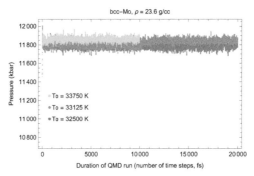

Figure 2. The same as in Figure 1 for the time evolution of pressure (in kbar; 10 kbar = 1 GPa): During melting, P increases from ~1180 GPa for the superheated state to ~1185 GPa for the liquid state at the corresponding melting point.

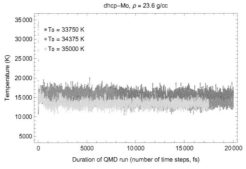

Figure 3. The same as in Figure 1 for double hexagonal close-packed (dhcp) Mo.

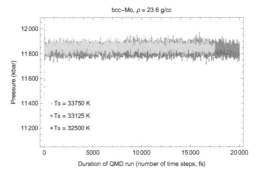

Figure 4. The same as in Figure 2 for dhcp-Mo.

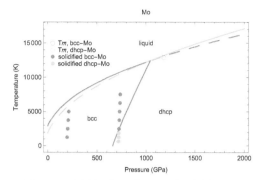

Figure 5. Ab initio phase diagram of molybdenum.

3. The Phase Diagram of W

Although the phase diagram of W has not been known in detail, there are several features of this phase diagram that have been firmly established. First, melting on the shock Hugoniot of W occurs at ~400 GPa and presumably at T ~12,000 K [27]; these values are extracted from the corresponding $P = P(U_p)$ and $T = T(U_p)$ dependences on the particle velocity U_p as those at $U_p \sim 2.5$ km/s at which melting occurs on the shock Hugoniot of W [27]. Second, the stability of bcc-W has been confirmed experimentally under room-T isothermal compression to 420 GPa [43] and to 500 GPa [44]. There is presently no experimental evidence for an s–s transition in W. However, there is compelling theoretical evidence for an s–s phase transition at high P. Calculations of the phonon spectra of both the bcc and face-centered cubic (fcc) structures of W show that bcc-W becomes mechanically unstable at pressures above 12 Mbar while fcc-W being mechanically unstable at low P becomes fully stabilized above 4 Mbar [45]. A very recent theoretical study [46] demonstrates that fcc-W becomes mechanically stable at ~450 GPa while bcc-W becomes mechanically unstable with increasing P; above 12 Mbar, bcc-W is unstable at low T but remains stable above 1000 K. Hence, an s–s phase transformation to another solid phase that is mechanically stable is expected to occur at $P \lesssim 12$ Mbar at low T and the s–s transition boundary is expected to have positive slope $(dT/dP > 0)$ since the bcc-W stability range widens with increasing T. Because fcc-W also becomes thermodynamically more favorable than bcc-W [46], fcc-W is one of the candidates for the high-P solid structure of W. In fact, the bcc-fcc s–s phase transition was predictied to occur at ~12 Mbar and calculations show that the bcc-fcc phase boundary does have positive slope at $P \geq 12$ Mbar [46]. However, in Reference [47], a transformation to a different solid structure, namely, dhcp, which is thermodynamically more stable than fcc, is predicted to occur, albeit at much lower pressure of 650 GPa. If a bcc–dhcp s–s phase transition does occur in W, just like in Mo, the two phase diagrams may look similar. As a matter of fact, as we demonstrate in what follows, the two phase diagrams are topologically equivalent.

We begin our theoretical study of the phase diagram of W with the calculation of the cold $(T = 0)$ free energies (i.e., enthalpies) of a number of different solid structures of W as a function of P.

3.1. Cold Enthalpies of Different Solid Structures of W

The calculations were based on density-functional theory (DFT) with the projector-augmented wave (PAW) [48] implementation and the generalized gradient approximation (GGA) for exchange-correlation energy in the form known as Perdew–Burke–Ernzerhof (PBE) [49]. All the calculations were done using VASP (Vienna Ab initio Simulation Package). Since the simulations were performed at high-PT conditions, we used accurate pseudopotentials where the semi-core 5s and 5p states were treated as valence states. Specifically, W was modeled with 14 valence electrons per atom (5s, 5p, 5d, and 6s orbitals). The valence electrons were represented with a plane-wave basis set with a cutoff energy of 500 eV, while the core electrons were represented by projector-augmented wave (PAW) pseudopotentials. The core radius (the largest value of RCUTs among those for each of the quantum orbitals) of this W pseudopotential is 2.3 a.u. or 1.217 Å. Since numerical errors in the calculations using VASP will remain almost negligible until the nearest neighbor distance reaches $2 \times \text{RCUT}/(1.25 \pm 0.05)$ [39], with this pseudopotential, one can study systems with densities up to $\sim$80 g/cm^3 (a pressure of $\sim$8000 GPa).

Cold enthalpies were calcualted using unit cells with very dense k-point meshes (e.g., $50 \times 50 \times 50$ for bcc-W) for high accuracy. In all the non-cubic cases, we first relaxed the structure to determine its unit cell parameters at each volume. Tight convergence criteria for the total energy (10^{-5} meV/atom) and structural relaxation (residual forces < 0.1 meV/Å and stresses ≤ 0.1 kbar) were employed.

Our calculated cold enthalpies of five different solid structures of W are shown in Figure 6. It is seen that, with inceasing P, a number of other solid structures become thermodynamically more favorable than bcc, but it is dhcp that does it first, at a pressure of 1060 GPa, and remains the most favorable solid structure of W at higher P. Hence, in the case of W, the s–s transition boundary is the bcc–dhcp one and its starting point is $(P, T) = (1060, 0)$.

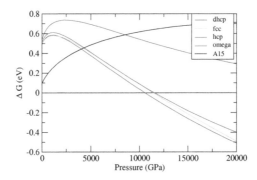

Figure 6. The $T = 0$ free energies of different solid structures of W (listed in the legend) from ab initio calculations using VASP (Vienna Ab initio Simulation Package): The free energy of bcc-W is taken to be identically zero.

3.2. Equations of State of bcc-W and dhcp-W

Next, we calculate the cold equations of state of both bcc-W and dhcp-W. For dhcp-W, we also determine the density dependence of the c/a ratio, where a and c are the lattice constants of the dhcp unit cell.

Our ab initio results on the cold equation of state (EOS) of bcc-W are described by the third-order Birch–Murnaghan (BM3) form:

$$P(\rho) = \frac{3}{2} B_0 \left(\left(\frac{\rho}{\rho_0}\right)^{7/3} - \left(\frac{\rho}{\rho_0}\right)^{5/3} \right) \left[1 + \frac{3}{4}(B_0' - 4) \left(\left(\frac{\rho}{\rho_0}\right)^{2/3} - 1 \right) \right], \tag{4}$$

where B_0 and B_0' are the values of the bulk modulus and its pressure derivative at the reference point $\rho = \rho_0$. In our case of bcc-W,

$$\rho_0 = 19.3 \text{ g/cm}^3, \quad B_0 = 314.1 \text{ GPa}, \quad B_0' = 4.07. \tag{5}$$

Since the $P = 0$ values of the density of W at $T = 0$ and 300 K differ by $\sim$0.3% (19.31 vs. 19.26 g/cc) and $T = 300$ K introduces a negligibly small thermal pressure correction, the $T = 0$ and $T = 300$ K isotherms can be described by the same values of B_0 and B_0'. Consequently, we can compare room-T isotherm data of References [44,50–53] to our $T = 0$ isotherm as determined with VASP. A comparison is shown in Figure 7. It is seen that our EOS is virtually identical to the experimental isotherm of

Reference [53], for which $B_0 = 317.5$ GPa and $B_0' = 4.05$ are very similar to ours, as well as to the theoretical calculations of Reference [51] in the so-called mean-field potential (MFP) approach.

For dhcp-W, the P dependence of the c/a ratio for the lattice constants of the dhcp unit cell is accurately described by

$$\frac{c}{a} = 3.2660 - 0.4912 \left(\frac{a}{a_0} \right)^3 + 0.7484 \left(\frac{a}{a_0} \right)^5, \tag{6}$$

where $a_0 = 2.8525$ Å coresponds to the ideal $(c/a = 4\sqrt{2/3} \approx 3.2660)$ unit cell of the same reference density ρ_0 as that for the corresponding cold EOS. Since the unit cell volume is $ca^2\sqrt{3}/8$ $(a^3/\sqrt{2}$ for the ideal structure), the above relation can be translated into the P dependence of c/a via the corresponding cold EOS of dhcp-W that we obtained. It is described by the BM3 form, Equation (4), with

$$\rho_0 = 18.6 \text{ g/cm}^3, \quad B_0 = 290.8 \text{ GPa}, \quad B_0' = 3.95. \tag{7}$$

Above 1000 GPa, the dhcp-W structure is virtually ideal. At the transition pressure of 1060 GPa, the two density values predicted by these EOSs are (in g/cm^3) 40.38 for bcc-W and and 40.53 for dhcp-W; hence, the bcc–dhcp transition corresponds to a small volume change of ~0.4%.

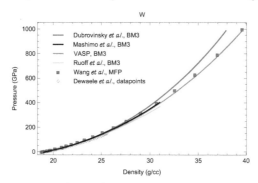

Figure 7. The $T = 0$ equation of state of bcc-W: our own ab initio calculations using VASP vs. experimental data (Dubrovinsky et al. [44]; Ruoff et al. [50]; Dewaele et al. [52]; and Mashimo et al. [53]) and other theoretical calculations (Wang et al. [51]).

4. Melting Curves of Different Solid Structures of W

We now discuss the calculation of the melting curves of different solid structures of W. For this calculation, we used the Z method, which is described in detail in References [9,36,54]. We calculated the melting curves of bcc-W as well as a number of other solid structures of W which have been mentioned in the literature in connection with transition metals: all the close-packed structures with different layer stacking (fcc, hcp, dhcp, thcp, and 9R), open structures (simple cubic, A15, and hex-ω), and different orthorhombic structures (Pnma, Pbca, Pbcm, Cmca, Cmcm, etc.). Again, in all the non-cubic cases, we relaxed the structure to determine its unit cell parameters; those unit cells were used for the construction of the corresponding supercells. We used systems of 400–500 atoms in each case.

4.1. Ab Initio Melting Curve of bcc-W

For the calculation of the melting curve of bcc-W (with the Z method), we used a 432-atom $(6 \times 6 \times 6)$ supercell with a single Γ-point. Full energy convergence (to $\lesssim 1$ meV/atom) was verified by performing short runs with $2 \times 2 \times 2$ and $3 \times 3 \times 3$ k-point meshes and by comparing their output with that of the run with a single Γ-point. For each of the five points listed in Table 1, we performed ten NVE runs (five values of V correspond to five values of densities in Table 1) of 10,000–20,000 time steps of 1 fs each, with an increment of the initial T of 250 K for the first datapoint and 625 K for the remaining four. Since the error in T is half of the increment [36], the T errors of our five values of T_m are within 5% each. The P errors are negligibly small: less than 1 GPa for the first point and 1–2 GPa for the remaining four. Hence, our melting results on bcc-W are very accurate.

Table 1. The five ab initio melting points of bcc-W, $(P_m, T_m \pm \Delta T_m)$, obtained from the Z method implemented with VASP.

Lattice Constant (Å)	Density (g/cm³)	P_m (GPa)	T_m (K)	ΔT_m (K)
3.350	16.240	-17.7	2800	125.0
3.050	21.519	90.6	6520	312.5
2.850	26.375	258	9840	312.5
2.675	31.897	543	13,760	312.5
2.535	37.479	947	18,070	312.5

The best fit to the five bcc-W melting points gives the melting curve of bcc-W in the Simon–Glatzel form (T_m in K, P in GPa):

$$T_m(P) = 3695 \left(1 + \frac{P}{41.8}\right)^{0.50}. \tag{8}$$

Its initial slope, $dT_m(P)/dP = 44.2$ K/GPa at $P = 0$, is in excellent agreement with 44 K/GPa from isobaric expansion measurements [26]. Both the five bcc-W melting points and the melting curve (Equation (8)) are shown in Figure 8 and compared to the experimental results in Reference [27,30], the calculated Hugoniot of W [55], and other theoretical calculations [6,56–58].

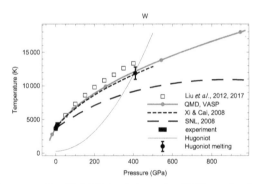

Figure 8. The melting curve of bcc-W: QMD simulations using VASP vs. other theoretical calculations (Xi an Cai [56]; SNL, 2008 [57]; and Liu et al., 2012 [6] and 2017 [58]), the low-pressure melting data (experiment in Reference [30]), and the experimental shock melting datapoint (Hugoniot melting [27]). The calculated Hugoniot is shown as a thin black curve.

4.2. Ab Initio Melting Curve of dhcp-W

The five melting points of dhcp-W that we obtained are listed in Table 2. The dhcp-W melting simulations were carried out exactly the same way as the bcc ones, namely, 10 runs per point, 10,000–20,000 time steps of 1 fs each per run, and an increment of the initial T of 300 K for the first datapoint and 625 K for the remaining four. Just as for bcc-W, a 432-atom ($6 \times 6 \times 3$) supercell was used to simulate each of the five dhcp-W melting points. Also, as for bcc-W, the T errors of our five values of T_m are within 5% each.

Table 2. The five ab initio melting points of dhcp-W, $(P_m, T_m \pm \Delta T_m)$, obtained from the Z method implemented with VASP: The lattice constant values correspond to the ideal dhcp structure.

Lattice Constant (Å)	Density (g/cm³)	P_m (GPa)	T_m (K)	ΔT_m (K)
2.90	17.702	8.1	3080	150.0
2.65	23.199	150	7380	312.5
2.40	31.230	510	13,110	312.5
2.25	37.902	970	18,000	312.5
2.15	43.440	1471	22,200	312.5

The best fit to the five dhcp-W points gives the melting curve of dhcp-W in the Simon–Glatzel form (T_m in K, P in GPa):

$$T_m(P) = 2640 \left(1 + \frac{P}{23.0}\right)^{0.51}. \tag{9}$$

Figures 9 and 10 demonstrate the time evolution of T and P, respectively, in the Z-method runs of the bcc-W melting point (P in GPa, T in K) $(P, T) = (947, 18, 070)$. Figures 11 and 12 demonstrate the same for the dhcp-W melting point $(970, 18, 000)$. These two points are chosen as examples and are shown in Figure 13 as open blue and green circles, respectively.

All the other solid structures that we considered melt below bcc. As an example, one of the fcc-W melting points, namely, $(P, T) = (942, 16,920)$, along with a short segment of the corresponding melting curve, are shown in Figure 13. The corresponding time evolution of T and P in the Z-method runs are shown in Figures 14 and 15, respectively.

The melting curves of bcc-W and dhcp-W cross each other at (P in GPa, T in K) $(P, T) = (1675, 23,680)$, which is the bcc–dhcp-liquid triple point. The choice of $T(P) = a(1060 - P)^b$, $0 < b < 1$ as a functional form for the bcc–dhcp phase boundary leads to
(iii) the bcc–dhcp solid–solid phase transition boundary:

$$T(P) = 73.2 (P - 1060)^{0.90}, \tag{10}$$

which crosses the triple point and lies within the bounds imposed by the solidification simulations using the inverse Z method. We discuss these inverse Z simulations in the following section.

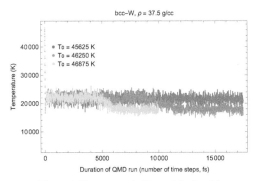

Figure 9. The same as in Figure 1 for bcc-W.

Although the rigorous derivation of the thermal equations of state of bcc-W and dhcp-W goes beyond the scope of this work, we note that the finite-T counterparts of the above two EOSs can be written approximately as $P(\rho, T) = P(\rho) + \alpha T$, where $\alpha_{bcc} = 7.3 \times 10^{-3}$ and $\alpha_{dhcp} = 6.8 \times 10^{-3}$. The resulting "approximate" thermal EOSs turn out to be quite accurate. For example, for the five bcc-W melting points in Table 1, the corresponding thermal EOS gives pressures of -17.8, 90.3, 258, 543, and 948, which are basically identical to those in the third column of Table 1. For the five dhcp-W melting points in Table 2, the five P values are 7.9, 150, 509, 970, and 1476, in excellent agreement with those in the third column of Table 2.

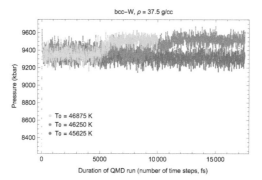

Figure 10. The same as in Figure 2 for bcc-W.

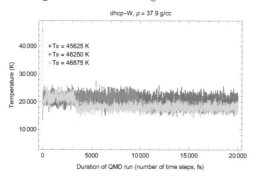

Figure 11. The same as in Figure 1 for dhcp-W.

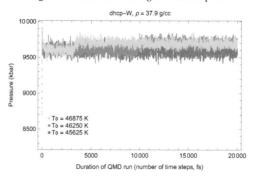

Figure 12. The same as in Figure 2 for dhcp-W.

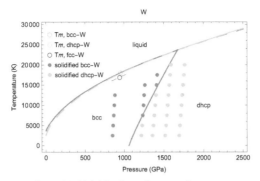

Figure 13. Ab initio phase diagram of tungsten.

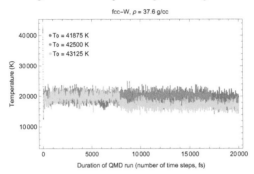

Figure 14. The same as in Figure 1 for fcc-W.

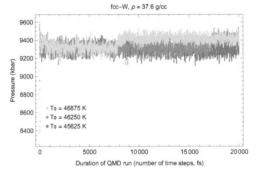

Figure 15. The same as in Figure 2 for fcc-W.

5. Inverse Z Solidification Simulations of Liquid W

To constrain the location of the bcc–dhcp solid–solid phase boundary on the P–T plane between the points $(P = 1060, T = 0)$ and $(P = 1675, T = 23,680)$, we carried out two sets of independent inverse Z runs [9] to solidify liquid W and to confirm that liquid W solidifies into bcc on one side of this boundary and into dhcp (or any other solid structure) on the other side, such that the location of this phase boundary may be constrained. Since the solidification kinetics is approximately governed by the factor $\exp\{\Delta F/T^*\}$, where $\Delta F \equiv F_l - F_s$ is the liquid–solid free-energy difference at the solidification temperature T^*, in the case of several energetically competitive solid phases, the most thermodynamically favorable solid phase has the largest ΔF and is therefore the fastest to solidify. Hence, the inverse Z method yields the most stable solid phase at a given (P, T).

For our inverse Z simulations, we used a computational cell of 512 atoms prepared by melting a $8 \times 8 \times 8$ solid simple cubic (sc) supercell, which would eliminate any bias towards solidification into bcc or any other solid structure (fcc, hcp, dhcp, etc.). We used sc unit cells of 2.0, 1.935, 1.915, 1.895, and 1.870 Å; the dimensions of bcc unit cells having the same volume as the sc ones are 2.520, 2.438, 2.413, 2.388, and 2.356 Å, respectively, which corresponds to the bcc-W pressures of ~870, 1225, 1355, 1505, and 1715 GPa and to slightly lower pressures for dhcp-W.

We carried out NVT simulations using the Nosé–Hoover thermostat with a timestep of 1 fs, with the initial T increment of 2500 K. Complete solidification typically was required from 15 to 25 ps or 15,000–25,000 timesteps. The inverse Z runs indicate that liquid W only solidifies into bcc at ~900 GPa in the whole temperature range from 0 to essentially the corresponding T_m. However, at ~1200–1400 GPa, it solidifies into bcc above the transition boundary in Figure 13, while below this boundary, it solidifies into another solid structure. The radial distribution functions (RDFs) of the final solid states are noisy; upon fast quenching of the seven structures (seven green bullets in Figure 13 in the 1200–1400 GPa range) to low T, where RDFs are more discriminating, and by comparing them to the RDFs of fcc, hcp, and dhcp, we conclude that dhcp is the closest strucure to those that liquid Mo solidies into below the transition boundary.

The RDFs of the solidified states at ~1200 GPa above the transition boundary are shown in Figure 16 and of those solidified below the transition boundary are shown in Figure 17. The 7500 K state virtually lies on the boundary. We tentatively assign it to bcc because it definitely has features of bcc (RDF peaks at $R \sim 55, 65$, and 85, etc.). At the same time, it certainly has some features that are both uncharacteristic of bcc (e.g., the disappearance of the bcc peak at $R \sim 95$) and characteristic of dhcp (peaks at $R \sim 90$ and 120, small peaks at $R \sim 100$ and 130, etc.). Most likely, this 7500 K state is bcc with some admixture of dhcp.

A few more comments are in order. The 17,500 K state at ~1250 GPa did not solidify, most likely for the reason of not being supercooled enough to intiate the solidification process [9]. Indeed, 17,500 K constitutes ~0.8 of the corresponding T_m of ~22,000 K (20% of supercooling) while, e.g., for the other set of points at ~900 GPa, the highest solidification T of 12,500 K constitutes ~0.75 of the corresponding T_m of ~17,000 K (25% of supercooling), which apparently allows for the solidification process to go through in this case.

The phase diagram of W to 2500 GPa is shown in Figure 13. It includes the two bcc and dhcp melting curves, the bcc–dhcp solid–solid phase transition boundary, and the results of the solidification of liquid W into final states of either solid bcc or solid dhcp using the inverse Z method.

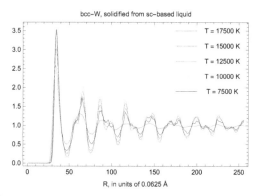

Figure 16. Radial distribution functions (RDFs) of the final states of the solidification of liquid W at ~1200 GPa at higher temperatures.

Both the Mo and W phase diagram figures, Figures 5 and 13, do not show the previous experimental DAC melting curves of Errandonea et al. [13,16] and are both flat, with the initial slope of ~7–8 K/GPa.

We did not include those in the two figures because we do not consider them to be relevant. The very recent experimental study by Hrubiak et al. [23] demonstrates that, on increasing T, compressed Mo undergoes a transformation that results in a texture (microstructure) change: large Mo grains become unstable at high T due to high atom mobility and reorganize into smaller crystalline grains. This transformation occurs below melting, and the pressure dependence of the transformation temperature is consistent with the previous DAC melting curve by Errandonea et al. [13]. Hence, most likely, Errandonea's curve is an intermediate DAC transition boundary, while our curve as well as Alfe's [41], which is basically identical to ours, is the actual melting curve of Mo.

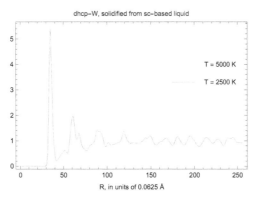

Figure 17. RDFs of the final states of the solidification of liquid W at ~1200 GPa at lower temperatures.

A behavior similar to that of Mo (a texture change) was recently observed in the high-PT melting experiments on V [25]. Hence, this phenomenon of a texture (microstructure) change can be common to a number of transition metals, including W. It is therefore natural to assume that, in the case of W, too, the corresponding Errandonea's melting curve is an intermediate DAC transition boundary while our curve is the actual melting curve of W.

6. Concluding Remarks

The phase diagrams of Mo and W, Group 6B partners in the periodic table, shown in Figures 5 and 13, respectively, are topologically equivalent. In both cases, the ambient bcc solid structure transforms into dhcp on increasing P. Only these two solid phases are confirmed as being present on the two phase diagrams, via ab initio QMD simulations using the Z methodology. Even the critical aspect ratio of the two phase diagrams is identical. Indeed, the ratio of the pressure values for the bcc–dhcp-liquid triple point and the $T = 0$ bcc–dhcp transition (it determines the curvatures and the locus of the bcc–dhcp transition boundary on the P–T plnae) is the same in both cases: 1045/660 = 1.583 for Mo and 1675/1060 = 1.580 for W. The two transition boundaries, Equations (3) and (10), are described by the same exponent 0.9 which further confirms the topological equivalence of the two phase diagrams.

It is interesting to note another case of the phase diagrams of Nb and Ta, group 5B partners in the periodic table, which appear to be topologically similar (look-alike) but not equivalent. In both cases, a bcc-Pnma solid–solid phase transition occurs [9,59], but in contrast to dhcp in Mo and W, the orthorhombic Pnma phase exists at high T only. In Ta, the bcc-Pnma-liquid triple point is at a pressure which is an order of magnitude higher than that in Nb [9,59]. Also, the bcc-Pnma transition T increases with P in Nb but decreases with T in Ta.

Inverse Z solidification simulations confirm the phase diagrams of Mo and W shown in Figures 5 and 13, respectively, and validate the corresponding bcc–dhcp solid–solid phase boundaries. Additional support for the existence of the high-P dhcp structure in both Mo and W comes from Reference [60]. The electronic wave functions of Mo and W and of their group 5B partner Cr are such that, with decreasing volume (increasing P), they become more spherically symmetric, which favors a higher atomic coordination number than 8 as that for bcc and, correspondingly, a more symmetric structure than bcc. It is well known that the highest possible coordination number is 12, and it corresponds to all of the hexagonal polytypes, namely, hcp (stacking sequence AB), fcc (ABC), dhcp ($ABAC$), triple-hcp ($ABCACB$), etc. It is therefore natural to expect that, at high P, both Mo and W and perhaps Cr as well will have one of the hexagonal polytypes as their equilibrium solid structure. Our study demonstrates that this polytype is in fact dhcp. It is interesting to note that, in Cr, as theoretical calculations show [61], hcp becomes energetically more favorable than bcc with decreasing volume (increasing P). Since different hexagonal polytypes are usually close to each other energetically, one can therefore expect them to be the high-P equilibrium solid structure of Cr. Thus, it is quite possible that the phase digram of Cr is topologically similar, if not equivalent, to those of both Mo and W. In contrast to both Mo and W, the study of Reference [62] shows that the electronic wave functions of V, Nb, and Ta are such that, with decreasing volume, they become less spherically symmetric, which favors a lower atomic coordination number than bcc's 8 and, correspondingly, a less symmetric structure than bcc, e.g., an orthorhombic one. This appears to be in agreement with the fact that, in both Nb and Ta, the high-P equilibrium solid structure is orthorhombic Pnma.

Fast recrystallization observed in HP-HT experiments on transition metals may imply either a microstructural transition (discussed above) at which the sample texture changes but its crystal structure remains the same or a true solid–solid phase transition. The former was recently proven to be the case in Mo [23], and it is very likely the case in V [25] as well as in W. The latter is apparently the case in Nb [59] and likely in Re as well [10]. In either case, the corresponding fast recrystallization lines have been misinterpreted as flat melting curves in the former laser-heating DAC experiments.

Author Contributions: All authors contributed equally to this work. All authors have read and agreed to the published version of the manuscript.

Funding: This research received no external funding.

Acknowledgments: The work was done under the auspices of the US DOE/NNSA. The QMD simulations were performed on the LANL clusters Pinto and Badger as parts of the Institutional Computing projects w17_molybdenum, w18_meltshear, and w19_thermoelast.

Conflicts of Interest: The authors declare no conflict of interest.

References

1. Belonoshko, A.B.; Burakovsky, L.; Chen, S.P.; Johansson, B.; Mikhaylushkin, A.S.; Preston, D.L.; Simak, S.I.; Swift, D.C. Molybdenum at high pressure and temperature: Melting from another solid phase. *Phys. Rev. Lett.* **2008**, *100*, 135701. [CrossRef] [PubMed]

2. Liu, Z.-L.; Cai, L.-C.; Chen, X.-R.; Jing, F.-Q. Molecular dynamics simulations of the melting curve of tantalum under pressure. *Phys. Rev. B* **2008**, *77*, 024103. [CrossRef]

3. Burakovsky, L.; Chen, S.P.; Preston, D.L.; Belonoshko, A.B.; Rosengren, A.; Mikhaylushkin, A.S.; Simak, S.I.; Moriarty, J.A. High-pressure–high-temperature polymorphism in Ta: Resolving an ongoing experimental controversy. *Phys. Rev. Lett.* **2010**, *104*, 255702. [CrossRef] [PubMed]

4. Haskins, J.; Moriarty, J.A.; Hood, R.Q. Polymorphism and melt in high-pressure tantalum. *Phys. Rev. B* **2012**, *86*, 224104. [CrossRef]

5. Hu, J.; Dai, C.; Yu, Y.; Liu, Z.; Tan, Y.; Zhou, X.; Tan, H.; Cai, L.; Wu, Q. Sound velocity measurements of tantalum under shock compression in the 10–110 GPa range. *J. Appl. Phys.* **2012**, *111*, 033511. [CrossRef]

6. Liu, C.M.; Chen, X.R.; Xu, C.; Cai, L.C.; Jing, F.Q. Melting curves and entropy of fusion of body-centered cubic tungsten under pressure. *J. Appl. Phys.* **2012**, *112*, 013518. [CrossRef]

7. Yao, Y.; Klug, D. Stable structures of tantalum at high temperature and high pressure. *Phys. Rev. B* **2013**, *88*, 054102. [CrossRef]

8. Pozzo, M.; Alfe, D. Melting curve of face-centered-cubic nickel from first-principles calculations. *Phys. Rev. B* **2013**, *88*, 024111. [CrossRef]

9. Burakovsky, L.; Chen, S.P.; Preston, D.L.; Sheppard, D.G. Z methodology for phase diagram studies: Platinum and tantalum as examples. *J. Phys. Conf. Ser.* **2014**, *500*, 162001. [CrossRef]
10. Burakovsky, L.; Burakovsky, N.; Preston, D.; Simak, S.I. Systematics of the third row transition metal melting: The hcp metals rhenium and osmium. *Crystals* **2018**, *8*, 243. [CrossRef]
11. Lazor, P.; Shen, G.; Saxena, S.K. Laser-heated diamond anvil cell experiments at high pressure: Melting curve of nickel up to 700 kbar. *Phys. Chem. Miner.* **1993**, *20*, 86. [CrossRef]
12. Kavner, A.; Jeanloz, R. High-pressure melting curve of platinum. *J. App. Phys.* **1998**, *83*, 7553. [CrossRef]
13. Errandonea, D.; Schwager, B.; Ditz, R.; Gessmann, C.; Boehler, R.; Ross, M. Systematics of transition-metal melting. *Phys. Rev. B* **2001**, *63*, 132104. [CrossRef]
14. Errandonea, D.; Somayazulu, M.; Häusermann, D.; Mao, H.K. Melting of tantalum at high pressure determined by angle dispersive x-ray diffraction in a double-sided laser-heated diamond-anvil cell. *J. Phys. Cond. Mat.* **2003**, *15*, 7635. [CrossRef]
15. Boehler, R.; Santamaría-Perez, D.; Errandonea, D.; Mezouar, M. Melting, density, and anisotropy of iron at core conditions: New X-ray measurements to 150 GPa. *J. Phys. Conf. Ser.* **2008**, *121*, 022018. [CrossRef]
16. Santamaría-Pérez, D.; Ross, M.; Errandonea, D.; Mukherjee, G.D.; Mezouar M.; Boehler, R. X-ray diffraction measurements of Mo melting to 119 GPa and the high pressure phase diagram. *J. Chem. Phys.* **2009**, *130*, 124509. [CrossRef]
17. Ruiz-Fuertes, J.; Karandikar, A.; Boehler, R.; Errandonea, D. Microscopic evidence of a flat melting curve of tantalum. *Phys. Earth Planet. Inter.* **2010**, *181*, 69. [CrossRef]
18. Yang, L.; Karandikar, A.; Boehler, R. Flash heating in the diamond cell: Melting curve of rhenium. *Rev. Sci. Instrum.* **2012**, *83*, 063905. [CrossRef]
19. Dewaele, A.; Mezouar, M.; Guignot, N.; Loubeyre, P. High melting points of tantalum in a laser-heated diamond anvil cell. *Phys. Rev. Lett.* **2010**, *104*, 255701. [CrossRef]
20. Errandonea, D. High-pressure melting curves of the transition metals Cu, Ni, Pd, and Pt. *Phys. Rev. B* **2013**, *87*, 054108. [CrossRef]
21. Anzellini, S.; Dewaele, A.; Mezouar, M.; Loubeyre, P.; Morard, G. Melting of iron at Earth's inner core boundary based on fast X-ray diffraction. *Science* **2013**, *340*, 464. [CrossRef] [PubMed]
22. Lord, O.T.; Wood, I.G.; Dobson, D.P.; Vočadlo, L.; Wang, W.; Thomson, A.R.; Wann, E.T.; Morard, G.; Mezouar, M.; Walter, M.J. The melting curve of Ni to 1 Mbar. *Earth Planet. Sci. Lett.* **2014**, *408*, 226. [CrossRef]
23. Hrubiak, R.; Meng, Y.; Shen, G. Microstructures define melting of molybdenum at high pressures. *Nat. Commun.* **2017**, *8*, 14562. [CrossRef] [PubMed]
24. Anzellini, S. *In situ* characterization of the high pressure–high temperature melting curve of platinum. *Sci. Rep.* **2019**, *9*, 13034. [CrossRef] [PubMed]
25. Errandonea, D.; MacLeod, S.G.; Burakovsky, L.; Santamaria-Perez, D.; Proctor, J.E.; Cynn, H.; Mezouar, M. Melting curve and phase diagram of vanadium under high-pressure and high-temperature conditions. *Phys. Rev. B* **2019**, *100*, 094111. [CrossRef]
26. Shaner, J.W.; Gathers, G.R.; Minichino, C. A new apparatus for thermophysical measurements above 2500 K. *High Temp.-High Pres.* **1976**, *8*, 425.
27. Hixson, R.S.; Fritz, J.N. Shock compression of tungsten and molybdenum. *J. Appl. Phys.* **1992**, *71*, 1721. [CrossRef]
28. Musella, M.; Ronchi, C.; Sheindlin, M. Dependence of the melting temperature on pressure up to 2000 bar in uranium dioxide, tungsten, and graphite. *Int. J. Thermophys.* **1999**, *20*, 1177. [CrossRef]
29. Kloss, A.; Hess, H.; Schneidenbach, H.; Grossjohann, R. Scanning the melting curve of tungsten by a submicrosecond wire-explosion experiment. *Int. J. Thermophys.* **1999**, *20*, 1199. [CrossRef]
30. Gustafson, P. Evaluation of the thermodynamic properties of tungsten. *Int. J. Thermophys.* **1985**, *6*, 395. [CrossRef]
31. Vereshchagin, L.F.; Fateeva, N.S. Melting curves of graphite, tungsten and platinum to 60 kbar. *Sov. Phys. JETP* **1969**, *28*, 597.
32. Gorecki, T. Vacancies and melting curve of metals at high pressure. *Z. Metallk.* **1977**, *68*, 231.
33. Gorecki, T. Vacancies and a generalised melting curve of metals. *High Temp.-High Press.* **1979**, *11*, 683.
34. Arblaster, J.W. What is the true melting point of osmium? *Platin. Met. Rev.* **2005**, *49*, 166. [CrossRef]
35. Zitserman, V. (JIHT, Moscow). Personal communication, 2019.

36. Burakovsky, L.; Burakovsky, N.; Preston, D.L. *Ab initio* melting curve of osmium. *Phys. Rev. B* **2015**, *92*, 174105.

37. Burakovsky, L.; Preston, D.L. Shear moduli of silicon and germanium in semiconducting and metallic phases. *Defect Diffus. Forum* **2002**, *210–212*, 43. [CrossRef]

38. Frohberg, M.G. Thirty years of levitation melting calorimetry-a balance. *Thermochem. Acta* **1999**, *337*, 7. [CrossRef]

39. Burakovsky, L.; Lushcer, D.J.; Preston, D.; Sjue, S.; Vaughan, D. Generalization of the unified analytic melt-shear model to multi-phase materials: Molybdenum as an example. *Crystals* **2019**, *9*, 86. [CrossRef]

40. Simon, F.; Glatzel, G. Bemerkungen zur Schmelzdruckkurve. *Z. Anorg. Allgem. Chem.* **1929**, *178*, 309. [CrossRef]

41. Cazorla, C.; Gillan, M.J.; Taioli, S.; Alfè, D. Melting curve and Hugoniot of molybdenum up to 400 GPa by *ab initio* simulations. *J. Phys. Conf. Ser.* **2008**, *121*, 012009. [CrossRef]

42. Shaner, J.W.; Gathers, G.R.; Minichino, C. Thermophysical properties of liquid tantalum and molybdenum. *High Temp.-High Press.* **1977**, *9*, 331.

43. Ruoff, A.L.; Xia, H.; Xia, Q. The effect of a tapered aperture on x-ray diffraction from a sample with a pressure gradient: Studies on three samples with a maximum pressure of 560 GPa. *Rev. Sci. Instrum.* **1992**, *63*, 4342. [CrossRef]

44. Dubrovinsky, L.; Dubrovinskaia, N.; Bykova, E.; Bykov, M.; Prakapenka, V.; Prescher, C.; Glazyrin, K.; Liermann, H.-P.; Hanfland, M.; Ekholm, M.; et al. The most incompressible metal osmium at static pressures above 750 gigapascals. *Nature* **2015**, *525*, 226. [CrossRef] [PubMed]

45. Einarsdotter, K.; Sadigh, B.; Grimvall, G.; Ozoliņš, V. Phonon instabilities in fcc and bcc tungsten. *Phys. Rev. Lett.* **1997**, *79*, 2073. [CrossRef]

46. Zhang, H.-Y.; Niu, Z.-W.; Cai, L.-C.; Chen, X.-R.; Xi, F. *Ab initio* dynamical stability of tungsten at high pressures and high temperatures. *Comput. Mater. Sci.* **2018**, *144*, 32. [CrossRef]

47. Ruoff. A.L.; Rodriguez, C.O.; Christensen, N.E. Elastic moduli of tungsten to 15 Mbar, phase transition at 6.5 Mbar, and rheology to 6 Mbar. *Phys. Rev. B* **1998**, *58*, 2998. [CrossRef]

48. Blöchl, P. Projector augmented-wave method. *Phys. Rev. B* **1994**, *50*, 17953. [CrossRef]

49. Perdew, J.P.; Burke, K.; Ernzerhof, M. Generalized gradient approximation made simple. *Phys. Rev. Lett.* **1996**, *77*, 3865. [CrossRef]

50. Ruoff, A.L.; Xia, H.; Luo, H.; Vohra, Y.K. Miniaturization techniques for obtaining static pressures comparable to the pressure at the center of the earth: X-ray diffraction at 416 GPa. *Rev. Sci. Instrum.* **1990**, *61*, 3830. [CrossRef]

51. Wang, Y.; Chen, D.; Zhang, X. Calculated equation of state of Al, Cu, Ta, Mo, and W to 1000 GPa. *Phys. Rev. Lett.* **2000**, *84*, 3220. [CrossRef]

52. Dewaele, A.; Loubeyre, P.; Mezouar, M. Equations of state of six metals above 94GPa. *Phys. Rev. B* **2004**, *70*, 094112. [CrossRef]

53. Mashimo, T.; Liu, X.; Kodama, M.; Zaretsky, E.; Katayama, M.; Nagayama, K. Effect of shear strength on Hugoniot-compression curve and the equation of state of tungsten (W). *J. Appl. Phys.* **2016**, *119*, 035904. [CrossRef]

54. Belonoshko, A.B.; Skorodumova, N.V.; Rosengren, A.; Johansson, B. Melting and critical superheating. *Phys. Rev. B* **2006**, *73*, 012201. [CrossRef]

55. McQueen, R.G.; Marsh, S.P. Equation of state for nineteen metallic elements from shock-wave measurements to two megabars. *J. Appl. Phys.* **1960**, *31*, 1253. [CrossRef]

56. Xi, F.; Cai, L. Theoretical study of melting curves on Ta, Mo, and W at high pressures. *Phys. B* **2008**, *403*, 2065. [CrossRef]

57. Carpenter, J.H.; Desjarlais, M.P.; Mattsson, A.E.; Cochrane, K.R. *A New Wide-Range Equation of State for Tungsten*; SNL Report SAND2008-1415C; American Physical Society: Washington, DC, USA, 2008.

58. Liu, C.M.; Xu, C.; Cheng, Y.; Chen, X.R.; Cai, L.C. Molecular dynamics studies of body-centered cubic tungsten during melting under pressure. *Chin. J. Phys.* **2017**, *55*, 2468. [CrossRef]

59. Errandonea, D.; Burakovsky, L.; Preston, D.L.; MacLeod, S.G.; Santamaría-Perez, D.; Chen, S.P.; Cynn, H.; Simak, S.I.; McMahon, M.L.; Proctor, J.E.; et al. Niobium at High Pressure and High Temperature: Combined Experimental and Theoretical Study. 2019, submitted.

60. Xie, Y.-Q.; Deng, Y.-P.; Liu, X.-B. Electroic structure and physical properties of pure Cr, Mo and W. *Trans. Nonferrous Met. Soc. China* **2003**, *13*, 5.
61. Pure Chromium. Available online: https://icme.hpc.msstate.edu/mediawiki/index.php/Pure_Chromium (accessed on 30 December 2019).
62. He, Y.; Xie, Y.-Q. Electroic structure and properties of V, Nb and Ta metals. *J. Cent. South Univ. Technol.* **2000**, *7*, 7. [CrossRef]

Article

Oxidation of High Yield Strength Metals Tungsten and Rhenium in High-Pressure High-Temperature Experiments of Carbon Dioxide and Carbonates

Raquel Chuliá-Jordán [1], David Santamaría-Pérez [1,*] , Tomás Marqueño [1], Javier Ruiz-Fuertes [2] and Dominik Daisenberger [3]

[1] Applied Physics Department - ICMUV - MALTA Consolider Team, University of Valencia, c/Dr. Moliner 50, 46100 Burjassot (Valencia), Spain; raquel.chulia@uv.es (R.C.-J.); tomasmarqueno@gmail.com (T.M.)
[2] DCITIMAC- MALTA Consolider Team, University of Cantabria, 39005 Santander, Spain; ruizfuertesj@unican.es
[3] Diamond Light Source, Didcot, Oxon OX11 0DE, UK; dominik.daisenberger@diamond.ac.uk
* Correspondence: david.santamaria@uv.es

Received: 24 November 2019; Accepted: 13 December 2019; Published: 17 December 2019

Abstract: The laser-heating diamond-anvil cell technique enables direct investigations of materials under high pressures and temperatures, usually confining the samples with high yield strength W and Re gaskets. This work presents experimental data that evidences the chemical reactivity between these refractory metals and CO_2 or carbonates at temperatures above 1300 °C and pressures above 6 GPa. Metal oxides and diamond are identified as reaction products. Recommendations to minimize non-desired chemical reactions in high-pressure high-temperature experiments are given.

Keywords: reactivity; tungsten; rhenium; carbon dioxide; carbonates; high-pressure high-temperature experiments

1. Introduction

Extreme high-pressure high-temperature (HP–HT) conditions are commonly encountered in nature. A primary source for acquiring knowledge of the Earth's interior is, for instance, generating similar pressure and temperature conditions in the laboratory, and studying how these conditions affect metals [1] or minerals [2,3]. A particular interesting problem is to unravel the fate of carbon in the deep mantle [2,4–6], for which HP–HT experiments involving carbon dioxide and carbonates need to be performed. However, the HP–HT chemistry of these compounds is largely unknown, which directly affects the proper design of such experiments.

High-pressure high-temperature physics and chemistry has greatly progressed in the last three decades, thanks to experimental developments. The diamond-anvil cell (DAC), for instance, has made possible to characterize in situ many materials at high pressure via spectroscopic and scattering techniques, due to the fact that diamond is transparent to many forms of electromagnetic radiation (visible, IR, X-ray) [2,7–9]. DACs are based on a simple design that consists of two opposed diamond anvils that compress a thin gasket which holds a tiny sample and a pressure transmitting medium. Experiments at pressures above 30 GPa require hard materials (rhenium, tungsten, diamond powder) with good mechanical properties to be used as gaskets [10,11]. Combined high-pressure high-temperature experiments are, however, less common, because they require the simultaneous use of cutting-edge techniques, such as DACs and resistive heaters, or high-power heating lasers, together with complementary characterization techniques [12].

In this context, new advances in static resistive-heated [13] and laser-heated diamond-anvil cells (LHDAC) technology [14], pyrometry and spectroradiometry diagnostic techniques [15], and the development of brighter X-ray sources [16] have extended the pressure–temperature regime accessible in HP-HT in situ studies, and allow giving new insights into materials' behavior (reactivity, phase transformations, properties) at these conditions. These experiments try to mimic planets' interiors, and can be therefore considered as windows to observe otherwise inaccessible parts of the Universe.

The thermodynamic stability of CO_2 and carbonates at ambient conditions may be significantly altered at high pressures and temperatures in the presence of rhenium or tungsten metal of the gasket. Several studies of the interaction of CO_2 with transition metal surfaces at low pressure are available in the literature [17,18]. Thus, for example, the overall CO_2 dissociative chemisorption on the Re (0001) surface increases below 140 K due to long time residence, whereas at higher temperatures, desorption is the dominant process [18]. The kinetics of the oxidation of tungsten by CO_2 at T > 2000 °C were also reported, and it was pointed out that the reactions of desorbed oxygen became important [19]. The dense CO_2 environment in diamond-anvil cell experiments, together with high temperatures, that could help to overcome the barriers to chemically interact with the Re and W gaskets, may result in undesired reaction products, as reported elsewhere for the case of rhenium [20].

In this work we review previous results on the chemical reactivity of molecular CO_2 in the presence of Re metal [21], providing new evidences of diamond formation, and show the occurrence of a redox reaction between tungsten metal and calcium silicate–carbonates under HP–HT conditions. Angle-dispersive powder X-ray diffractometry (XRD) data of in situ and recovered samples in contact with tungsten during compression and laser-heating (T > 1400 K) show the presence of scheelite-type calcium tungstate (after tilleyite decomposition and carbonate reduction). These results give an overview of the reactivity of these gasket materials at extreme conditions, and emphasize the need of meticulousness while preparing suitable sample chamber configurations to try to avoid non-desired chemical reactions in carbon dioxide and carbonates above that temperature.

2. Materials and Methods

High-pressure high-temperature X-ray diffraction (XRD) measurements were performed using symmetric and Boehler–Almax diamond-anvil cells (DACs). The chemical interaction between tungsten and carbonates was observed while laser-heating natural silicate–carbonate samples [6,22,23]. Natural tilleyite mineral (specimen YPM MIN 041104, Yale Peabody Museum) from the Crestmore quarry was analyzed by energy-dispersive X-ray spectroscopy, and only traces (<1 wt%) of Al and K were found apart the Ca, Si, C and O atoms present in the ideal $Ca_5(Si_2O_7)(CO_3)_2$ formulae. HP–HT experiments of tilleyite were performed in a DAC with diamonds of 250 μm culet diameter, using a W gasket preindented to a thickness of 35 μm. A tilleyite sample contained within the W pressure chamber was heated using a high-power diode-pumped fiber laser with wavelength of 1.07 μm and a Cu disk as an internal heater, creating a 30 μm-diameter stable hotspot comparable in size with the heater. Copper was chosen as this heater material because of its known difficulty in forming carbonates [24] and silicates [25]. In this experiment, the pressure chamber was ~50 μm in diameter, and hot Cu diffused, forming a thin layer that covered the whole pressure chamber. Even though the hotspot was centered, the heat was conducted to the gasket (W)-sample (silicate–carbonate) boundary and a chemical reaction occurred. Here, we report the results of tilleyite compressed to 9 GPa, which already converted into post-tilleyite [6], and laser-heated above 1500 °C.

Experimental details of the HP–HT Re + CO_2 measurements were already reported elsewhere [21]. In short, rhenium disks of 30 μm diameter and 5–12 μm thickness were placed at the center of a ~100 μm diameter hole of tungsten gasket. These metallic samples are good absorbers of the infrared radiation of the heating laser described above. High-purity CO_2 was loaded in the DAC at room temperature using a gas loading apparatus. Pure SiO_2 was used to thermally isolate the diamond anvils. Two runs were carried out: (1) 8 GPa–1300 °C and (2) 24 GPa–1400 °C. In this experiment, the W gasket was not in contact to Re heater, and consequently, no W oxidation was observed.

Pressures were measured before and after heating by the ruby fluorescence method [26]. We also used the equations of state of the metals, Cu [27] and Re [28], as secondary pressure gauges, with a pressure difference with respect to that of the ruby standard of less than 2%. Temperatures were measured fitting the Planck's radiation function to the emission spectra (in the range 550–800 nm) assuming wavelength-independent emissivities [29].

Angle-dispersive X-ray diffraction data here presented were collected in two different synchrotron beamlines: GSECARS at the Advanced Photon Source for the Re+CO_2 experiments, and I15 at the Diamond Light Source for silicate–carbonate heating experiments. At GSECARS, incident X-rays had a wavelength of 0.3344 Å and the beam was focused to ~3 × 3 μm^2 spot. At the I15 beamline of the Diamond Light Source we used an incident monochromatic wavelength of 0.4246 Å focused to $10 \times 10\ \mu m^2$. No rotation of the sample was done during XRD pattern collection. Detectors calibration, correction of distortion and integration to conventional 2 θ-intensity data were carried out with the Dioptas software [30]. The indexing and refinement of the powder patterns were performed using the, Unitcell [31], Fullprof [32] and Powdercell [33] program packages.

3. Results and Discussion

Suggested gasket materials to reach tens of gigapascals or megabar pressures include high yield strength metals such as rhenium or tungsten.

3.1. W oxidation

Oxidation of tungsten produced by different routes has been widely studied in the past [34]. These studies concluded that elevated temperatures would enhance the oxidation rate through the formation of highly volatile tungsten oxides, the oxidation kinetics of this metal being quite complex. For instance, a large number of tungsten oxides and phases have been determined [34]. Tungsten metal, in spite of many attractive properties such as good HT strength and high electrical resistivity, easily oxidizes at ambient pressure.

This low resistance to oxidation was also patent while studying the crystallochemical behavior of natural calcium silicate–carbonate minerals at HP–HT conditions. We heated up the W gasket as described in the "Methods" section, and we characterized the heated area afterwards through XRD mapping. Tilleyite $Ca_5(Si_2O_7)(CO_3)_2$ was compressed to 9 GPa and laser-heated above 1500 °C, and the XRD patterns of the recovered sample located close to the pressure chamber walls revealed that a chemical reaction had occurred (see Figure 1 and Table 1).

Table 1. List of phases identified at ambient conditions in the recovered sample after compression and laser-heating the silicate-carbonate tilleyite $Ca_5(Si_2O_7)(CO_3)_2$ in the presence of tungsten.

Compound	Space Group (Nr.)	Lattice Parameters (Å)
Scheelite-$CaWO_4$	$I4_1/a$ (88)	$a = 5.2404(10)$ $c = 11.374(4)$
bcc-W	$Im\bar{3}m$ (229)	$a = 3.163(3)$
fcc-Cu	$Fm\bar{3}m$ (225)	$a = 3.614(3)$
Quartz-SiO_2 (distorted)	$C222$ (21)	$a = 4.996$ $b = 8.668$ $c = 5.46$
B1-CaO	$Fm\bar{3}m$ (225)	$a = 4.818$
Aragonite-$CaCO_3$	$Pmcn$ (62)	$a = 4.962$ $b = 7.969$ $c = 5.743$
Diamond -C	$Fd\bar{3}m$ (227)	$a = 3.567$

Note that the temperature at the reaction's location was not measured, but it must be considerably lower than that at the center of the hotspot [35]. The most intense diffraction peaks correspond to scheelite-type $CaWO_4$, but peaks corresponding to bcc-W (gasket), fcc-Cu (laser absorber), quartz-SiO_2 [36], B1-type CaO and aragonite-$CaCO_3$ are also visible.

These compounds are compatible with silicate–carbonate decomposition and a subsequent redox reaction involving tungsten and calcium carbonate. Then, the following sequence of reactions:

$$\text{Decomposition: } Ca_5(Si_2O_7)(CO_3)_2 \rightarrow 2 \cdot CaCO_3 + 2 \cdot SiO_2 + 3 \cdot CaO$$

$$\text{Redox reaction: } 2 \cdot W + 3 \cdot CaCO_3 \rightarrow 2 \cdot CaWO_4 + CaO + 3 \cdot C$$

could eventually give rise to the oxidation of tungsten, from W^0 to W^{+6}, and the reduction of carbon, from C^{+4} to C^0. Note, however, that there exists a previous HP–HT study carried out by Liu and Lin [37] where they reported the decomposition of tilleyite at 4 GPa into $CaSiO_3$ wollastonite, Ca_2SiO_4 larnite and $CaCO_3$ aragonite. Of these phases, only aragonite was found in our recovered sample, and for this reason, this decomposition path was not considered. Figure 1 shows the diffraction image with the trace of diamond formation (((111) reflection), which could not be discerned in the XRD integrated pattern. The chemical reactivity of calcium carbonate under reducing conditions was also reported at 5.5 GPa and 1350 °C when chemically interacted with iron metal, forming Ca-ferrites and graphite [38]. These redox reactions could take place through a previous decomposition of $CaCO_3$ into CaO and CO_2, or CaO, O_2 and C, but the temperature values cannot be compared, since the latter decomposition was reported to occur at approximately 3200 °C between 9 and 21 GPa [39]. In fact, Chepurov et al. mentioned that they detected $CaWO_4$ at the contact between $CaCO_3$ and the W-foil used to minimize the contact of Fe with the Pt-ampoule in the multi-anvil HP–HT experiments [38]. It should be also stressed that, in the case at hand, the reaction between tilleyite and tungsten could alternatively occur as a one-step process with W acting as catalyst.

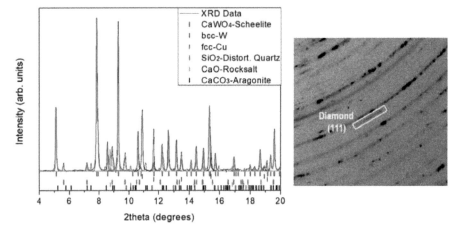

Figure 1. (Right) LeBail fit of the X-Ray diffractometry (XRD) pattern of quenched recovered sample obtained close to the W gasket after compression (8 GPa) and heating (>1500° C) of tilleyite. Wavelength: 0.4246 Å. Color profiles and vertical marks correspond to the different phases found: scheelite-$CaWO_4$ (red), bcc-W (blue, gasket), fcc-Cu (green, heater), distorted quartz-SiO_2 (magenta), B1-CaO (orange) and aragonite–$CaCO_3$ (black). (Left) Diamond (111) reflection was also found at $2\theta = 11.835°$ after inspection of the XRD image.

3.2. Re Oxidation

Oxidation of rhenium in oxygen/air occurs above 420 °C at ambient pressure, and takes place by the formation of volatile oxides, but information available on heterogeneous equilibria in the rhenium–oxygen system is not always consistent. Thus, in a previous work [40], the oxidation reaction has been examined at 427 °C with X-ray photoelectron and ultra-violet photoelectron spectroscopies. ReO_2, ReO_3 and Re_2O_7 were found on the Re surface [40].

In the temperature range 600 – 1300 °C, Re_2O_7 was identified by X-ray examination as the only volatile [41] whereas, at higher temperatures (1500-1900 °C), ReO_3 was the primary oxide present in the condensed vapor deposits [42]. When mixtures of Re metal and ReO_3 in different gross compositions are heated up, a stable intermediary ReO_2 phase is formed [43]. However, differences in the thermodynamic characteristics of sublimation, decomposition and disproportionation of these phases are significant [44]. The reasons for literature discrepancies could be the Re surface state and different oxidizing conditions. All in all, like tungsten, rhenium has a limited use as an HT structural material due to its low resistance to oxidation.

At high pressures (from 8 to 25 GPa), no Re reactivity has been reported with solid or fluid O_2 up to 1000 °C [45]. In the presence of supercritical fluid $H_2O–O_2$ mixtures, however, Re undergoes a series of reactions to form perrhenic acid ($HReO_4$) and its hydrates at pressures as low as 0.5 GPa and at room temperature [46]. From a practical perspective for high pressure research, it is important to know the potential chemical reactivity between the Re gasket and solid and fluid samples. Re + CO_2 interaction at HP and HT was recently studied by our group. Our results demonstrate the existence of carbon dioxide reduction reaction in the presence of Re metal at high pressures (8 and 24 GPa) and high temperatures (above 1200 °C), obtaining orthorhombic β-ReO_2 as high-pressure high-temperature reaction product (see Figure 2 and Table 2).

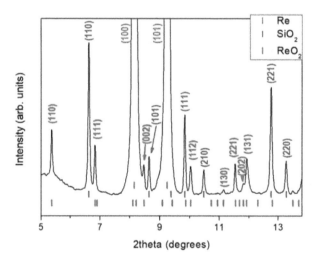

Figure 2. In situ XRD pattern of compressed (~24 GPa) laser-heated (1300 °C) rhenium in the presence of CO_2. Wavelength: 0.3344 Å. Silica was used as thermal isolating material. At these P–T conditions, CO_2 is liquid and, consequently, no diffraction signal of this phase was observed. Above 1250 °C, the diffraction peaks of orthorhombic β-ReO_2 were clearly visible. Green, blue and red vertical ticks correspond to hcp-Re, SiO_2-stishovite and β-ReO_2, respectively.

Table 2. List of phases identified in the in situ XRD pattern of compressed (~24 GPa) laser-heated (1300 °C) rhenium in the presence of CO_2.

Compound	Space Group (Nr.)	Lattice Parameters (Å)
β-ReO_2	*Pbcn* (60)	$a = 4.669(3)$ $b = 5.556(3)$ $c = 4.524(2)$
Stishovite-SiO_2	$P4_2/mnm$ (136)	$a = 4.0929(11)$ $c = 2.639(2)$
hcp-Re	$P6_3/mmc$ (194)	$a = 2.717$ $c = 4.413$

This compound, with Re(IV) oxidation state, is stable and solid at temperatures between 1250 and 1500 °C while compressed at 24 GPa, in comparison to that occurring in vacuum conditions, where ReO_2 dissociates at 850-1050 °C into $Re_2O_7(g)$ and metallic Re [42,47].

This rhenium oxide still exists at 50 GPa after heating above 1600 °C (see Figure 3 and Table 3), together with diamond. In fact, its stability field extends above 1 Mbar, as shown in a recent publication [48]. No Re_2O_7 or ReO_3 were found during our experiments, β-ReO_2 being recovered after pressure and temperature quenching [21].

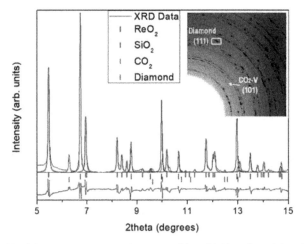

Figure 3. LeBail fit of the in situ XRD pattern of compressed (~50 GPa) laser-heated (1600 °C) rhenium in the presence of CO_2. Wavelength: 0.3344 Å. Silica was used as thermal isolating material. Experimental data are depicted as a solid black line, and calculated XRD profiles of β-ReO_2, SiO_2 stishovite, CO_2–V and diamond are represented as red, blue, dark yellow and magenta lines, respectively. The full difference profile is represented at the bottom as a green line. Inset: XRD image showing the (111) diffraction peak of diamond, after CO_2 reduction.

Table 3. List of phases identified in the in situ XRD pattern of compressed (~50 GPa) laser-heated (1600 °C) rhenium in the presence of CO_2. The lattice parameters of the structures recovered at ambient conditions can be found in Reference. [21].

Compound	Space Group (Nr.)	Lattice parameters (Å)
β-ReO_2	*Pbcn* (60)	$a = 4.571(3)$ $b = 5.492(2)$ $c = 4.462(2)$
Stishovite-SiO_2	$P4_2/mnm$ (136)	$a = 4.027(2)$ $c = 2.607(2)$
CO_2-V	$I\bar{4}2d$ (122)	$a = 3.563$ $c = 5.905$
Diamond -C	*Fd-3m* (227)	$a = 3.453$

From these results, we can infer that the following reaction is taking place: $Re + CO_2 \rightarrow ReO_2 + C(d)$. This is in agreement with the observation of (111) reflections of diamond in XRD patterns, as shown in the CCD image depicted in the inset of Figure 3. β-ReO_2 was quenched both in temperature and in pressure, and recovered at ambient conditions. Thus, the existence of rhenium dioxide through such a wide P–T range illustrates the increasing thermal stability with pressure of this oxidation state IV phase to disproportionation to rhenium metal and higher oxidation state oxides.

4. Conclusions

These results allow drawing an important conclusion: Refractory metals rhenium and tungsten chemically interact with CO_2 and carbonates at moderate high temperatures (T > 1300 °C), even at high pressure. These redox reactions give rise to the formation of metal oxide species and diamond. This fact suggests that extreme care should be taken in high-pressure high-temperature experiments involving the aforementioned oxidized carbon species in order to avoid non desired chemical reactions. Our experience suggests following some methodological recommendations to rule out the possibility of chemical interaction in laser-heated diamond-anvil cell experiments:

(i)   Center the laser hotspot in the pressure chamber, place it as far as possible from the gasket.
(ii) If heating with a Nd-doped YAG laser, λ = 1.06 μm, do not use Re or W as an internal absorber of the laser radiation. As seen above, they easily react with the sample above 1300 °C. Pay special attention when using chemically inert Au as a heater. In the absence of a proper sample containment, this low melting temperature metal [49] would disperse across the pressure chamber and could eventually reach the Re or W gasket. A small and perfectly centered heater foil will minimize this possibility. Something similar occurs with soft Cu [49] or Pt [50,51].
(iii) A convenient approach is heating the optically transparent samples with a CO_2 laser, λ = 10.6 μm. This wavelength equates to a photon energy of the same order of magnitude as lattice phonons in covalent crystals, and it is usually absorbed by minerals and transparent oxides [52]. Note, however, that the focal size of the laser spot in this case is larger than in the case of solid-state lasers [12], making the heating of the gasket metal more likely. Steering and focusing of the beam becomes then critical. The use of gold-coated or diamond gaskets could help in preventing potential metal oxidation products.

From the data obtained, it follows that the decomposition of carbon dioxide and carbonates occurs at relatively low temperatures under reducing conditions, due to chemical interaction with refractory metals.

Author Contributions: Conceptualization, D.S.-P.; laser-heating DAC XRD experiments, D.S.-P., T.M., J.R.-F. and D.D.; Formal analysis, R.C.-J. and D.S.-P.; Writing, R.C.-J. and D.S.-P.; Review, all the authors.

Funding: This work was supported by the Spanish Ministerio de Ciencia, Innovación y Universidades (MICINN) under Grants PGC 2018-097520-A-I00 and RED2018-102612-T (MALTA Consolider), and by the Generalitat Valenciana Grant PROMETEO/2018/123. D. S-P. acknowledges MINECO for a Ramón y Cajal (RyC-2014-15643).

Acknowledgments: Authors thank the APS and Diamond synchrotrons for beamtime allocation at GSECARS and I15 lines and the Mineralogy and Meteoritic Division of the Peaboby Museum for the provision of natural samples.

Conflicts of Interest: The authors declare no conflict of interest.

References

1.   Boehler, R.; Santamaria-Perez, D.; Errandonea, D.; Mezouar, M. Melting, density, and anisotropy of iron at core conditions: New x-ray measurements to 150 GPa. *J. Phys. Conf. Ser.* **2008**, *121*, 022018. [CrossRef]
2.   Boulard, E.; Pan, D.; Galli, G.; Liu, Z.X.; Mao, W.L. Tetrahedrally coordinated carbonates in Earth's lower mantle. *Nat. Commun.* **2015**, *6*, 6311. [CrossRef] [PubMed]
3.   Litasov, K.; Othani, E. The solidus of carbonated eclogite in the system CaO-Al$_2$O$_3$-MgO-SiO$_2$-Na$_2$O-CO$_2$ to 32 GPa and carbonatite liquid in the deep mantle. *Earth Planet. Sci. Lett.* **2010**, *295*, 115. [CrossRef]
4.   Dasgupta, R. Ingassing, storage, and outgassing of terrestrial carbon through geologic time. *Rev. Mineral. Geochem.* **2013**, *75*, 183. [CrossRef]
5.   Stagno, V.; Ojwang, D.O.; McCammon, C.A.; Frost, D.J. The oxidation state of the mantle and the extraction of carbon from Earth's interior. *Nature* **2013**, *493*, 84. [CrossRef]
6.   Santamaria-Perez, D.; Ruiz-Fuertes, J.; Peña-Alvarez, M.; Chulia-Jordan, R.; Marqueño, T.; Zimmer, D.; Gutierrez-Cano, V.; MacLeod, S.; Gregoryanz, E.; Popescu, C.; et al. Post-tilleyite, a dense calcium silicate-carbonate phase. *Sci. Rep.* **2019**, *9*, 7898. [CrossRef]

7. Dunstan, D.J.; Spain, I.L. Technology of diamond anvil high-pressure cell. I. Principles, design and construction. *J. Phys. E Sci. Instrum.* **1989**, *22*, 913. [CrossRef]
8. Santamaria-Perez, D.; Kumar, R.S.; Santos-Garcia, A.J.D.; Errandonea, D.; Chulia-Jordan, R.; Saez-Puche, R.; Rodriguez-Hernandez, P.; Muñoz, A. High-pressure transition to the post-barite phase in $BaCrO_4$ hashemite. *Phys. Rev. B* **2012**, *86*, 094116. [CrossRef]
9. Aguado, F.; Santamaria-Perez, D. Structure of Earth's interior. In *An Introduction to High-Pressure Science and Technology*; CRC Press: Boca Raton, FL, USA, 2016.
10. Lavina, B.; Dera, P.; Meng, Y. Synthesis and microdiffraction at extreme pressures and temperatures. *J. Vis. Exp.* **2013**, *80*, 50613. [CrossRef]
11. Zou, G.; Ma, Y.; Mao, H.K.; Hemley, R.J.; Gramsch, S.A. A diamond gasket for the laser-heated diamond anvil cell. *Rev. Sci. Instrum.* **2001**, *72*, 1298. [CrossRef]
12. Salamat, A.; Fischer, R.A.; Briggs, R.; McMahon, M.I.; Petitgirard, S. In situ synchrotron X-ray diffraction in the laser-heated diamond anvil cell: Melting phenomena and synthesis of new materials. *Coord. Chem. Rev.* **2014**, *278*, 15–30. [CrossRef]
13. Santamaria-Perez, D.; Marqueño, T.; MacLeod, S.; Ruiz-Fuertes, J.; Daisenberger, D.; Chulia-Jordan, R.; Errandonea, D.; Jorda, J.L.; Rey, F.; McGuire, C.; et al. Structural evolution of CO_2-filled pure silica LTA zeolite under high-pressure high-temperature conditions. *Chem. Mater.* **2017**, *29*, 4502–4510. [CrossRef]
14. Boehler, R.; Chopelas, A. A new approach to laser-heating in high-pressure mineral physics. *Geophys. Res. Lett.* **1991**, *18*, 1147. [CrossRef]
15. Benedetti, L.R.; Loubeyre, P. Temperature gradients, wavelength-dependent emissivity and accuracy of high temperatures measured in a LHDAC. *High Press. Res.* **2004**, *24*, 423–445. [CrossRef]
16. Schultz, E.; Mezouar, M.; Crichton, W.; Bauchau, S.; Blattmann, G.; Andrault, D.; Fiquet, G.; Boehler, R.; Rambert, N.; Sitaud, B.; et al. Double-sided laser heating system for in situ HP-HT monochromatic XRD at the ESRF. *High Press. Res.* **2005**, *25*, 71–83. [CrossRef]
17. Freund, H.J.; Roberts, M.W. Surface chemistry of carbon dioxide. *Surf. Sci. Rep.* **1996**, *25*, 225–273. [CrossRef]
18. Peled, H.; Asscher, M. Dissociative chemisorption of CO_2 on Re(0001) single crystal surface. *Surf. Sci.* **1987**, *183*, 201–215. [CrossRef]
19. Walsh, P.N.; Quets, J.M.; Graff, R.A.; Ladd, I.R. Kinetics of the oxidation of tungsten by CO_2 at high temperatures. *J. Chem. Phys.* **1967**, *46*, 3571–3576. [CrossRef]
20. Santamaria-Perez, D.; McGuire, C.; Mahkluf, A.; Kavner, A.; Chulia-Jordan, R.; Rey, F.; Pellicer-Porres, J.; Martinez-Garcia, D.; Rodriguez-Hernandez, P.; Muñoz, A. Strongly-driven $Re + CO_2$ redox reaction at high-pressure and high-temperature. *Nat. Commun.* **2016**, *7*, 13647. [CrossRef]
21. Santamaria-Perez, D.; McGuire, C.; Kavner, A.; Chulia-Jordan, R.; Pellicer-Porres, J.; Martinez-Garcia, D.; Doran, A.; Kunz, M.; Rodriguez-Hernandez, P.; Muñoz, A. Exploring the chemical reactivity between carbon dioxide and three transition metals (Au, Pt, and Re) at high pressures and temperatures. *Inorg. Chem.* **2016**, *55*, 10793. [CrossRef]
22. Grice, J.D. The structure of spurrite, tileyite and scawtite, and relationships to other silicate-carbonate minerals. *Can. Mineral.* **2005**, *43*, 1489–1500. [CrossRef]
23. Santamaria-Perez, D.; Ruiz-Fuertes, J.; Marqueño, T.; Pellicer-Porres, J.; Chulia-Jordan, R.; MacLeod, S.; Popescu, C. Structural behavior of natural silicate-carbonate spurrite mineral, $Ca_5(SiO_4)_2(CO_3)$, under high-pressure high-temperature conditions. *Inorg. Chem.* **2018**, *57*, 98–105. [CrossRef] [PubMed]
24. Isaacs, T. The mineralogy and chemistry of nickel carbonates. *Mineral. Mag.* **1963**, *33*, 663–678. [CrossRef]
25. Otto, H.H.; Meibohm, M. Crystal structure of copper polysilicate, $Cu[SiO_3]$. *Z. Kristallogr.* **1999**, *214*, 558–565. [CrossRef]
26. Mao, K.K.; Xu, J.; Bell, P.M. Calibration of the Ruby Pressure Gauge to 800-Kbar under Quasi-Hydrostatic Conditions. *J. Geophys. Res.* **1986**, *91*, 4673–4676. [CrossRef]
27. Dewaele, A.; Loubeyre, P.; Mezouar, M. Equations of state of six metals above 94 GPa. *Phys. Rev. B* **2004**, *70*, 094112. [CrossRef]
28. Jeanloz, R.; Godwal, B.K.; Meade, C. Static Strength and Equation of State of Rhenium at Ultra-High Pressures. *Nature* **1991**, *349*, 687–689. [CrossRef]
29. Santamaria-Perez, D.; Ross, M.; Errandonea, D.; Mukherjee, G.D.; Mezouar, M.; Boehler, R. X-ray diffraction measurements of Mo melting to 119 GPa and the high pressure phase diagram. *J. Chem. Phys.* **2009**, *130*, 124509. [CrossRef]

30. Prescher, C.; Prakapenka, V.B. DIOPTAS: A Program for Reduction of Two-Dimensional X-Ray Diffraction Data and Data Exploration. *High Press. Res.* **2015**, *35*, 223–230. [CrossRef]
31. Holland, T.J.B.; Redfern, S.A.T. Unit cell refinement from powder diffraction data: The use of regression diagnostics. *Mineral. Mag.* **1997**, *61*, 65–77. [CrossRef]
32. Rodriguez-Carvajal, J. Recent Advances in Magnetic-Structure Determination by Neutron Powder Diffraction. *Phys. B* **1993**, *192*, 55–69. [CrossRef]
33. Nolze, G.; Kraus, W. Powdercell 2.0 for Windows. *Powder Diffr.* **1998**, *13*, 256–259.
34. Ivanov, V.Y.; Nechiporenko, Y.P.; Yefimenko, L.N.; Yurchenko, M.I. *High Temperature Oxidation Protection of Tungsten*; NASA Translation of "Zashchita Vol'frama ot Okisleniya pri Vysokikh Temperaturakh" Atom Press: Moscow, Russia, 1968.
35. Boehler, R. High-pressure experiments and phase diagram of lower mantle and core materials. *Rev. Geophys.* **2000**, *38*, 221–245. [CrossRef]
36. Marqueño, T.; Santamaria-Perez, D.; Ruiz-Fuertes, J.; Chulia-Jordan, R.; Jorda, J.L.; Rey, F.; McGuire, C.; Kavner, A.; MacLeod, S.; Daisenberger, D.; et al. An ultrahigh CO_2-loaded silicalite zeolite: Structural stability and physical properties at high pressures and temperaturas. *Inorg. Chem.* **2018**, *57*, 6447–6455. [CrossRef]
37. Liu, L.G.; Lin, C.C. High-pressure phase transformations of carbonates in the system CaO-MgO-SiO_2-CO_2. *Earth Planet. Sci. Lett.* **1995**, *134*, 297–305. [CrossRef]
38. Chepurov, A.I.; Sonin, V.M.; Zhimulev, E.I.; Chepurov, A.A.; Tomilenko, A.A. On the formation of element carbon during decomposition of $CaCO_3$ at high P-T parameters under reducing conditions. *Dokl. Earth Sci.* **2011**, *441*, 1738–1741. [CrossRef]
39. Bayarjargal, L.; Shumilova, T.G.; Friedrich, A.; Winkler, B. Diamond formation from $CaCO_3$ at high pressure and temperature. *Eur. J. Mineral.* **2010**, *22*, 29–34. [CrossRef]
40. Alnot, M.; Ehrhardt, J. A study of oxygen-rhenium interactions in various pressure and temperature conditions. *J. Chim. Phys.* **1982**, *79*, 735–739. [CrossRef]
41. Phillips, W.L., Jr. The rate of oxidation of rhenium at elevated temperatures in air. *J. Less Common Met.* **1963**, *5*, 97–100. [CrossRef]
42. Chou, T.C.; Joshi, A.; Packer, C.M. Oxidation behavior of rhenium at high temperatures. *Scr. Metall. Mater.* **1993**, *28*, 1565–1570. [CrossRef]
43. Magneli, A. Studies on rhenium oxides. *Acta Scand.* **1957**, *11*, 28–33. [CrossRef]
44. Shcheglov, P.A.; Drobot, D.V. Heterogeneous equilibria in the rhenium-oxygen system. *Russ. J. Phys. Chem.* **2006**, *80*, 1819–1825. [CrossRef]
45. Santoro, M.; Gregoryanz, E.; Mao, H.K.; Hemley, R.J. New phase diagram of oxygen at high pressures and temperatures. *Phys. Rev. Lett.* **2004**, *93*, 265701–265704. [CrossRef] [PubMed]
46. Chellappa, R.S.; Somayazulu, M.; Hemley, R.J. Rhenium reactivity in H_2O-O_2 supercritical mixtures at high pressures. *High Press. Res.* **2009**, *29*, 792–799. [CrossRef]
47. Haynes, W.M. *Physical Constants of Inorganic Compounds in CRC Handbook of Chemistry and Physics*, 97th ed.; CRC Press: Boca Raton, FL, USA, 2016.
48. Dziubek, K.F.; Ende, M.; Scelta, D.; Bini, R.; Mezouar, M.; Garbarino, G.; Miletich, R. Crystalline polymeric carbón dioxide stable at megabar pressures. *Nat. Commun.* **2018**, *9*, 3148. [CrossRef]
49. Hieu, H.K.; Ha, N.N. High pressure melting curves of silver, gold and copper. *AIP Adv.* **2013**, *3*, 112125. [CrossRef]
50. Kavner, A.; Jeanloz, R. High-pressure melting curve of platinum. *J. Appl. Phys.* **1998**, *83*, 7553–7559. [CrossRef]
51. Anzellini, S.; Monteseguro, V.; Bandiello, E.; Dewaele, A.; Burakovsky, L.; Errandonea, D. In situ characterization of the high pressure—High temperature melting curve of platinum. *Sci. Rep.* **2019**, *9*, 13034. [CrossRef]
52. Smith, D.; Smith, J.S.; Childs, C.; Rod, E.; Hrubiak, R.; Shen, G.; Salamat, A. A CO_2 laser heating system for in situ high pressure-temperature experiments at HPCAT. *Rev. Sci. Instrum.* **2018**, *89*, 083901. [CrossRef]

Article

Fe Melting Transition: Electrical Resistivity, Thermal Conductivity, and Heat Flow at the Inner Core Boundaries of Mercury and Ganymede

Innocent C. Ezenwa [†] and Richard A. Secco *

Department of Earth Sciences, University of Western Ontario London, London, ON N6A5B7, Canada
* Correspondence: secco@uwo.ca
† Now at Institute for Planetary Materials, Okayama University, Misasa, Tottori 682-0193, Japan.

Received: 10 June 2019; Accepted: 12 July 2019; Published: 15 July 2019

Abstract: The electrical resistivity and thermal conductivity behavior of Fe at core conditions are important for understanding planetary interior thermal evolution as well as characterizing the generation and sustainability of planetary dynamos. We discuss the electrical resistivity and thermal conductivity of Fe, Co, and Ni at the solid–liquid melting transition using experimental data from previous studies at 1 atm and at high pressures. With increasing pressure, the increasing difference in the change in resistivity of these metals on melting is interpreted as due to decreasing paramagnon-induced electronic scattering contribution to the total electronic scattering. At the melting transition of Fe, we show that the difference in the value of the thermal conductivity on the solid and liquid sides increases with increasing pressure. At a pure Fe inner core boundary of Mercury and Ganymede at ~5 GPa and ~9 GPa, respectively, our analyses suggest that the thermal conductivity of the solid inner core of small terrestrial planetary bodies should be higher than that of the liquid outer core. We found that the thermal conductivity difference on the solid and liquid sides of Mercury's inner core boundary is ~2 W(mK)$^{-1}$. This translates into an excess of total adiabatic heat flow of ~0.01–0.02 TW on the inner core side, depending on the relative size of inner and outer core. For a pure Fe Ganymede inner core, the difference in thermal conductivity is ~7 W(mK)$^{-1}$, corresponding to an excess of total adiabatic heat flow of ~0.02 TW on the inner core side of the boundary. The mismatch in conducted heat across the solid and liquid sides of the inner core boundary in both planetary bodies appears to be insignificant in terms of generating thermal convection in their outer cores to power an internal dynamo suggesting that chemical composition is important.

Keywords: melting transition; Fe; electrical resistivity; thermal conductivity; high pressure; heat flow; thermal and chemical convection

1. Introduction

The processes of magnetic field generation and sustainability in planetary bodies depend on the composition and thermal state of their cores. Among the rocky planetary bodies with an active dynamo, Mercury has the weakest internally generated magnetic field, with a surface field strength of ~0.3 µT or ~1% compared with the Earth's field. Though a possible remnant crustal magnetization has been suggested [1], a self-sustained dynamo in Mercury's Fe core is the most plausible source of its global magnetic field [2]. A recent study suggests that a double-diffusive convective regime operates, where both thermal and compositional convection drive the system [3]. Earth-based radar measurements of subtle deviations from the mean resonant spin rate of Mercury demonstrate that Mercury's mantle is decoupled from its liquid or partially molten core [4–7]. This supports earlier assertions that Mercury has a molten outer core [8,9]. Recent geodetic constraints on the interior of

Mercury from the MESSENGER spacecraft are consistent with a high degree of internal differentiation and a solid inner core with a radius of 0.4–0.7 times the outer core radius [10]. While the role of a solid inner core and its contribution to chemical composition convection in a liquid outer core was recognized long ago for Mercury [9], the possibility of Mercury's weak surface magnetic field resulting from dynamo action in a thin shell geometry has been shown more recently [11]. For the case of Ganymede, although remnant magnetization cannot be completely ruled out as the source of its magnetic field, magneto-convection in its core like that in the Earth has been suggested [12]. However, dynamo action in Ganymede differs from that of the Earth due to the presence of the strong nearby Jovian magnetic field. Thus, the magneto-hydrodynamic equation is variant under the transformation of $B \rightarrow -B$ with a directional preference for the self-generated field which could lead to a non-reversing magnetic field unlike the reversing nature of the geomagnetic field [13].

Convection in a terrestrial planetary core can arise from heat transport in excess of the conducted heat (i.e., by thermal convection) or from exsolution and precipitation of core components (i.e., by chemical convection) such as Fe at the inner core boundary, ICB [14], or SiO_2 [15], or MgO [16] at the core-mantle boundary as suggested for Earth. Recent studies have both challenged [17] and supported [18] the MgO precipitation model. There is continuing debate about the relative contributions of thermal vs. chemical convection throughout the thermal and chemical evolution of terrestrial-like planetary cores [19–27]. In a purely thermally driven core, as expected in the early stages of core evolution where a solid inner core and chemical convection are absent, thermal convection is the only source of energy to power the dynamo. Thus, knowing the relative contribution of thermal conduction and thermal convection to thermal transport in the core is essential to understanding the source of energy of a core-generated magnetic field, inner core age, and thermal evolution of the core.

The contribution of conductive heat flow in the core requires the thermal conductivity of core material to be known. Thermal conductivity for metals can be approximated using the Wiedemann-Franz relation if values of electrical resistivity of Fe at high pressure *(P)* and temperature *(T)* conditions are known. This approach is often adopted [19–21] over direct measurement of thermal conductivity due to the enormous challenges in maintaining a well-controlled *T*-gradient in a small sample at very high *T* and *P* conditions [22].

Much recent attention in attempts to determine core conductive heat flow is focused on Earth. The electrical resistivity of the Earth's core was estimated to be 350–450 μΩcm from analysis of low *P* static and high *P* dynamic shock compression data [23,24], leading to calculated values of core thermal conductivity of 30–50 W(mK)$^{-1}$ that are generally consistent with the only experimental measurements of thermal conductivity made on Fe at core *P,T* in the diamond anvil cell [25]. However, theoretical [26,27] and experimental investigations [19,20] have suggested a much lower core resistivity (and thermal conductivity values greater than 90 W(mK)$^{-1}$) for the outer core because of the effect of resistivity saturation at high *T*.

Theoretical investigation by Wagle and Steinle-Neumann [28] used a thermodynamic model and the Ziman approximation to determine the resistivity of solid and liquid Fe up to core *P* and *T* conditions. They found a decreasing resistivity change ($\rho_{liq} - \rho_{sol}$) on melting with increasing *P*. From their experimental resistivity data on *hcp* Fe at high *P* and room *T* in the diamond anvil cell (DAC) Gomi et al. [19] asserted that Fe resistivity at core conditions is close to saturation and therefore the resistivity change on melting should be negligible. From their DAC measurements, Ohta et al. [20] reported, ~20% change in Fe resistivity on melting from the *fcc* Fe phase at 51 GPa. However, lower pressure measurements in the multi-anvil press of the *T*-dependent electrical resistivity of Co up to 5 GPa [29], Ni up to 9 GPa [30], and Fe up to 12 GPa [21] demonstrated an increasing change of resistivity on melting with increasing *P*. This lower pressure regime is relevant for thermal transport at the ICB in the small planetary bodies Mercury and Ganymede.

From these multi-anvil studies, the resistivity of liquid Co and Ni along their respective *P*-dependent melting boundaries remained invariant while Fe showed a decreasing trend of resistivity below the δ-γ-liquid triple point at ~5 GPa but then remains constant above the triple point *P*. Although experimental investigation of electrical resistivity of the α, γ, ε phases of Fe at combined static high *P* and *T* conditions have been made [20,21,31–33], its behavior through the melting transition is still contentious, hence, a detailed discussion is needed.

Generally, for the *3d* ferromagnetic metals Fe, Co, and Ni, the weak interaction of *d* electrons gives rise to an ordered magnetic state characterized by different numbers of electrons with up and down spins. Since the electronic state of a metal can be probed through the investigation of its electrical resistivity, and since electronic state and magnetism in a metal are interwoven [34], electrical resistivity can also provide information about the magnetic state of these metals. We discuss qualitatively our observation of the effect of decreasing magnon-induced electron scattering with increasing *P* on the *T*-dependent electrical resistivity of these metals at the solid–liquid transition. In addition, we discuss the possible implications of this behavior on the thermal conductivity and heat flow at the ICB of Mercury and Ganymede.

2. Electronic Scattering in Ferromagnetic Metals

For unfilled *d* band transition metals, *s–d* scattering dominates over normal *s–s* electron scattering as *T* increases due to the high density of *d*-band states. This is generally understood in Mott's *s–d* scattering model [35]. For diamagnetic metals at 1 atm, those with filled *d*-bands (e.g., Cu, Ag, and Au), the combined results from many studies show that their *T*-dependent resistivity in the solid state follows a linear dependence on *T* [36]. Similarly, for some paramagnetic metals at 1 atm (e.g., Pt, Pd, etc), their resistivity follows a near-linear dependence on *T* [37,38]. However, for the ferromagnetic metals, resistivity follows a T^2-dependence below the Curie point and *T*-dependence above the Curie point [39–43]. With increasing *T*, the increasing phonon and spin-disorder induced scattering (magnon-induced scattering) of the highly mobile *s* conduction electrons into unfilled *d*-band states leads to decreased mobility of *s* electrons and higher resistivity. Below the Curie *T*, electron scattering is caused by a combination of phonon- and magnon-induced scattering, as well as a contribution from the asymmetry of the Fermi surface (Mott, 1964). Above the Curie *T* in the paramagnetic state, paramagnon-induced scattering tends toward a constant value while the phonon-induced scattering continues to increase with increasing *T* and therefore controls the *T*-dependent resistivity trend. Even if only qualitatively known, the relative contribution of the different scattering mechanisms is important for our study.

Probing band structure effects through resistivity investigation of the ferromagnetic metals under *P* and *T* conditions may provide an understanding of the complex electron scattering mechanisms which can occur due to topological features of the Fermi surfaces, Fermi level position, and energy gap between the spin sub-bands (δE_{ex}). Experimental studies mapping the Fermi surfaces of Fe, Co, and Ni have been accomplished primarily through the use of de Hass-van Alphen (dHvA) oscillatory effects [44] along with magnetoresistance investigations that have confirmed the existence of a complicated open orbit topology of the Fermi surfaces of these metals [45–47]. In *3d* ferromagnetic metals, magnetism is largely caused by electrons in the high density of states *3d* bands at the Fermi level. Angle-resolved photoemission studies demonstrated that the decrease in δE_{ex} above the Curie *T* for Fe, Co, and Ni is due to the energy of the spin-down sub-band shifting ~2–3 times faster than the spin-up sub-band [47–49]. Interestingly, values of δE_{ex} for Fe, Co, and Ni, and the population of the *3d*-band at ambient conditions correlate with the magnitude of the abrupt change in the electrical resistivity on melting as shown in Figure 1. Ni has the highest number of *3d* electrons (least number of unoccupied *3d* states), lowest value of δE_{ex}, and it has the greatest change in resistivity on melting as shown in Figure 1. Fe has the least number of *3d* electrons (highest number of unoccupied *3d* states), highest value of δE_{ex}, and it has the smallest change in resistivity [38] on melting. This implies that Fe, having the highest number of unoccupied *3d* states with the greatest contribution of *s–d* electron scattering induced by phonons and magnons, should show a smaller change in the resistivity on melting. The small change in the resistivity on melting can thus be explained by the extensive pre-melting scattering relative to the additional scattering arising from atomic structural change on melting. Conversely, just prior to melting, Ni has the least contribution of scattering from phonon- and magnon-induced s–d electron scattering and therefore shows a larger jump in resistivity arising from the relatively larger scattering contribution on melting due to the effect of atomic structural change.

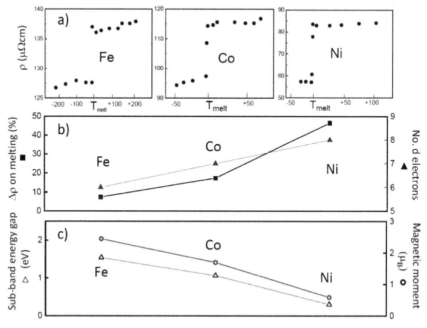

Figure 1. Data of 1 atm of Fe, Co, and Ni. (a) Resistivity discontinuity [38] on melting (note the differences in resistivity scale for Fe, Co, and Ni, whose melting T's are 1809 K, 1768 K, and 1728 K, respectively); (b) resistivity discontinuity on melting and number of occupied d-electron band, and; (c) sub-band energy gap and magnetic moments [34,50].

Focusing on the magnitude of resistivity of Fe, Co, and Ni in the solid state just prior to melting, the T-dependent resistivity of paramagnetic Fe above the Curie T is similar in trend to the T-dependent resistivity of paramagnetic Pd as phonon-induced scattering dominates in both cases, as shown in Figure 2. On the other hand, an x-ray magnetic circular dichroism study [51] showed that the net magnetic moment of Fe decreases with increasing P and vanishes at ~18 GPa at ambient T while both Ni and Co remain ferromagnetic to well over 100 GPa. The increasing population of d-band electrons due to s–d hybridization with increasing P [52–54] will lead to termination of magnetism. It is expected that the relative change in the positions of s and d bands in Fe, Co, and Ni with increasing P control the rate of d-band population and loss/retention of magnetism. Theoretical investigation demonstrated that the non-spin state of Fe is the most energetically favored electronic state at high P [51]. Through P-induced reduction of magnetism and tendency toward spin disorder saturation above the Curie T, these two effects combine to reduce or eliminate the contribution of paramagnon-induced electron scattering in the T-dependent resistivity region of ferromagnetic metals at high P.

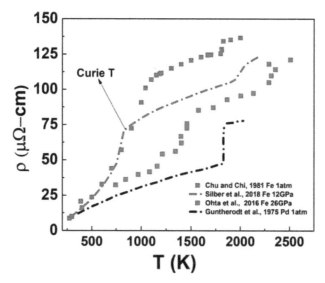

Figure 2. *T*-dependent electrical resistivity of Fe at different *P* compared with Pd at 1 atm.

3. Results and Discussion

3.1. Electrical Resistivity and Thermal Conductivity at the Melting Transition

As shown in Figure 3, recent experimental investigations of the *T*-dependence of resistivity of Co [29], Ni [30], and Fe [21] at high *P* demonstrate that the effect of *P* on resistivity is greater in the high *T* region (*T*-dependent resistivity) above the Curie *T* than in the low *T* region (T^2-dependent resistivity) below the Curie *T*. This suggests that magnon-induced scattering is less sensitive to *P* than is scattering caused by simple phonon scattering or phonon scattering that results in *s*-electrons being scattered into *d*-states. This appears intuitively expected as phonon scattering or phonon-induced *s–d* scattering arise from atomic vibration whereas magnon-induced scattering is operative at the electronic level.

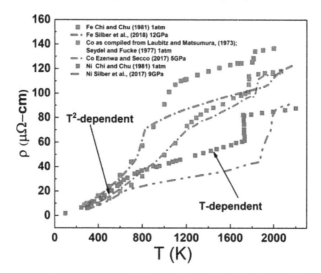

Figure 3. Temperature dependence of resistivity of solid and liquid Fe, Co, and Ni at 1 atm and at various high pressures.

The *P*-dependence of liquid resistivity of Co and Ni along the melting boundary appears constant up to 5 GPa and 9 GPa, respectively, with values of resistivity on melting $\left(\rho_{liq}\right)$ comparable to their corresponding values at 1 atm. The resistivity of Fe on melting decreases up to 5 GPa as it melts from the *bcc* phase but then resistivity on melting remains constant up to 12 GPa as it melts from the *fcc* phase. With a constant value of resistivity on the melting boundary, ρ_{liq}, and a decreasing value of solid resistivity just before melting $\left(\rho_{sol}\right)$ with increasing *P*, $\rho_{liq} - \rho_{sol}$ increases with increasing *P* up to the maximum pressures investigated in these studies as shown in Figure 4. Although, these data for Fe show an increasing $\rho_{liq} - \rho_{sol}$ with increasing *P* in this low *P* range, theoretical calculation [28] up to core *P* and *T* show that the $\rho_{liq} - \rho_{sol}$ for Fe melting from the *hcp* phase decreases with increasing *P*. Further experimental work is needed at higher *P* to assess the trend of $\rho_{liq} - \rho_{sol}$ for Fe shown here within the context of a larger pressure range.

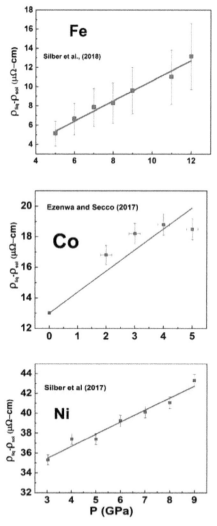

Figure 4. The difference in electrical resistivity value of solid and liquid Fe, Co, and Ni at the melting transition with increasing *P*. The least squares fits of $\rho_{liq} - \rho_{sol}$ vs. *P* for Fe, Co, and Ni are, respectively, $(11.33 \pm 3.19) + 0.74\,P$, $(12.99 \pm 0.0016) + (1.35 \pm 0.6)\,P$, and $(31.84 \pm 0.61) + (1.21 \pm 0.1)\,P$.

Theoretical calculations demonstrate that *d*-resonance scattering dominates the electrical resistivity of unfilled *d*-band liquid metals [55–58]. Experimental study using a flash heating technique in the DAC showed the electrical resistivity of Pt along its high *P* melting boundary is constant [59]. The constancy in the liquid resistivity on the melting boundary may be understood based on the expectation that increasing *P* brings the Fermi level closer to the *d*-resonance site, hence, decreasing conduction electron mobility and increasing resistivity. However, increasing *P* also decreases phonon amplitudes and thus phonon-induced scattering which decreases resistivity. The combined antagonistic effects of *P* on these scattering mechanisms on melting could compensate each other in such a way that resistivity remains constant along the melting boundary, especially in closed packed structures [29,60]. Upon loss of, or reduction in, paramagnon-induced electron scattering at high *P* and *T* conditions, one might infer that the *T*-dependence of resistivity of ferromagnetic metals Fe, Co, and Ni in the solid state could eventually, at high enough *P*, mimic that of paramagnetic metals such as Pt and Pd, at 1 atm [36,37] or perhaps Cu, Ag, Au, and Zn [61–64], where there is a constant paramagnon-induced scattering contribution to its *T*-dependent resistivity.

3.2. Heat Flow at the Inner Core Boundaries of Mercury and Ganymede

The electronic thermal conductivity, k_e, at planetary inner core conditions can be estimated using the Wiedemann-Franz law $\left(k_e = \frac{LT}{\rho}\right)$ where L is the Lorenz number. The total thermal conductivity of metals is dominated by electronic thermal conductivity [65] and one can reasonably assume they are similar in value. Mercury is thought to have a solid inner and liquid outer core with the *P* and *T* conditions at the ICB of ~5 GPa [6] and (1800–2000) K [66], respectively. Parameter values are provided in Table 1. For a pure Fe core in Mercury, using measured resistivity and melting *T* data of Fe at 5 GPa by Silber et al. [21] and the Sommerfeld value ($L_o = 2.445 \times 10^{-8}$ WΩ/K^2) of the Lorenz number [67], we compute a value of k_e of 39 W(mK)$^{-1}$ for the solid just before melting and 37 W(mK)$^{-1}$ for the liquid side. The errors on these calculated values are mainly due to the errors on the experimentally derived values of ρ_{sol} and ρ_{liq} which are ~5% [21] and the *T* at the ICB; however, we used the same value of 1880 K to calculate both values. The difference in the calculated k_e values suggest ~5% difference in thermal conductivity across the ICB of a pure Fe core in Mercury. While the choice of Lorenz number may also be debated, a single value of *L* is appropriate for calculating k_e on both sides of the ICB which are at a single set of *P,T* conditions. An *L* value different than the one used here will not change the relative values of k_e across the ICB. For Ganymede with P_{ICB} of ~9 GPa [68] and using measured Fe resistivity and melting *T* data by Silber et al. [21], we calculate k_e on the solid side of the ICB in a pure Fe core to be 46 W(mK)$^{-1}$ and on the liquid side to be 39 W(mK)$^{-1}$, a difference of more than 7%. For Mercury and Ganymede, this analysis suggests that their thermal conductivity on the solid side of their ICB is likely to be higher than on the liquid side of their ICB, but only marginally when errors are considered. This difference is likely to be higher in Ganymede with P_{ICB} of ~9 GPa compared with Mercury with P_{ICB} of ~5 GPa.

The heat flow (*Q*) along the adiabatic *T* gradient in a liquid outer core can be expressed as:

$$Q_{cond} = k\left(\frac{dT}{dz}\right)_{adiabatic} = k\frac{\alpha g T}{C_p} \tag{1}$$

where $\left(\frac{dT}{dz}\right)_{adiabatic}$ is the adiabatic *T* gradient and α, g, and C_p are thermal expansion, gravitational acceleration, and heat capacity at constant *P*, respectively. Heat flow transported away from the inner core that exceeds the conducted heat flow in the liquid outer core is transported by thermal convection, which in turn is available for driving a dynamo. Here, we concentrate on the heat transport across the solid and the liquid sides of the ICB of Mercury and Ganymede. At the solid side of Mercury ICB of ~5 GPa, we use a melting *T* for Fe at ~5 GPa of 1880 K [21] and we adopt an average value of 8.9×10^{-5} K^{-1} for α from the range of values $(6.4–11.4) \times 10^{-5}$ K^{-1} estimated at the top of Mercury's core by Secco [67], a value of 4.0 ms^{-2} for *g* [69], a value of 39 W(mK)$^{-1}$ for k_e, and for C_p a value of

835 J/Kg/K which is assumed independent of P and T [70], we calculate a value of 31 mWm^{-2} for the heat flow conducted down the adiabat on the solid side of Mercury ICB. To calculate the total adiabatic heat flow on the solid side of the ICB, we use a total core radius of 2004 km [69] along with the recently obtained estimates of inner core radius of 0.4–0.7 times the outer core radius [10]. These values yield a total adiabatic heat flow of 0.25–0.77 TW. On the liquid side of Mercury ICB, using the calculated k_e of 37 W(mK)$^{-1}$ and keeping other quantities constant, we calculate a value of 30 mWm^{-2} for the conducted heat flow and a range of total adiabatic heat flow of 0.24–0.75 TW. This analysis suggests that for a pure Fe core in Mercury, the difference in the heat conducted along the adiabat across the ICB is small and in the range of 0.01–0.02 TW and likely too small to generate significant thermal convection in the liquid outer core.

Table 1. Parameter values for Mercury and Ganymede used in heat flow calculations *.

Parameter	Mercury	Ref.	Ganymede	Ref.
P_{ICB}—pressure at ICB	5 GPa	[6]	9 GPa	[68]
T_{ICB}—temperature at ICB	1880 K	[66]	2200 K	[71]
L_o—Lorenz number	2.445 × 10^{-8} WΩ/K^2	[67]	2.445 × 10^{-8} WΩ/K^2	[67]
$k_{e\ solid}$—electronic thermal conductivity of solid	39 W/m K		46 W/m K	
$k_{e\ liquid}$—electronic thermal conductivity of solid	37 W/m K		39 W/m K	
a—thermal expansion	8.9 × 10^{-5} K^{-1}	[67]	4.8 × 10^{-5} K^{-1}	[72]
g—gravitational acceleration	4.0 m/s^2	[69]	4.36 m/s^2	
C_P—specific heat	835 J/kg K	[70]	835 J/kg K	[70]
$Q_{cond\ solid}$—conducted heat on solid side of ICB	31 mW/m^2		23 mW/m^2	
$Q_{cond\ liquid}$—conducted heat on liquid side of ICB	30 mW/m^2		19 mW/m^2	
r_{ICB}—radius of ICB	800–1400 km	[10,69]	650 km	[73]
total adiabatic heat flow on solid side of ICB	0.25–0.77 TW		0.12 TW	
total adiabatic heat flow on liquid side of ICB	0.24–0.75 TW		0.10 TW	

* Values in table without references are calculated in this study.

We calculate the heat flow on the solid side of a pure Fe core in Ganymede where P_{ICB} is taken as ~9 GPa and T_{ICB} ~2200 K [71]. We determine k_e on the solid and liquid side of Ganymede's ICB to be 46 W(mK)$^{-1}$ and 39 W(mK)$^{-1}$, respectively. The size of Ganymede inner core, r, is not well determined, however, its core size has been estimated to lie between 650–900 km [73] and we assume an ICB radius of 650 km in our calculations. We estimate gravity $g(r)$ by $4\pi G\rho_c r$ where, G is the gravitational constant, ρ_c is core density ~8000 kg/m^2 [12] to be 4.36 m/s^2. From the research of Jeanloz [72], we determine α_{Fe} at 9 GPa to be 4.8 × 10^{-5} K^{-1}. The melting T of Fe at ~9 GPa is ~1990 K [21]. Using these parameters in equation 1, we estimate the heat flow on the solid and liquid sides of Ganymede's ICB to be 23 mW/m^2 and 19 mW/m^2, respectively. For an inner core radius of ~650 km, this yields a total adiabatic heat flow of ~0.12 TW and ~0.10 TW conducted on the solid and liquid side of Ganymede ICB, respectively. This analysis shows that the larger thermal conductivity difference on the solid and liquid sides of Ganymede's ICB of ~7 W(mK)$^{-1}$ compared to Mercury only causes a difference of ~0.02 TW in the heat flow conducted along its adiabat, which is similar to the value for Mercury.

4. Conclusions

The T variation of the electrical resistivity of solid and liquid Fe, Co, and Ni through the melting transition at high P was discussed using experimentally measured data from previous studies. These findings were examined on the basis of reduction of magnon-induced electron scattering (quadratic dependence on T) at high P and T. The scattering of s-electrons to d-states in Fe, Co, and Ni above their Curie T can be related to the increasing phonon-induced scattering to empty d-states (linear dependence on T) and the diminishing relative effect of constant magnon-induced scattering. Relative increases of resistivity on melting in these three metals are self-consistently interpreted within this model. The k_e of solid and liquid at the onset of melting was calculated using the Wiedemann-Franz law with the Sommerfeld value of Lorenz number. These analyses suggest that the thermal conductivity of the solid inner core of small terrestrial planetary bodies could be higher than the liquid outer core. Analysis of the thermal conductivity difference on the solid and liquid side of a pure Fe Mercury and

Ganymede inner core were performed. We found that the thermal conductivity difference on the solid and liquid sides of Mercury's ICB at ~5 GPa is ~2 W(mK)$^{-1}$, which translates into a difference in total adiabatic heat flow of ~0.01–0.02 TW, depending on the size of the inner core relative to the outer core. For a pure Fe Ganymede inner core at ~9 GPa, the difference in thermal conductivity is ~7 W(mK)$^{-1}$, corresponding to difference in total adiabatic heat flow of ~0.02 TW across its ICB. The cores of both planetary bodies appear to have a difference in conducted heat across their ICB that is insignificant in terms of generating thermal convection to power an internal dynamo suggesting that chemical composition is important.

Author Contributions: I.C.E. conceptualized the work and carried out heat flow analyses; I.C.E. and R.A.S. contributed equally to the interpretations and writing of the manuscript.

Funding: RAS acknowledges the Natural Sciences and Engineering Research Council of Canada (NSERC) (grant number 2018-05021) for research funding.

Acknowledgments: The authors thank Wenjun Yong for comments on the manuscript. ICE thanks Takashi Yoshino for the opportunity to continue research on transport properties applied to planetary bodies at the Institute for Planetary Materials, Okayama University, Misasa, Japan.

Conflicts of Interest: The authors declare no conflict of interest.

References

1. Aharonson, O.; Zuber, M.T.; Solomon, S.C. Crustal remanence in an internally magnetized non-uniform shell: A possible source for Mercury's magnetic field? *Earth Planet. Sci. Lett.* **2004**, *218*, 261–268. [CrossRef]

2. Christensen, U.R. A deep dynamo generating Mercury's magnetic field. *Nature* **2006**, *444*, 1056–1058. [CrossRef] [PubMed]

3. Takahashi, F.; Shimizu, H.; Tsunakawa, H. Mercury's anomalous magnetic field caused by a symmetry-breaking self-regulating dynamo. *Nat. Commun.* **2019**, *10*, 208. [CrossRef] [PubMed]

4. Margot, J.L.; Peale, S.J.; Jurgens, R.F.; Slade, M.A.; Holin, I.V. Large longitude libration of Mercury reveals a molten core. *Science* **2007**, *316*, 710–714. [CrossRef]

5. Margot, J.L.; Peale, S.J.; Solomon, S.C.; Hauck, S.A.; Ghigo, F.D.; Jurgens, R.F.; Yseboodt, M.; Giorgini, J.D.; Padovan, S.; Campbell, D.B. Mercury's moment of inertia from spin and gravity data. *J. Geophys. Res. Planets* **2012**, *117*. [CrossRef]

6. Hauck, S.A.; Margot, J.L.; Solomon, S.C.; Phillips, R.J.; Johnson, C.L.; Lemoine, F.G.; Perry, M.E. The curious case of Mercury's internal structure. *J. Geophys. Res. Planets* **2013**, *118*, 1204–1220. [CrossRef]

7. Stark, A.; Oberst, J.; Preusker, F.; Peale, S.J.; Margot, J.L.; Phillips, R.J.; Neumann, G.A.; Smith, D.E.; Zuber, M.T.; Solomon, S.C. First MESSENGER orbital observations of Mercury's librations. *Geophys. Res. Lett.* **2015**, *42*, 7881–7889. [CrossRef]

8. Ness, N.F.; Behannon, K.W.; Lepping, R.P.; Whang, Y.C.; Schatten, K.H. Magnetic field observations near Mercury: Preliminary results from Mariner 10. *Science* **1974**, *185*, 151–160. [CrossRef]

9. Stevenson, D.J.; Spohn, T.; Schubert, G. Magnetism and thermal evolution of the terrestrial planets. *Icarus* **1983**, *54*, 466–489. [CrossRef]

10. Genova, A.; Goossens, S.; Mazarico, E.; Lemoine, F.G.; Neumann, G.A.; Kuang, W.; Sabaka, T.J.; Hauck, I.I.; Smith, D.E.; Solomon, S.C.; et al. Geodetic evidence that Mercury has a solid inner core. *Geophys. Res. Lett.* **2019**, *46*, 3625–3633. [CrossRef]

11. Stanley, S.; Bloxham, J.; Hutchison, W.E.; Zuber, M.T. Thin shell dynamo models consistent with Mercury's weak observed magnetic field. *Earth Planet. Sci. Lett.* **2005**, *234*, 27–38. [CrossRef]

12. Kimura, J.; Nakagawa, T.; Kurita, K. Size and compositional constraints of Ganymede's metallic core for driving an active dynamo. *Icarus* **2009**, *202*, 216–224. [CrossRef]

13. Schubert, G.; Zhang, K.; Kivelson, M.G.; Anderson, J.D. The magnetic field and internal structure of Ganymede. *Nature* **1996**, *384*, 544–545. [CrossRef]

14. Verhoogen, J. *Energetics of the Earth*; National Academy Press: Washington, DC, USA, 1980.

15. Hirose, K.; Morard, G.; Sinmyo, R.; Umemoto, K.; Hernlund, J.; Helffrich, G.; Labrosse, S. Crystallization of silicon dioxide and compositional evolution of the Earth's core. *Nature* **2017**, *543*, 99. [CrossRef] [PubMed]

16. O'Rourke, J.G.; Stevenson, D.J. Powering Earth's dynamo with magnesium precipitation from the core. *Nature* **2016**, *529*, 387. [CrossRef] [PubMed]

17. Du, Z.; Jackson, C.; Bennett, N.; Driscoll, P.; Deng, J.; Lee, K.M.; Greenberg, E.; Prakapenka, V.B.; Fei, Y. Insufficient energy from MgO exsolution to power early geodynamo. *Geophys. Res. Lett.* **2017**, *44*, 11376–11381. [CrossRef]

18. Badro, J.; Aubert, J.; Hirose, K.; Nomura, R.; Blanchard, I.; Borensztajn, S.; Siebert, J. Magnesium partitioning between Earth's mantle and core and its potential to drive an early exsolution geodynamo. *Geophys. Res. Lett.* **2018**, *45*, 13240–13248. [CrossRef]

19. Gomi, H.; Ohta, K.; Hirose, K.; Labrosse, S.; Caracas, R.; Verstraete, M.J.; Hernlund, J.W. The high conductivity of iron and thermal evolution of the Earth's core. *Phys. Earth Planet. Inter.* **2013**, *224*, 88–103. [CrossRef]

20. Ohta, K.; Kuwayama, Y.; Hirose, K.; Shimizu, K.; Ohishi, Y. Experimental determination of the electrical resistivity of iron at Earth's core conditions. *Nature* **2016**, *534*, 95–98. [CrossRef] [PubMed]

21. Silber, R.E.; Secco, R.A.; Yong, W.; Littleton, J.A. Electrical resistivity of liquid Fe to 12 GPa: Implications for heat flow in cores of terrestrial bodies. *Sci. Rep.* **2018**, *8*, 10758. [CrossRef] [PubMed]

22. Williams, Q. The thermal conductivity of Earth's core: A key geophysical parameter's constraints and uncertainties. *Annu. Rev. Earth Planet. Sci.* **2018**, *46*, 47–66. [CrossRef]

23. Stacey, F.D.; Anderson, O.L. Electrical and thermal conductivities of Fe–Ni–Si alloy under core conditions. *Phys. Earth Planet. Inter.* **2001**, *124*, 153–162. [CrossRef]

24. Stacey, F.D.; Loper, D.E. A revised estimate of the conductivity of iron alloy at high pressure and implications for the core energy balance. *Phys. Earth Planet. Inter.* **2007**, *161*, 13–18. [CrossRef]

25. Konôpková, Z.; McWilliams, R.S.; Gómez-Pérez, N.; Goncharov, A.F. Direct measurement of thermal conductivity in solid iron at planetary core conditions. *Nature* **2016**, *534*, 99.

26. Pozzo, M.; Davies, C.; Gubbins, D.; Alfe, D. Thermal and electrical conductivity of iron at Earth's core conditions. *Nature* **2012**, *485*, 355–358. [CrossRef]

27. De Koker, N.; Steinle-Neumann, G.; Vlček, V. Electrical resistivity and thermal conductivity of liquid Fe alloys at high P and T, and heat flux in Earth's core. *Proc. Natl. Acad. Sci. USA* **2012**, *109*, 4070–4073. [CrossRef] [PubMed]

28. Wagle, F.; Steinle-Neumann, G. Electrical resistivity discontinuity of iron along the melting curve. *Geophys. J. Int.* **2017**, *213*, 237–243. [CrossRef]

29. Ezenwa, I.C.; Secco, R.A. Invariant electrical resistivity of Co along melting boundary. *Earth Planet. Sci. Lett.* **2017**, *474*, 120–127. [CrossRef]

30. Silber, R.E.; Secco, R.A.; Yong, W. Constant electrical resistivity of Ni along the melting boundary up to 9 GPa. *J. Geophys. Res. Solid Earth* **2017**, *122*, 5064–5081. [CrossRef]

31. Secco, R.A.; Schloessin, H.H. The electrical resistivity of solid and liquid Fe at pressures up to 7 GPa. *J. Geophys. Res. Solid Earth* **1989**, *94*, 5887–5894. [CrossRef]

32. Deng, L.; Seagle, C.; Fei, Y.; Shahar, A. High pressure and temperature electrical resistivity of iron and implications for planetary cores. *Geophys. Res. Lett.* **2013**, *40*, 33–37. [CrossRef]

33. Pommier, A. Influence of sulfur on the electrical resistivity of a crystallizing core in small terrestrial bodies. *Earth Planet. Sci. Lett.* **2018**, *496*, 37–46. [CrossRef]

34. Landrum, G.A.; Dronskowski, R. The orbital origins of magnetism: From atoms to molecules to ferromagnetic alloys. *Angew. Chem. Int. Ed.* **2000**, *39*, 1560–1585. [CrossRef]

35. Mott, N.F. The electrical conductivity of transition metals. *Proc. R. Soc. Lond. A Math. Phys. Eng. Sci.* **1936**, *153*, 699–717.

36. Matula, R.A. Electrical resistivity of copper, gold, palladium, and silver. *J. Phys. Chem. Ref. Data* **1979**, *8*, 1147–1298. [CrossRef]

37. Powell, R.W.; Tye, R.P. The promise of platinum as a high temperature thermal conductivity reference material. *Br. J. Appl. Phys.* **1963**, *14*, 662. [CrossRef]

38. Güntherodt, H.J.; Hauser, E.; Künzi, H.U.; Müller, R. The electrical resistivity of liquid Fe, Co, Ni and Pd. *Phys. Lett. A* **1975**, *54*, 291–292. [CrossRef]

39. Laubitz, M.J.; Matsumura, T. Transport properties of the ferromagnetic metals. i Cobalt. *Can. J. Phys.* **1973**, *51*, 1247–1256. [CrossRef]

40. Laubitz, M.J.; Matsumura, T.; Kelly, P.J. Transport properties of the ferromagnetic metals. II. Nickel. *Can. J. Phys.* **1976**, *54*, 92–102. [CrossRef]

41. Seydel, U.; Fucke, W. Sub-microsecond pulse heating measurements of high temperature electrical resistivity of the 3d-transition metals iron, cobalt, and nickel. *Chem. Inf.* **1977**, *32*, 994–1002. [CrossRef]
42. Chu, T.K.; Chi, T.C. McGram-Hill/CICADAS data series on materials properties. In *Properties of Selected Ferrous Aloying Elements, Vol. III-1*; Touloukian, Y.S., Ho, C.Y., Eds.; McGraw-Hill: New York, NY, USA, 1981.
43. Campbell, I.A.; Fert, A. Chapter 9 Transport properties of ferromagnets. In *Handbook of Ferromagnetic Materials*; Wohlfarth, E.P., Ed.; Elsevier: Amsterdam, The Netherlands, 1982; pp. 747–804.
44. Gold, A.V. Review paper: Fermi surfaces of the ferromagnetic transition metals. *J. Low Temp. Phys.* **1974**, *16*, 3–42. [CrossRef]
45. Coleman, R.V.; Morris, R.C.; Sellmyer, D.J. Magnetoresistance in Iron and Cobalt to 150 kOe. *Phys. Rev. B* **1973**, *8*, 317–331. [CrossRef]
46. Angadi, M.A.; Fawcett, E.; Rasolt, M. High field magnetoresistance and quantum oscillations in iron whiskers. *Can. J. Phys.* **1975**, *53*, 284–298. [CrossRef]
47. Himpsel, F.J.; Heimann, P.; Eastman, D.E. Band structure measurements and multi-electron effects (satellites) for nearly-filled d-band metals: Fe, Co, Ni, Cu, Ru, and Pd. *J. Appl. Phys.* **1981**, *52*, 1658–1663. [CrossRef]
48. Eastman, D.E.; Himpsel, F.J.; Knapp, J.A. Experimental exchange-split energy-band dispersions for Fe, Co, and Ni. *Phys. Rev. Lett.* **1980**, *44*, 95–98. [CrossRef]
49. Korenman, V.; Prange, R.E. Local-band-theory analysis of spin-polarized, angle-resolved photoemission spectroscopy. *Phys. Rev. Lett.* **1984**, *53*, 186–189. [CrossRef]
50. Chikazumi, S.; Graham, C.D. *Physics of Ferromagnetism 2e (No. 94)*; Oxford University Press: Oxford, UK, 2009.
51. Iota, V.; Klepeis, J.H.P.; Yoo, C.S.; Lang, J.; Haskel, D.; Srajer, G. Electronic structure and magnetism in compressed 3d transition metals. *Appl. Phys. Lett.* **2007**, *90*, 042505. [CrossRef]
52. Ross, M.; Boehler, R.; Errandonea, D. Melting of transition metals at high pressure and the influence of liquid frustration: The late metals Cu, Ni, and Fe. *Phys. Rev. B* **2007**, *76*, 184117. [CrossRef]
53. McMahan, A.K.; Albers, R.C. Insulating nickel at a pressure of 34 TPa. *Phys. Rev. Lett.* **1982**, *49*, 1198. [CrossRef]
54. McMahan, A.K. Pressure effects on the electronic structure of 4f and 5f materials. *J. Less Common Met.* **1989**, *149*, 1–11. [CrossRef]
55. Gaspari, G.D.; Gyorffy, B.L. Electron-phonon interactions, d resonances, and superconductivity in transition metals. *Phys. Rev. Lett.* **1972**, *28*, 801–805. [CrossRef]
56. Brown, J.S. D resonance calculation of the resistivity and thermopower of liquid Ni and Pd. *J. Phys. F Met. Phys.* **1973**, *3*, 1003–1007. [CrossRef]
57. Shvets, V.T. Influence of sd hybridization of the electrical conductivity of liquid transition metals. *Theor. Math. Phys.* **1982**, *53*, 1040–1046. [CrossRef]
58. Shvets, V.T.; Savenko, S.; Datsko, S. Perturbation theory for electrical resistivity of liquid transition metals. *Condens. Matter Phys* **2002**, *5*, 511–522. [CrossRef]
59. McWilliams, R.S.; Konôpková, Z.; Goncharov, A.F. A flash heating method for measuring thermal conductivity at high pressure and temperature: Application to Pt. *Phys. Earth Planet. Inter.* **2015**, *247*, 17–26. [CrossRef]
60. Silber, R.E.; Secco, R.A.; Yong, W.; Littleton, J.A. Heat flow in Earth's core from invariant electrical resistivity of Fe-Si on the melting boundary to 9 GPa: Do light elements matter? *J. Geophys. Res. Solid Earth* **2019**, *124*. [CrossRef]
61. Ezenwa, I.C.; Secco, R.A.; Yong, W.; Pozzo, M.; Alfè, D. Electrical resistivity of solid and liquid Cu up to 5GPa: Decrease along the melting boundary. *J. Phys. Chem. Solids* **2017**, *110*, 386–393. [CrossRef]
62. Littleton, J.A.; Secco, R.A.; Yong, W. Decreasing electrical resistivity of silver along the melting boundary up to 5 GPa. *High Press. Res.* **2018**, *38*, 99–106. [CrossRef]
63. Berrada, M.; Secco, R.A.; Yong, W. Decreasing electrical resistivity of gold along the melting boundary up to 5 GPa. *High Press. Res.* **2018**, *38*, 367–376. [CrossRef]
64. Ezenwa, I.C.; Secco, R.A. Constant electrical resistivity of Zn along the melting boundary up to 5 GPa. *High Press. Res.* **2017**, *37*, 319–333. [CrossRef]
65. Klemens, P.G.; Williams, R.K. Thermal conductivity of metals and alloys. *Int. Met. Rev.* **1986**, *31*, 197–215. [CrossRef]
66. Breuer, D.; Rueckriemen, T.; Spohn, T. Iron snow, crystal floats, and inner-core growth: Modes of core solidification and implications for dynamos in terrestrial planets and moons. *Prog. Earth Planet. Sci.* **2015**, *2*, 39. [CrossRef]

67. Secco, R.A. Thermal conductivity and Seebeck coefficient of Fe and Fe-Si alloys: Implications for variable Lorenz number. *Phys. Earth Planet. Inter.* **2017**, *265*, 23–34. [CrossRef]
68. Sohl, F.; Spohn, T.; Breuer, D.; Nagel, K. Implications from Galileo observations on the interior structure and chemistry of the Galilean satellites. *Icarus* **2002**, *157*, 104–119. [CrossRef]
69. Rivoldini, A.; Van Hoolst, T. The interior structure of Mercury constrained by the low-degree gravity field and the rotation of Mercury. *Earth Planet. Sci. Lett.* **2013**, *377*, 62–72. [CrossRef]
70. Desai, P.D. Thermodynamic properties of iron and silicon. *J. Phys. Chem. Ref. Data* **1986**, *15*, 967–983. [CrossRef]
71. Hussmann, H.; Sotin, C.; Lunine, J.I. Interiors and evolution of icy satellites. In *Treatise on Geophysics*; Schubert, G., Ed.; Elsevier: Amsterdam, The Netherlands, 2007; Volume 10.
72. Jeanloz, R. Properties of iron at high pressures and the state of the core. *J. Geophys. Res. Solid Earth* **1979**, *84*, 6059–6069. [CrossRef]
73. Schubert, G.; Anderson, J.D.; Spohn, T.; McKinnon, W.B. Interior composition, structure and dynamics of the Galilean satellites. *Jupit. Planet Satell. Magnetos.* **2004**, *1*, 281–306.

Article

Stepwise Homogeneous Melting of Benzene Phase I at High Pressure

Ravi Mahesta [1] and **Kenji Mochizuki** [1,2,*]

[1] Department of Chemistry and Materials, Faculty of Textile Science and Technology, Ueda 386-8567, Japan; mahestaravi@gmail.com

[2] Institute for Fiber Engineering, Shinshu University, Ueda 386-8567, Japan

* Correspondence: mochizuki@shinshu-u.ac.jp

Received: 26 April 2019; Accepted: 27 May 2019; Published: 28 May 2019

Abstract: We investigate, using molecular dynamics simulations, the spontaneous homogeneous melting of benzene phase I under a high pressure of 1.0 GPa. We find an apparent stepwise transition via a metastable crystal phase, unlike the direct melting observed at ambient pressure. The transition to the metastable phase is achieved by rotational motions, without the diffusion of the center of mass of benzene. The metastable crystal completely occupies the whole space and maintains its structure for at least several picoseconds, so that the phase seems to have a local free energy minimum. The unit cell is found to be unique—no such crystalline structure has been reported so far. Furthermore, we discuss the influence of pressure control on the melting behavior.

Keywords: benzene phase I; homogeneous melting; Ostwald's step rule; molecular dynamics simulation; high pressure; metastable phase

---

1. Introduction

Benzene is a renowned small and simple molecule. Regardless, its polymorphism [1–3] and local liquid structure [4–6] have been extensively explored until now. Five crystalline phases of pure benzene have been reported so far experimentally [7–14], and computational studies predicted many other potential crystalline structures [1,15–18]. Although the phase diagram shows the most stable phase at a given condition, there can be rich intermediate phases on the transition pathway from the initial to the finally prevailed phase [19–23]. Ostwald argued that a phase transition can proceed via an intermediate metastable phase due to the reduction in the surface energy of nucleation (Ostwald's step rule) [24] in contrast to the direct nucleation described in the classical nucleation theory [25]. Indeed, a metastable phase or structure has been shown to play a key role in the melting of colloids [26], copper and aluminum [27], and ice [28]. Further, under high pressures, the melting pathway can be much more complicated, because of the competition between potential energy, entropy, and packing effect (negligible at ambient pressure).

The existence of intermediate metastable phase in the melting of benzene phase I is still controversial, although the phase transition dynamics of benzene has been investigated [29–31]. Tohji et al. claimed the existence of a premelting stage at 10 K below melting point using powder X-ray diffraction [30]. Furthermore, they predicted that a plastic crystal transiently appears in the melting. This premelting stage was later disproven by the investigation over the complete temperature range of 4 to 280 K [31]. Very recently, we performed the molecular dynamics (MD) simulations for the homogeneous melting of benzene phase I crystal near the limit of superheating and statistically demonstrated that there is no intermediate transient state between the crystal and liquid phases [32]. While these studies have been conducted under ambient pressure, the melting dynamics under high pressure have not been explored.

In this study, we perform MD simulations of the homogeneous melting of benzene phase I at a high pressure of 1.0 GPa and observe the stepwise transition via a metastable crystal phase. We show that the formation of the metastable phase is achieved by rotational motions, and the unit cell structure differs from any other crystalline phases reported so far.

2. Methods

2.1. Force Field

Benzene was modelled with full atomistic detail using CHARM22 [33,34]. This force field reproduces the crystal structure of benzene phase I [29,35] and the physical properties of liquid benzene, such as density and self-diffusion coefficient [36]. The intermolecular nonbonded interactions were described by Lennard–Jones plus Coulomb potential. The intermolecular interactions were truncated at 1.20 nm. The Lennard–Jones parameters for cross-interactions were obtained using Lorentz–Berthelot mixing rules and the long-range Coulombic interactions were evaluated using particle-mesh Ewald algorithm [37].

2.2. MD Simulations

MD simulations were performed using GROMACS 2019 package [38], in which the equations of motion are integrated using leapfrog algorithm with a time step of 2 fs. The temperature T for equilibration was controlled using a Berendsen thermostat [39] with a damping constant of 1.0 ps, while a Nose-Hoover thermostat [40,41] was used for the production runs. The pressure p was isotropically controlled by a Berendsen barostat [39] for both equilibration and production runs, with a damping constant of 2.0 ps. The pressure was set to 1.0 GPa for all the simulations. The periodic boundary condition was applied in all three directions.

To evaluate the influence of pressure control on the melting behavior, we performed some additional simulations using anisotropic pressure control, where the box dimensions and the box angles can change.

2.3. Crystals

The structure of benzene phase I was obtained from Cambridge Structural Database (refcode: BENZEN) [42]. Phase I is the most stable crystalline phase up to 1.2 GPa [9,14]. The unit cell consists of four benzene molecules with space group *Pbca*. The lattice parameters are $a = 0.7460$ nm, $b = 0.9666$ nm, and $c = 0.7034$ nm at 270 K [11]. Lattice axes a, b, and c corresponded to x, y, and z axes, respectively, in this study.

For the spontaneous homogeneous melting, we replicated the unit cell to manufacture an approximately cubic system with a box size of 5.21 nm × 4.78 nm × 4.84 nm. The number of molecules, N, was 980. To equilibrate the crystal structure, energy minimization was first carried out using the steepest descent algorithm. Subsequently, an isochoric isothermal (*NVT*) MD simulation was performed at 270 K for 500 ps. Thereafter, an isobaric isothermal (*NpT*) MD simulation for 5 ns at 270 K and 0.1 MPa was performed, followed by a 5 ns *NpT* MD simulation at 270 K and 1.0 GPa. Furthermore, an *NpT* MD simulation at 500 K and 1.0 GPa was performed for 5 ns. The final configuration obtained was used as the initial structure for the following heating simulations.

To determine the equilibrium melting temperature we used the two-phase method [43–46]. First, we developed two identical boxes of benzene phase I crystal containing 448 molecules with a box size of 2.68 nm × 6.45 nm × 2.86 nm. We randomly erased 48 molecules from one of these crystal structures, and we melted it by performing a short *NVT* MD simulation at 800 K. Then, we joined these crystal and liquid boxes. The box dimension was 2.68 nm × 13.5 nm × 2.86 nm. To equilibrate the solid-liquid coexistence structure, energy minimization was first carried out using the steepest descent algorithm. Subsequently, an *NVT* MD simulation was performed at 200 K for 20 ps. Thereafter, the system was gradually heated up by performing each 20 ps *NpT* MD simulations at two temperatures of 200 and

400 K and at 1.0 GPa, and the equilibrated configuration was used for the production runs at 1.0 GPa. The configuration equilibrated at 200 K and 1.0 GPa was gradually decompressed from 1.0 to 0.0 GPa in increments of 0.2 GPa at 200 K by performing each 20 ps NpT MD simulations. The systems were further equilibrated at 400 K for 0.8 GPa, at 300 K for 0.6 and 0.4 GPa, and at 200 K for 0.2 and 0.0 GPa by performing each 20 ps NpT MD simulation, then the obtained configurations were used for the production runs at each pressure.

3. Results and Discussion

We initially evaluated the equilibrium melting temperature T_m of benzene phase I at 1.0 GPa in our model using the two-phase method [43–46]. The initial configuration for production runs was the coexistence between liquid and phase I crystal, as shown in Figure 1A. We performed each 10 ns NpT MD simulation at 430K, 435K, and 440 K and at 1.0 GPa. Figure 1B depicts the time evolution of the volume for these runs. At T_m, the phase I crystal and liquid exhibit the same stability in free energy, and the total volume does not change. On the other hand, the volume increases at $T > T_m$ due to melting, while it decreases at $T < T_m$. The results show that T_m is 435 ± 5 K at 1.0 GPa for our computational model, which is approximately 20 K lower than the experimental estimation of 457 K [9]. We also computed T_m at 0.0, 0.2, 0.4, 0.6, 0.8, 1.0, and 1.2 GPa in the same manner. Figure 1C shows that the melting curve for our computational model qualitatively reproduces the experimental result [9,47].

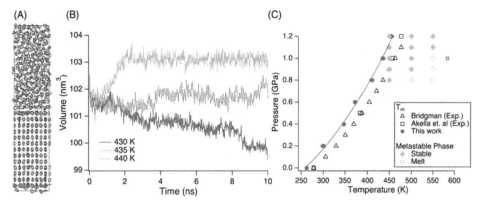

Figure 1. Two-phase method for T_m: (**A**) initial configuration consisting of liquid and phase I structures, (**B**) time evolution of the total volumes in NpT MD simulations at 1.0 GPa and at three different temperatures (430, 435 and 440 K), (**C**) the resulting melting curve (T_m) for the phase I (blue filled circles) with the experimental data by Bridgman (open triangles) [47] and by Akella et. al (open squares) [9]. The panel C also presents the homogeneous melting temperature (red sharp), and the stable (filled diamonds) and unstable (open diamonds) conditions for the metastable phase, estimated in this study.

Benzene phase I, equilibrated at 500 K, was heated up in the increments of 10 K at 1.0 GPa, and an NpT MD simulation was performed for 10 ns at each T. We first observed the melting of crystal at 590 K. This result indicates that the limit of superheating achieved by our heating rate (1 K/ns) is 590 K, which is considerably higher than the equilibrium melting temperature of 435 K for our computational model. The large superheating for homogeneous melting has also been reported for other materials both experimentally [26,48,49] and computationally [28,50–55], because of the absence of an apparent trigger, such as surfaces and impurities, to initiate the disordering. To observe the melting under moderate conditions, we focused on the melting at 584 K, heated up from 580 K. At 584 K, 6 of 95 independent trajectories melted within 10 ns when we gave different momenta to the same configuration.

Figure 2A,B present the time evolution of potential energy (PE) and density for a typical melting trajectory at 584 K. We find a plateau between 490 and 510 ps (shown by green shade), where both the

PE and density remain constant. The preservation of these properties over a time range implies the existence of a metastable phase. Figure 3A–C depicts the molecular structures at each stage of 100, 490, and 575 ps, respectively. In contrast to the disordered liquid structure in Figure 3C, layered structures along the *y* axis are depicted in Figure 3A,B. However, their molecular orientations are clearly different. The same structure as that in Figure 3B was also observed in all the other melting trajectories at 584 K, although the structure sometimes did not fill the whole space of the simulation box and its lifetime was in the order of several 10 picoseconds.

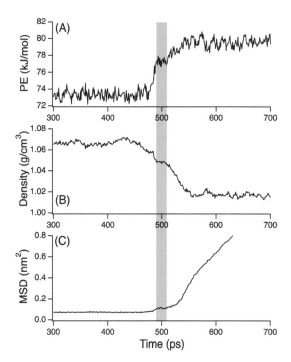

Figure 2. Time dependence of (**A**) PE, (**B**) density, and (**C**) MSD for the melting of benzene phase I at 584 K. The time range when the metastable phase is formed is shaded in green.

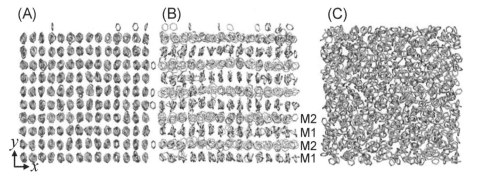

Figure 3. Molecular structures of (**A**) phase I (100 ps), (**B**) metastable phase (490 ps), and (**C**) liquid phase (575 ps). The two distinct layers, M1 and M2, and the *y* axis are shown in (**B**).

Although the metastable structure exists for a very short time in the melting pathway at 584 K, the same structure might be (meta) stable at different pT conditions. To explore such a pT region, we performed NpT MD simulations for each 10 ns at pressures ranging from 0.8 to 1.2 GPa and at temperatures ranging from 450 to 550 K. At a given pT condition, we gave 10 different momenta to the structure obtained at 490 ps in the melting trajectory (Figure 3B). The metastable structure was found to be preserved for 10 ns in the all trajectories at several thermodynamic conditions: 450 K and 0.8, 0.9, 1.0, 1.1, and 1.2 GPa; 500 K and 0.9, 1.0, 1.1, and 1.2 GPa; 550 K and 1.1 and 1.2 GPa. At the other conditions, the metastable structure melted into liquid. These stable and unstable T-p conditions for the metastable structure are plotted in Figure 1C. These results imply that the metastable structure can be a "phase" rather than a transiently appearing fragile structure. Its free energy is likely to be between phase I and liquid benzene, in accordance with Ostwald's step rule [23].

In contrast to the short lifetime of the metastable phase, the waiting times for the phase transition from phase I to the metastable phase range from hundreds of picoseconds to several nanoseconds at 584 K. The significant difference in their lifetimes indicates that the free energy barrier from the metastable phase to liquid is much lower than that from phase I to the metastable phase at 584 K.

To characterize the self-diffusion over melting, we computed the mean square displacement (MSD) of center of mass (COM) of benzene from the initial configuration at 584 K ($t = 0$ ps), using the expression of $\langle |\vec{r}(t) - \vec{r}(0)| \rangle^2$ (Figure 2C). The time-dependent MSD shows a small hump at around 490 ps, arising from the re-organization of molecules to form the metastable phase. The difference in the MSD of COM is negligibly small (0.04 nm^2). The MSD does not change between 490 and 510 ps, where the PE and density are also preserved. After 510 ps, the MSD gradually increases, during which a liquid nucleus forms and erodes the metastable phase. Finally, the MSD increases rapidly after 530 ps, at which point the liquid structure covers approximately half the space of the simulation box. Figure 4 shows the radial distribution functions (RDFs) between COMs of benzene molecules in the three phases. The results show that the metastable phase maintains a long-range order structure in comparison with the liquid phase. Although the metastable phase exhibits broader peaks than phase I, the peaks in the two phases are located at a similar radial distance. This indicates that there is only a minor change in the position of COM between phase I and the metastable phases. Hence, the analyses of MSD and RDF demonstrate that the transition from phase I to the metastable phase arises from the rotational motion, and COM in the metastable phase exhibits an ordered structure.

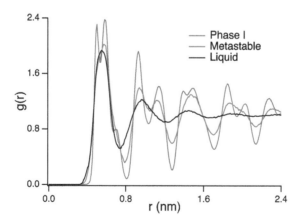

Figure 4. RDFs, g(r), of the COM of benzene molecule for phase I, metastable phase, and liquid phase, computed in the time range 100–150 ps, 490–510 ps, and 9000–9050 ps, respectively.

We next characterize the molecular orientation in the metastable phase. Figure 3B demonstrates that benzene molecules in every other layer of the metastable phase (we name them M1 and M2 layers) tend to have the same alignment. We project the probability density of the normal vector on a sphere using HEALPix algorithm [56–58], in which the order parameter (θ, φ) is used to define the vector orientation (Figure 5A). In phase I, benzene molecules are most likely to orient at $(90°, 40°)$ and $(90°, 140°)$, as shown in Figure 5B. It is to be noted that the molecular orientations of $(90°, 40°)$ and $(90°, 140°)$ are identical to $(90°, -140°)$ and $(90°, -40°)$, respectively. The molecular orientation in phase I is slightly different from that obtained experimentally [11] (see the green circles in Figure 5B); the difference probably arises from the limitations of the force field or the effect of high pressure. In the metastable phase, there are distinct orientations in both M1 and M2 layers (Figure 5C,D). More specifically, the molecules in the M1 layer are most likely to orient at $(90°, 0°)$ and $(90°, 180°)$ (or $(90°, -180°)$), while those in the M2 layer are most likely to orient at $(90°, 90°)$ and $(90°, -90°)$. These preferential orientations are obviously different from those for phase I. These results indicate that all molecules change the orientations in the transition from phase I to the metastable phase. In the previous study, we have shown that at ambient pressure the flipping motion played an important role in the formation of the critical nucleus [32]. Hence, the rotational motion triggers the melting transition of phase I at both the pressures. Further, it is primarily the high pressure that induces the formation of the metastable phase in the melting dynamics, because there is no intermediate metastable phase at ambient pressure.

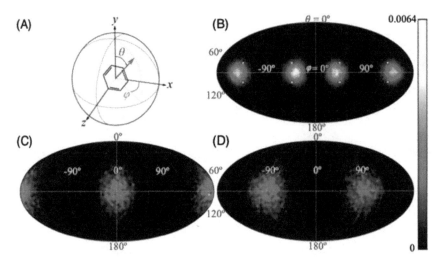

Figure 5. Probability density of the normal benzene vector (**A**) whose orientation is defined by angles θ and φ. The normalized distributions for (**B**) phase I (100–120 ps), (**C**) M1, and (**D**) M2 layers in the metastable phase (490–510 ps) are depicted using HEALPix, such that each pixel represents an equal area on the spherical surface. The experimentally obtained orientation for phase I [11] is shown in green circles.

Furthermore, we find that the probability densities in the metastable phase are more broadly distributed than phase I. The loose orientation results in the increase of conformational entropy, which may compensate for the increase in PE (Figure 2A) and stabilize the metastable phase in term of free energy. Thus, based on the preferential orientation in each layer and the negligible diffusivity of COM, we conclude that the metastable phase is a crystal.

Using the same method, we evaluated the probability density for the two-fold axis of the benzene molecule, which is in parallel to the molecular plane and determined its most probable orientation. The average box size was calculated between 490 and 510 ps. Based on the repeating structural pattern, we determined that the unit cell is tetragonal and contains two molecules with a dimension of $0.51 \text{ nm} \times 0.51 \text{ nm} \times 0.96 \text{ nm}$ (Figure 6). The space group is $P4_2nm$. Interestingly, as far as we know, the obtained unit cell structure is different from any other crystalline structures of benzene reported so far [1,7–18]

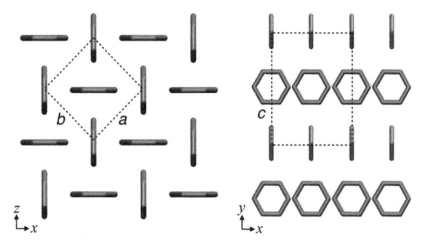

Figure 6. Schematic representation of the metastable phase and its unit cell (dashed black lines). Every other layer along the *y* axis has the same color code. The *x*, *y*, and *z* for the simulation box and the *a*, *b*, and *c* parameters for the unit cell are shown.

It is worth investigating the influence of pressure control on the melting behavior. In the above simulations, we used the isotropic pressure control, where the dimension of box was isotropically changed. For comparison, we performed some *NpT* MD simulations using anisotropic pressure control, which allows the deformation of the simulation box. First, we performed 10 independent *NpT* MD simulations at 500 K for each 10 ns under the anisotropic pressure control, initiated from the metastable structure obtained under the isotropic pressure control. In all the trajectories, the metastable structure transformed into a different crystal form—phase III with line defect [16]. The formed phase III structure was preserved for 10 ns. Note that on the *pT* phase diagram, the phase III locates at very high pressures over 4 GPa and does not have the phase boundary with phase I [9,14]. Our results imply that the phase III can be a metastable phase between phase I and liquid. Second, we heated up the phase I crystal under the anisotropic pressure control until it melted at 580K. However, we did not observe a stepwise homogeneous melting via phase III at the limit of superheating. The phase III may appear on the melting pathway at lower temperatures. These comparisons indicate that the metastable structure that we found in this study is likely to appear under the isotropic pressure control, which can be realized by the recently developed experimental techniques [59–61].

4. Summary

We carried out the MD simulations to investigate the homogeneous melting of benzene phase I under a high pressure of 1.0 GPa. We found a stepwise change in the PE, density, and MSD in the melting trajectories. The metastable phase preserves the layered structure as phase I but the molecular orientations are obviously different from those of phase I. The formation of the metastable phase is attributable to the flipping motion of benzene rather than the diffusion of COM. Although the flipping motion has been previously shown to lead the formation of the critical nucleus, no intermediate metastable phase was observed at ambient pressure. Furthermore, the probability density of the normal vector demonstrates the preferential orientation in the metastable phase. Their orientations are more relaxed than those in phase I, indicating the gain of conformational entropy compensates the increase of potential energy. We further determined the unit cell and found that such a crystalline structure has not been reported so far.

Author Contributions: Conceptualization, K.M.; molecular simulations, R.M.; writing, K.M. and R.M.; funding acquisition, K.M.

Funding: We acknowledge the support from the Japan Society for the Promotion of Science (KAKENHI 18K19060) and Inoue Foundation for Science (Inoue Science Research Award).

Acknowledgments: We thank M. Matsumoto for determining the unit cell. A part of calculations was performed at the Research Center for Computational Science in Okazaki, Japan.

Conflicts of Interest: The authors declare no conflict of interests.

References

1. Raiteri, P.; Martoňák, R.; Parrinello, M. Exploring Polymorphism: The Case of Benzene. *Angew. Chem. Int. Ed.* **2005**, *44*, 3769–3773. [CrossRef] [PubMed]
2. Chanyshev, A.D.; Litasov, K.D.; Rashchenko, S.V.; Sano-Furukawa, A.; Kagi, H.; Hattori, T.; Shatskiy, A.F.; Dymshits, A.M.; Sharygin, I.S.; Higo, Y. High-Pressure–High-Temperature Study of Benzene: Refined Crystal Structure and New Phase Diagram up to 8 GPa and 923 K. *Cryst. Growth Des.* **2018**, *18*, 3016–3026. [CrossRef]
3. Wen, X.-D.; Hoffmann, R.; Ashcroft, N.W. Benzene under High Pressure: a Story of Molecular Crystals Transforming to Saturated Networks, with a Possible Intermediate Metallic Phase. *J. Am. Chem. Soc.* **2011**, *133*, 9023–9035. [CrossRef]
4. Headen, T.F.; Howard, C.A.; Skipper, N.T.; Wilkinson, M.A.; Bowron, D.T.; Soper, A.K. Structure of pi-pi interactions in aromatic liquids. *J. Am. Chem. Soc.* **2010**, *132*, 5735–5742. [CrossRef] [PubMed]
5. Yoshida, K.; Fukuyama, N.; Yamaguchi, T.; Hosokawa, S.; Uchiyama, H.; Tsutsui, S.; Baron, A.Q.R. Inelastic X-ray scattering on liquid benzene analyzed using a generalized Langevin equation. *Chem. Phys. Lett.* **2017**, *680*, 1–5. [CrossRef]
6. Nagasaka, M.; Yuzawa, H.; Mochizuki, K.; Rühl, E.; Kosugi, N. Temperature-Dependent Structural Changes in Liquid Benzene. *J. Phys. Chem. Lett.* **2018**, *9*, 5827–5832. [CrossRef]
7. Pravica, M.; Grubor-Urosevic, O.; Hu, M.; Chow, P.; Yulga, B.; Liermann, P. X-ray Raman Spectroscopic Study of Benzene at High Pressure. *J. Phys. Chem. B* **2007**, *111*, 11635–11637. [CrossRef]
8. Piermarini, G.J.; Mighell, A.D.; Weir, C.E.; Block, S. Crystal Structure of Benzene II at 25 Kilobars. *Science* **1969**, *165*, 1250–1255. [CrossRef]
9. Akella, J.; Kennedy, G.C. Phase Diagram of Benzene to 35 kbar. *J. Chem. Phys.* **1971**, *55*, 793–796. [CrossRef]
10. Block, S.; Weir, C.E.; Piermarini, G.J. Polymorphism in benzene, naphthalene, and anthracene at high pressure. *Science* **1970**, *169*, 586–587. [CrossRef] [PubMed]
11. Cox, E.G.; Smith, J.A.S. Crystal Structure of Benzene at −3 °C. *Nature* **1954**, *173*, 75. [CrossRef]
12. Cansell, F.; Fabre, D.; Petitet, J. Phase transitions and chemical transformations of benzene up to 550 °C and 30 GPa. *J. Chem. Phys.* **1993**, *99*, 7300–7304. [CrossRef]
13. Ciabini, L.; Santoro, M.; Bini, R.; Schettino, V. High pressure crystal phases of benzene probed by infrared spectroscopy. *J. Chem. Phys.* **2001**, *115*, 3742–3749. [CrossRef]
14. Thiéry, M.M.; Léger, J.M. High pressure solid phases of benzene. I. Raman and x-ray studies of C_6H_6 at 294 K up to 25 GPa. *J. Chem. Phys.* **1988**, *89*, 4255–4271. [CrossRef]
15. Schneider, E.; Vogt, L.; Tuckerman, M.E. Exploring polymorphism of benzene and naphthalene with free energy based enhanced molecular dynamics. *Acta Crystallogr. B Struct. Sci. Cryst. Eng. Mater.* **2016**, *72*, 542–550. [CrossRef]
16. Yu, T.-Q.; Tuckerman, M.E. Temperature-accelerated method for exploring polymorphism in molecular crystals based on free energy. *Phys. Rev. Lett.* **2011**, *107*, 015701. [CrossRef] [PubMed]
17. Van Eijck, B.P.; Spek, A.L.; Mooij, W.T.M.; Kroon, J. Hypothetical Crystal Structures of Benzene at 0 and 30 kbar. *Acta Crystallogr. B* **1998**, *54*, 291–299. [CrossRef]
18. Shoda, T.; Yamahara, K.; Okazaki, K.; Williams, D.E. Molecular packing analysis: prediction of experimental crystal structures of benzene starting from unreasonable initial structures. *J. Mol. Struct. THEOCHEM* **1994**, *313*, 321–334. [CrossRef]
19. Yagasaki, T.; Matsumoto, M.; Tanaka, H. Phase Diagrams of TIP4P/2005, SPC/E, and TIP5P Water at High Pressure. *J. Phys. Chem. B* **2018**, *122*, 7718–7725. [CrossRef] [PubMed]
20. Jacobson, L.C.; Hujo, W.; Molinero, V. Amorphous precursors in the nucleation of clathrate hydrates. *J. Am. Chem. Soc.* **2010**, *132*, 11806–11811. [CrossRef]

21. Wolde, P.R.T.; Frenkel, D. Enhancement of Protein Crystal Nucleation by Critical Density Fluctuations. *Science* **1997**, *277*, 1975–1978. [CrossRef]
22. Hirata, M.; Yagasaki, T.; Matsumoto, M.; Tanaka, H. Phase Diagram of TIP4P/2005 Water at High Pressure. *Langmuir* **2017**, *33*, 11561–11569. [CrossRef]
23. Mochizuki, K.; Himoto, K.; Matsumoto, M. Diversity of transition pathways in the course of crystallization into ice VII. *Phys. Chem. Chem. Phys.* **2014**, *16*, 16419–16425. [CrossRef]
24. Ostwald, W. Studien über die Bildung und Umwandlung fester Körper. *Z. Phys. Chem.* **1897**, *22*, 289. [CrossRef]
25. Karthika, S.; Radhakrishnan, T.K.; Kalaichelvi, P. A Review of Classical and Nonclassical Nucleation Theories. *Cryst. Growth Des.* **2016**, *16*, 6663–6681. [CrossRef]
26. Wang, Z.; Wang, F.; Peng, Y.; Zheng, Z.; Han, Y. Imaging the homogeneous nucleation during the melting of superheated colloidal crystals. *Science* **2012**, *338*, 87–90. [CrossRef]
27. Samanta, A.; Tuckerman, M.E.; Yu, T.-Q.; Weinan, E. Microscopic mechanisms of equilibrium melting of a solid. *Science* **2014**, *346*, 729–732. [CrossRef]
28. Mochizuki, K.; Matsumoto, M.; Ohmine, I. Defect pair separation as the controlling step in homogeneous ice melting. *Nature* **2013**, *498*, 350–354. [CrossRef]
29. Shah, M.; Santiso, E.E.; Trout, B.L. Computer simulations of homogeneous nucleation of benzene from the melt. *J. Phys. Chem. B* **2011**, *115*, 10400–10412. [CrossRef]
30. Tohji, K.; Murata, Y. X-Ray Diffraction Study of the Melting of Benzene. *Jpn. J. Appl. Phys.* **1982**, *21*, 1199–1204. [CrossRef]
31. Craven, C.J.; Hatton, P.D.; Howard, C.J.; Pawley, G.S. The structure and dynamics of solid benzene. I. A neutron powder diffraction study of deuterated benzene from 4 K to the melting point. *J. Chem. Phys.* **1993**, *98*, 8236–8243. [CrossRef]
32. Mochizuki, K. Computational Study on Homogeneous Melting of Benzene Phase I. *Crystals* **2019**, *9*, 84. [CrossRef]
33. MacKerell, A.D.; Bashford, D.; Bellott, M.; Dunbrack, R.L.; Evanseck, J.D.; Field, M.J.; Fischer, S.; Gao, J.; Guo, H.; Ha, S.; et al. All-atom empirical potential for molecular modeling and dynamics studies of proteins. *J. Phys. Chem. B* **1998**, *102*, 3586–3616. [CrossRef]
34. MacKerell, A.D.; Wiorkiewicz-Kuczera, J.; Karplus, M. An all-atom empirical energy function for the simulation of nucleic acids. *J. Am. Chem. Soc.* **1995**, *117*, 11946–11975. [CrossRef]
35. Nemkevich, A.; Bürgi, H.-B.; Spackman, M.A.; Corry, B. Molecular dynamics simulations of structure and dynamics of organic molecular crystals. *Phys. Chem. Chem. Phys.* **2010**, *12*, 14916–14929. [CrossRef] [PubMed]
36. Fu, C.-F.; Tian, S.X. A Comparative Study for Molecular Dynamics Simulations of Liquid Benzene. *J. Chem. Theory Comput.* **2011**, *7*, 2240–2252. [CrossRef]
37. Darden, T.; York, D.; Pedersen, L. Particle mesh Ewald: An N·log(N) method for Ewald sums in large systems. *J. Chem. Phys.* **1993**, *98*, 10089–10092. [CrossRef]
38. Berendsen, H.J.C.; van der Spoel, D.; van Drunen, R. GROMACS: A message-passing parallel molecular dynamics implementation. *Comput. Phys. Commun.* **1995**, *91*, 43–56. [CrossRef]
39. Berendsen, H.J.C.; Postma, J.P.M.; van Gunsteren, W.F.; DiNola, A.; Haak, J.R. Molecular dynamics with coupling to an external bath. *J. Chem. Phys.* **1984**, *81*, 3684–3690. [CrossRef]
40. Hoover, W.G. Canonical dynamics: Equilibrium phase-space distributions. *Phys. Rev. A Gen. Phys.* **1985**, *31*, 1695–1697. [CrossRef] [PubMed]
41. Nosé, S. A molecular dynamics method for simulations in the canonical ensemble. *Mol. Phys.* **2002**, *100*, 191–198. [CrossRef]
42. Bacon, G.E.; Curry, N.A.; Wilson, S.A. A crystallographic study of solid benzene by neutron diffraction. *Proc. R. Soc. Lon. A* **1964**, *279*, 98–110.
43. Fernández, R.G.; Abascal, J.L.F.; Vega, C. The melting point of ice Ih for common water models calculated from direct coexistence of the solid-liquid interface. *J. Chem. Phys.* **2006**, *124*, 144506. [CrossRef]
44. Belonoshko, A.B. Molecular Dynamics of MgSiO$_3$ perovskite at high pressures: Equation of state, structure, and melting transition. *Geochim. Chosmochim. Acta* **1994**, *58*, 4039–4047. [CrossRef]
45. Belonoshko, A.B.; Ahuja, R.; Johansson, B. Quasi-Ab initio molecular dynamic study of Fe melting. *Phys. Rev. Lett.* **2000**, *84*, 3638–3641. [CrossRef] [PubMed]

46. Belonoshko, A.B.; Arapan, S.; Martonak, R.; Rosengen, A. MgO phase diagram from first principles in a wide pressure-temperature range. *Phys. Rev. B* **2010**, *81*. [CrossRef]
47. Bridgman, P.W. The Technique of High Pressure Experimenting. *Proc. Am. Acad. Arts Sci.* **1914**, *49*, 627. [CrossRef]
48. Mo, M.Z.; Chen, Z.; Li, R.K.; Dunning, M.; Witte, B.B.L.; Baldwin, J.K.; Fletcher, L.B.; Kim, J.B.; Ng, A.; Redmer, R.; et al. Heterogeneous to homogeneous melting transition visualized with ultrafast electron diffraction. *Science* **2018**, *360*, 1451–1455. [CrossRef] [PubMed]
49. Wang, F.; Zhou, D.; Han, Y. Melting of Colloidal Crystals. *Adv. Funct. Mater.* **2016**, *26*, 8903–8919. [CrossRef]
50. Bai, X.-M.; Li, M. Nature and extent of melting in superheated solids: Liquid-solid coexistence model. *Phys. Rev. B* **2005**, *72*. [CrossRef]
51. Stegailov, V. Homogeneous and heterogeneous mechanisms of superheated solid melting and decay. *Comput. Phys. Commun.* **2005**, *169*, 247–250. [CrossRef]
52. Delogu, F. Mechanistic Aspects of Homogeneous and Heterogeneous Melting Processes. *J. Phys. Chem. B* **2006**, *110*, 12645–12652. [CrossRef] [PubMed]
53. Burakovsky, L.; Burakovsky, N.; Cawkwell, M.J.; Preston, D.L.; Errandonea, D.; Simak, S.I. Ab initio phase diagram of iridium. *Phys. Rev. B* **2016**, *94*.
54. Errandonea, D.; MacLeod, S.G.; Ruiz-Fuertes, J.; Burakovsky, L.; McMahon, M.I.; Wilson, C.W.; Ibañez, J.; Daisenberger, D.; Popescu, C. High-pressure/high-temperature phase diagram of zinc. *J. Phys. Condens. Matter* **2018**, *30*, 295402. [CrossRef]
55. Burakovsky, L.; Burakovsky, N.; Preston, D.; Simak, S. Systematics of the Third Row Transition Metal Melting: The HCP Metals Rhenium and Osmium. *Crystals* **2018**, *8*, 243. [CrossRef]
56. Gorski, K.M.; Hivon, E.; Banday, A.J.; Wandelt, B.D.; Hansen, F.K.; Reinecke, M.; Bartelmann, M. HEALPix: A Framework for High-Resolution Discretization and Fast Analysis of Data Distributed on the Sphere. *Astrophys. J.* **2005**, *622*, 759–771. [CrossRef]
57. Kovačević, T.; Reinhold, A.; Briesen, H. Identifying Faceted Crystal Shape from Three-Dimensional Tomography Data. *Cryst. Growth Des.* **2014**, *14*, 1666–1675. [CrossRef]
58. Wise, P.K.; Ben-Amotz, D. Interfacial Adsorption of Neutral and Ionic Solutes in a Water Droplet. *J. Phys. Chem. B* **2018**, *122*, 3447–3453. [CrossRef] [PubMed]
59. Alexeev, A.D.; Revva, V.N.; Alyshev, N.A.; Zhitlyonok, D.M. True triaxial loading apparatus and its application to coal outburst prediction. *Int. J. Coal Geol.* **2004**, *58*, 245–250. [CrossRef]
60. Jonsson, H.; Gråsjö, J.; Nordström, J.; Johansson, N.; Frenning, G. An apparatus for confined triaxial testing of single particles. *Powder Technol.* **2015**, *270*, 121–127. [CrossRef]
61. Jonsson, H.; Frenning, G. Investigations of single microcrystalline cellulose-based granules subjected to confined triaxial compression. *Powder Technol.* **2016**, *289*, 79–87. [CrossRef]

Article

Pressure Effects on the Optical Properties of NdVO$_4$

Enrico Bandiello [1,*], Josu Sánchez-Martín [1], Daniel Errandonea [1] and Marco Bettinelli [2]

[1] Departamento de Física Aplicada-ICMUV, Universitat de València, Calle Dr. Moliner 50, 46100 Burjassot, Spain; sanmarj2@alumni.uv.es (J.S.-M.); daniel.errandonea@uv.es (D.E.)

[2] Laboratorio Materiali Luminescenti, Dipartimento di Biotecnologie, Università di Verona, and INSTM, UdR Verona, Strada Le Grazie 15, 37134 Verona, Italy; marco.bettinelli@univr.it

* Correspondence: enrico.bandiello@uv.es

Received: 25 April 2019; Accepted: 4 May 2019; Published: 6 May 2019

Abstract: We report on optical spectroscopic measurements in pure NdVO$_4$ crystals at pressures up to 12 GPa. The influence of pressure on the fundamental absorption band gap and Nd^{3+} absorption bands has been correlated with structural changes in the crystal. The experiments indicate that a phase transition takes place between 4.7 and 5.4 GPa. We have also determined the pressure dependence of the band-gap and discussed the behavior of the Nd^{3+} absorption lines under compression. Important changes in the optical properties of NdVO$_4$ occur at the phase transition, which, according to Raman measurements, corresponds to a zircon to monazite phase change. In particular, in these conditions a collapse of the band gap occurs, changing the color of the crystal. The changes are not reversible. The results are analyzed in comparison with those deriving from previous studies on NdVO$_4$ and related vanadates.

Keywords: vanadate; zircon; high pressure; band gap; phase transition; optical absorption

---

1. Introduction

Trivalent metal orthovanadates have been constantly studied during the last decade because of their optical properties, which makes them appropriate for many different technological applications. Such applications go from photocatalytic water purification and hydrogen production to scintillators, phosphors, solid-state lasers, magnetoelectric, and medical applications [1–6]. Neodymium vanadate (NdVO$_4$) is one of the most promising orthovanadates for the development of technological applications [1–6], for which an accurate knowledge of NdVO$_4$ optical properties is a fundamental requisite. Several studies have been carried out on the optical properties of NdVO$_4$ [6–11], mainly focused on the study of the band-gap energy and the optical absorptions associated to internal transitions of Nd^{3+} ions from ^{4}I$_{9/2}$ to ^{4}G$_{5/2}$, ^{4}F$_{7/2}$, ^{4}F$_{5/2}$, and ^{4}F$_{3/2}$ [12]. Discrepancies on the value of the fundamental band gap (E_{gap}) can be found in the literature [6–11], with values of the E_{gap} ranging from 2.95 to 3.72 eV. Consequently, new studies are needed to accurately determine this parameter, which is crucial for the understanding of the properties of this wide gap semiconductor.

A second focus of research on NdVO$_4$ has been the influence of pressure on its properties and crystal structure. Numerous studies have been carried out, reporting the existence of a phase transition around 5–6 GPa [13–16]. Interestingly, it has been found that the crystal structure of the high-pressure (HP) phase strongly depends on the conditions of the experiment (namely, hydrostatic or non-hydrostatic conditions). On the other hand, a second phase transition has been found to occur at 18–20 GPa, according to Raman and X-ray diffraction (XRD) experiments [13,14]. In contrast, optical-absorption experiments located this transition at 12 GPa. These results also suggest that additional studies of the pressure effects on the optical properties of NdVO$_4$ are required. This could give not only information about the pressure dependence of the band gap but also on the structural stability. In particular, the study of the behavior under compression of the internal absorption transitions of Nd^{3+} ions, which strongly depend on their local environment, can be useful to characterize the HP phase transitions [17,18].

Here we report a systematic study of the optical properties on $NdVO_4$ in the ultraviolet (UV)–visible (VIS) range. The studies have been carried on single crystals growth by the flux growth method. The maximum pressure achieved is 12 GPa. Our results confirm that $NdVO_4$ is a wide-gap semiconductor with $E_{gap} = 3.72(2)$ eV. The pressure dependence of E_{gap} is reported and discussed, as well as the pressure dependence of the absorption bands associated to internal transition of Nd^{3+}. We demonstrate the existence of only one phase transition in the pressure range covered by the experiment. This transition is non-reversible and causes a band-gap collapse, along with important changes in the absorption bands associated to Nd^{3+}, as shown and discussed in the following.

2. Materials and Methods

$NdVO_4$ single crystals, up to 10 mm long, were obtained by the flux growth method [19–21]. A 1:1.923 mol. mixture of V_2O_5 and PbO (99%, Carlo Erba, Val-De-Reuil, France, and 99.9%, Sigma Aldrich, St. Louis, MO, USA, respectively) plus borax ($Na_2B_4O_7 \cdot 10H_2O$, > 99%, JT Baker, Waltham, MA, USA) was used as the flux. Borax was added as a flux modifier, in order to increase the size of the crystals [22]. Nd_2O_3 (0.7394 g, purity 99.99%, Sigma Aldrich, St. Louis, MO, USA) was added as the crystal precursor. All the reagents were in form of fine powders. The flux composition was 21.3082 g of PbO, 9.1235 g of V_2O_5, and 1.25 g of $Na_2B_4O_7 \cdot 10H_2O$. Platinum crucibles were filled with the flux and the precursor, sealed with a platinum lid, and put in a programmable oven. To dehydrate the mixture, the temperature was ramped up at 0.6 °C/hour and kept at 250 °C for 2 hours. Afterwards, the temperature was kept fixed for 15 hours at 1300 °C, after a gradual increment at a rate of 105 °C/hour. During this step, the flux melts and acts as a solvent for Nd_2O_3. The temperature was then slowly decreased until 950 °C (−1.8 °C/hour), in order to promote the formation of the crystals by precipitation and spontaneous nucleation. The crucible was then removed from the oven, rapidly reversed (to facilitate the recovery of the crystals) and allowed to cool down to room temperature. Subsequently, the platinum lid was removed and the crucible was immersed in hot nitric acid (1.5 M), which was continuously stirred and renewed multiple times until the complete dissolution of the flux. Finally, the clean crystals were washed with deionized water and recovered using a paper filter.

In order to characterize the growth of $NdVO_4$ crystals we performed energy-dispersive X-ray spectroscopy (EDXS) in a transmission electron microscopy (TEM, FEI Company, Hillsboro, OR, USA) operated at 200 kV (TECNAI G2 F20 S-TWIN, Waltham, MA, USA), powder XRD measurements using a Rigaku D/Max diffractometer (Tokyo, Japan) with Cu-K_α radiation (λ = 1.5406 Å), and Raman measurements in backscattering geometry with a Jobin Yvon THR1000 single spectrometer (Kyoto, Japan) equipped with an edge filter, a 10X objective (Mitutoyo, Kawasaky, Japan) with a numerical aperture of 0.28, and a thermoelectric-cooled multichannel charge-coupled device (CCD) detector, using a 632.8 nm HeNe laser (10 mW power).

Optical-absorption experiments were accomplished in the UV–VIS range using the optical set-up previously described by Segura et al. [23], which is equipped with Cassegrain objectives (6X magnification and 0.60 numerical aperture, Edmund Optics, Mainz Germany) and an UV–VIS spectrometer (USB4000-UV–VIS - Ocean Optics, Duiven, Netherlands). Optical absorption spectra were obtained from the transmittance of the sample, measured using the sample-in sample-out method [24]. For these measurements, platelets of size 60 × 60 μm and approximately 5 μm thick were cleaved along the {110} plane. The experiments were carried out both at ambient and at high pressure. The HP experiments were performed using a diamond anvil cell (DAC, Chervin-type, Sorbonne University, Paris, France) equipped with IIA-type diamonds with a culet size of 480 μm. The crystals were loaded in a 200 μm hole of a steel gasket, pre-indented to a thickness of 50 μm. As the pressure medium, we employed a 16:3:1 methanol–ethanol–water mixture, which remains quasi-hydrostatic up to 10.5 GPa [25]. The $NdVO_4$ crystals were carefully located at the center of the gasket hole to avoid bridging [26]. Small ruby chips were loaded next to the sample for pressure determination [27] with an accuracy of 0.05 GPa. The full width half maximum of R_1 and R_2 lines of ruby fluorescence and the splitting between them confirm the quasi-hydrostatic conditions of the experiments [28].

3. Results and Discussion

3.1. Ambient-Pressure Characterization

The obtained single crystals are transparent with a strong violet hue. The growth is preferentially along the *c*-axis, with size of the order of $5 \times 1 \times 0.2$ mm (see inset in Figure 1). EDXS shows the presence of only Nd, O, and V, indicating that the grown crystals are free from impurities. According to EDXS analysis, the ratio of Nd to V is found to be ~1, confirming the formation of stoichiometric $NdVO_4$. Results from powder XRD measurements are shown in Figure 1 together with the results of the structural refinement and residuals. The phase purity of $NdVO_4$ has been confirmed by the Rietveld refinement, along with the expected zircon-type crystal structure (space group $I4_1/amd$) [29]. A total number of 31 reflections have been observed and all of them have been used in the refinement. Nd and V are located at high-symmetry atomic positions 4a (0, 3/4, 1/8) and 4b (0, 1/4, 3/8), respectively. This facilitates the determination of the positions of oxygen atoms, which were established to be at 16 h (0, 0.4292(5), 0.2055(5)). The unit-cell parameters are $a = 7.3311(8)$ Å and $c = 6.4359(7)$ Å. According to these results, the zircon structure is composed of regular VO_4 tetrahedral units with a V-O distance of 1.7076(6) Å and NdO_8 dodecahedra with two different Nd-O distances, 2.4082(9) and 2.5001(9) Å. These results are in full agreement with the literature [4,29]. The small residuals in Figure 1 and the R-factors of the refinement ($R_P = 5.8\%$ and $R_{WP} = 7.11\%$) fully support the assignment of the zircon-type structure.

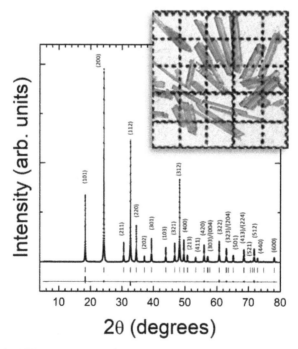

Figure 1. Powder XRD pattern measured in $NdVO_4$. The dots are the experiment. The solid lines the refinement and residual. Ticks indicate the positions of Bragg peaks. The most intense peaks are labeled with their indexes. A picture of the grown crystals is show in the inset, showing their size and transparency. For reference, the squares are 1×1 mm.

Results from the Raman experiment at ambient pressure are shown in Figure 2. Group theory predicts 12 Raman active modes for the zircon-type structure, whose irreducible representation is $2A_{1g} + 4B_{1g} + B_{2g} + 5E_g$ [30]. All the expected Raman active modes could be identified in our experiment. Notice that the two broadest peaks (those at ~240 and 378 cm^{-1}) are doublets consistent of modes very close in frequency (see Table 1). No other modes than those representative of zircon-type NdVO$_4$ have been observed. This supports that only the stable polymorph of NdVO$_4$ is present in our crystals, a conclusion that agrees with the results of XRD experiments. The wavenumbers of the different modes are summarized in Table 1. They agree with those reported by Santos et al. [30], Nguyen et al. [31], and Panchal et al. [13]. The mode assignment was done according to these previous works.

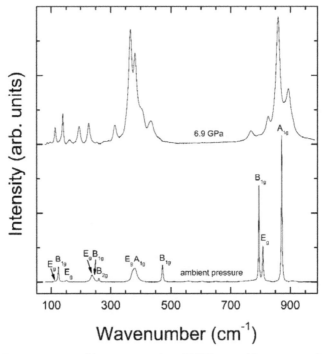

Figure 2. Raman spectrum of low-pressure zircon NdVO$_4$ at ambient pressure (bottom) and high-pressure NdVO$_4$ at 6.9 GPa (top). The modes of the zircon phase are identified.

Table 1. Frequencies (ω) in cm^{-1} of the Raman-active phonon modes at room conditions in NdVO$_4$.

Mode	ω	Mode	ω	Mode	ω
E_g	113(1)	B_{1g}	244(1)	B_{1g}	472(1)
B_{1g}	124(1)	B_{2g}	261(1)	B_{1g}	795(1)
E_g	151(1)	E_g	374(1)	E_g	808(1)
E_g	238(1)	A_{1g}	381(1)	A_{1g}	871(1)

Results from optical-absorption experiments are shown in Figure 3. The absorption spectrum has an abrupt absorption edge at low wavelengths plus eight absorption bands, named as A_1 to A_8. These bands resemble those of the transitions between the levels of Nd^{3+}, while the strong low-wavelength absorption corresponds to the fundamental band gap of NdVO$_4$, with the typical Urbach-tail shape of this family of compounds [32,33]. This can be clearly seen in the inset of the figure, where the absorption is plotted as a function of the photon energy (E). In particular, it closely resembles the absorption edges of the direct band gap of zircon-type YVO$_4$, LuVO$_4$, and YbVO$_4$ [11]. By assuming that the absorption coefficient (α) of NdVO$_4$ obeys an Urbach-tail energy dependence, $\alpha(E) = \alpha_0 e^{-(Egap-E)/Eu}$, we have been able to explain the observed strong low-wavelength (high-energy) absorption edge. In the equation, E_u is the Urbach energy, which is related to the steepness of the absorption edge. In the inset of Figure 3 we show the goodness of this dependence fit for the high-energy part of the absorption spectrum (red line). The best fit has been obtained with E_{gap} = 3.72(2) eV and E_u = 0.081(5) eV. The value of the band-gap agrees with the results reported by Panchal et al. [11] and the value of E_u is comparable with the values reported for related oxides [34].

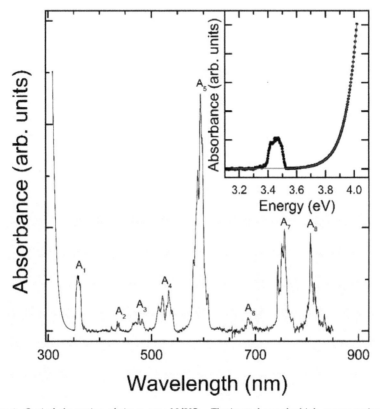

Figure 3. Optical-absorption of zircon-type NdVO$_4$. The inset shows the high-energy region as a function of energy including the fir to the step fundamental absorption described in the text (red line).

The rest of the absorption features observed in our experiment (A$_1$ to A$_8$ in Figure 3) can be correlated to transitions between 4f^3 levels of Nd^{3+}. All the absorption bands are composed by several overlapping peaks, due to the expected broadening at room temperature. They are centered at 359, 439, 474, 528, 593, 688, 753, and 817 nm. They likely correspond to transitions from ^{4}I$_{9/2}$ to ^{4}G$_{11/2}$, ^{4}G$_{9/2}$, ^{4}G$_{7/2}$, ^{4}F$_{9/2}$, ^{4}G$_{5/2}$, ^{4}F$_{7/2}$, ^{4}F$_{5/2}$, and ^{4}F$_{3/2}$, respectively [35]. The features at long wavelengths are the strongest and have been previously reported [6–10]. Among the others, only the absorption band centered at 359 nm has been previously reported, even if mistakenly assigned to the fundamental band-gap of NdVO$_4$ [8]. This leads to the underestimation of the band gap, with values from ranging 2.95 to 3.5 eV [6–10]. The band gap of NdVO$_4$ measured in our experiment is larger (3.72 eV), compatible with the band gap of rare-earth vanadates, and its value is determined by the classical field splitting of the VO$_4$$^{3-}$ ion [26], the top of the valence band and bottom of the conduction band being dominated by V 3d and O 2p states. A possible cause for the aforementioned underestimation of the band-gap can be the fact that the previous studies were carried out on nanoparticles and nanowires and not on single crystals. The diffusion of light through nanomaterials used in previous experiments, along with the poorer crystallinity, would lead to a larger Urbach energy, making the fundamental absorption edge less steep and causing it to merge with the highest energy absorption band of Nd^{3+}, therefore leading to an underestimation of E_{gap}.

3.2. High-Pressure Studies

We will now present the results from HP experiments. In Figure 4, we show the high-energy portion of the absorption spectrum, which we will use first to discuss the effect of pressure on the band gap. Afterwards, we will focus on the influence of pressure on the absorptions related to Nd^{3+}. These results will be correlated with the information known on the HP behavior of the crystal structure and Raman modes of $NdVO_4$ [13–16]. The first phenomenon we observe is that the absorption edge gradually blue-shifts from ambient pressure to 4.7 GPa. In particular, we found that in this range of pressure E_{gap} increases from 3.72(2) eV to 3.79(2) eV. The shape of the absorption edge does not change with pressure, which indicates that the band gap remains a direct one. This result is in good agreement with the report by Panchal et al. [11]. In Figure 5, we present the pressure dependence of E_{gap} and compare it with the literature [11]. Both experiments show a linear increase of E_{gap}, which is a consequence of the increase of the repulsion of bonding and anti-bonding states as the V-O distances of the VO_4 tetrahedron are reduced upon compression [13].

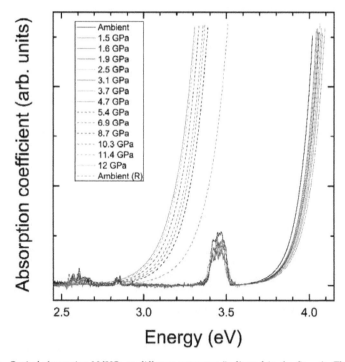

Figure 4. Optical-absorption $NdVO_4$ at different pressures (indicated in the figure). The spectra assigned to the low(high)-pressure phase are represented with solid (dashed) lines.

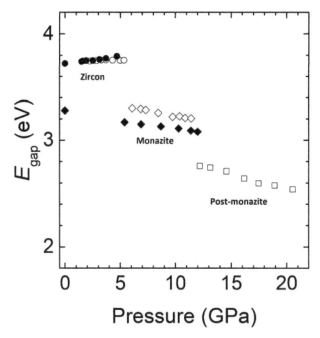

Figure 5. Pressure dependence of the band-gap energy of NdVO$_4$. Circles (diamonds) correspond to the zircon (monazite) phase. Squares correspond to the post-monazite phase. Results from the present study are shown with solid symbols and results from the literature [11] are shown with empty symbols.

At 5.4 GPa we found an abrupt decrease of E_{gap}. This can be seen in Figure 4 by the sudden red-shift of the absorption edge when moving from 4.7 to 5.4 GPa. This corresponds to a decrease of the band gap from 3.79(2) eV to 3.17(2) eV. This fact is indicative of the occurrence of a phase transition. From 5.4 up to 12 GPa, the absorption edge gradually red-shifts, with no indication of the existence of a second phase transition. Upon decompression, the transition is found to be irreversible in our experiment, being E_{gap} = 3.26(2) eV at ambient pressure in the recovered sample. According to the literature, the observed transition can be either from zircon to scheelite or monazite [13–16]. In order to clarify this issue, we have carried out a Raman experiment at 6.9 GPa. The results are shown in Figure 2. For the scheelite phase, according to group theory, 13 Raman-active modes are expected (3A$_g$ + 5B$_g$ + 5E$_g$) [36]. For the monazite phase, 36 Raman-active modes are expected (18A$_g$ + 18B$_g$) [14]. The difference in number of modes is very noticeable in the high-frequency region, where scheelite has a very characteristic feature with only three modes. In our experiment, the Raman spectrum has manifestly more than thirteen peaks. In addition, it very much resembles the Raman spectrum associated to monazite-type NdVO$_4$ by Panchal et al. [14]. Therefore, the changes observed in the absorption spectrum are likely to be triggered by the occurrence of the zircon–monazite transition. This is a first-order reconstructive transition that involves a volume collapse and an increase in the coordination number of Nd from 8 to 9. In the HP monazite phase, Nd is surrounded by nine oxygens forming NdO$_9$ pentagonal interpenetrating tetrahedral units, while in zircon Nd has eight neighboring oxygens forming NdO$_8$ dodecahedral units.

From the absorption spectra shown in Figure 4 we obtained the pressure dependence of E_{gap}, using the same procedure to determine it at ambient pressure. The results are shown in Figure 5 and compared with those reported by Panchal et al. [11]. For the zircon-type phase we obtained a very similar linear pressure dependence. In our case dE_{gap}/dP = 11.2(9) meV/GPa. In the figure, the abrupt collapse of the band gap from 4.7 to 5.4 GPa can be seen, which is associated to the zircon–monazite transition. This transition involves a collapse of the volume, a lowering of the crystal symmetry, and a distortion of the VO_4 tetrahedron [16]. These changes of the crystalline structure are reflected in the electronic structure of $NdVO_4$, producing the above described collapse of E_{gap}. Notice that the variation of the electronic band gap is much more rapid and sharp than the changes observed in the Raman and XRD experiments, suggesting that optical experiments at the intrinsic absorption edge are more accurate for the determination of the transition pressures. This is because the electronic properties are much more sensitive than the structural and vibrational properties to the influence of pressure. A small change in the crystal structure can induce large changes in the band structure and consequently in the electronic properties. In our case, the band-gap collapse at the transition is 0.1 eV larger than in previous experiments [11].

After the phase transition, the band gap follows a linear behavior but red-shifts under compression with a rate dE_{gap}/dP = 19.5(9) meV/GPa. The HP monazite phase remains stable up to 12 GPa, the highest pressure covered by our experiments. Notice that we did not find evidence of the second transition reported by Panchal et al. [11]. However, we observed the formation of cracks in the crystal at 12 GPa, which allowed the diffusion of light and precluded the performance of accurate experiments at higher pressure. We consider the formation of cracks a possible precursor effect of the second transition reported by Panchal et al. [11] at 11. 4 GPa [37], which could be triggered at a lower pressure than in our experiment because of the use of a thicker crystal, causing the bridging of the crystal between diamonds [23]. Interestingly, we found that after releasing pressure, our samples remain in the HP phase. This allows us to determine the band gap of the metastable monazite-type polymorph at ambient pressure, with E_{gap} = 3.26(2) eV. This value is probably more suitable for many photocatalytic applications than that of zircon-type $NdVO_4$. The red-shift of the band gap in the HP phase is consistent with the results obtained for monazite-type $LaVO_4$ [18]. The explanation for the closing of the gap with pressure in monazite (in contrast to the opening of the gap in zircon) comes from the contribution of Nd 4f and 5d orbitals to the bottom of the conduction band in monazite $NdVO_4$. These states are much more localized than O 2p and V 3d states. As a consequence, the bottom of the conduction band does not shift with pressure, but the top of the valence band moves toward higher energies, causing the observed band gap decrease.

We will comment now on the absorption bands related to Nd^{3+}. In Figure 4, it can be seen that at the phase transition the absorption band around 359 nm (A_1 in Figure 3) assigned to $^4I_{9/2}$ to $^4G_{11/2}$ disappears because the band gap of the HP phase is at a lower energy (longer wavelength). In the low-pressure zircon phase, the effect of pressure on the A_1–A_8 absorption bands is qualitatively the same for all of them. We will therefore focus on bands A_4 and A_5 ($^4I_{9/2}$ to $^4F_{9/2}$ and $^4I_{9/2}$ to $^4F_{5/2}$ transitions), two of the better resolved bands, to illustrate the pressure behavior of internal Nd^{3+} transitions. Their pressure dependence is shown in Figure 6. The absorption bands consist of several multiplets caused by the Stark splitting. In the low-pressure phase (pressure ≤ 4.7 GPa), there are no qualitative changes as pressure increases. Basically, the absorption bands show the typical red shift which can be attributed to an increase of the covalence in $NdVO_4$ [37]. The position of each of the identified peaks as function of pressure is presented in Figure 7. The behavior is qualitatively similar to related oxides doped with Nd^{3+} or other lanthanides [17,18,38,39]. In particular, our results suggest that the increase of the crystal field interaction with pressure in $NdVO_4$ is qualitatively similar to that of isomorphic Nd-doped YVO_4 [38].

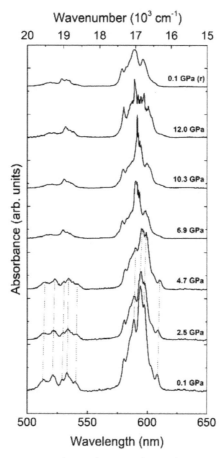

Figure 6. Absorbance relative to the $^4I_{9/2}$ to $^4G_{11/2}$ and $^4I_{9/2}$ to $^4F_{9/2}$ transitions measured at room temperature at different pressures. The spectrum collected after pressure release is marked with (r). The spectra are plotted as function of wavenumber (top) and wavelength (bottom). They have been vertically shifted to facilitate comparison.

It is noteworthy that, for pressures below the phase transition, some band splitting and merging is present in the 1.5–3.8 GPa range. This may be caused by the well-established distortion of the NdO_8 tetrahedron under high-pressure conditions [16]. For pressures higher than 4.7 GPa, the absorption bands are totally different, confirming the phase transition. The optical spectroscopy associated to Nd^{3+} is thus quite sensitive to pressure, proving also to be a very efficient tool for the detection of phase transitions. Again, the changes induced by pressure in the absorption spectra are not reversible upon decompression (see Figure 4). The changes of the absorption bands are consistent with the coordination change of Nd^{3+} at the zircon-monazite transition. In the HP monazite phase Nd^{3+} is coordinated by nine oxygen atoms forming a distorted polyhedron. As in the low-pressure phase, in the HP phase the absorption bands also mainly red shift under compression. However, the pressure shift is slightly reduced, which is caused by the decrease of the compressibility due to the density increase associated to the transition [16]. To conclude, we would like to add that in both phases we observed the splitting of some of the multiplets. We believe this is the results of enhancement of exchange interaction in Nd^{3+}-Nd^{3+} pairs as pressure increases [18,40].

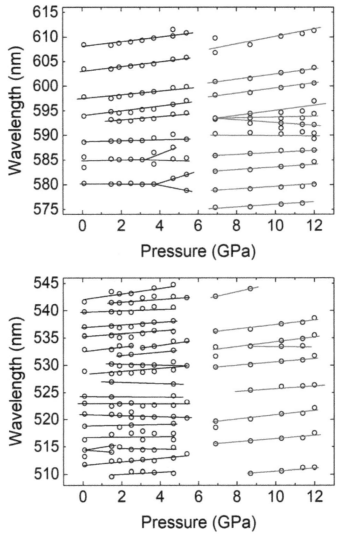

Figure 7. Peak positions observed under pressure increase. Lines are a guide for the eye.

4. Conclusions

In this study, single crystals of $NdVO_4$ where synthetized and characterized by XRD, UV–VIS absorption, and Raman spectroscopy at ambient conditions and at high pressure (up to 12 GPa). In particular, optical spectroscopy allowed a precise determination of the band-gap energy of $NdVO_4$ (3.72(2) eV), which is found to be larger than that reported in previous studies. We compared the pressure evolution of the fundamental band-gap and of the absorption bands associated with transitions of Nd^{3+} energy levels. The shape of the fundamental absorption edge does not change with pressure and a phase transition around 5 GPa is evidenced by a sudden collapse of the band gap ($\Delta E_g = -0.69$ eV). These HP experiments are thus an accurate tool for establishing the band gap. On the other hand, the absorption bands of Nd^{3+} show a more gradual evolution but the phase transition is equally put in evidence by their reduced number in the high-pressure phase. The band-gap collapse near 5 GPa can be correlated to the well-known zircon–monazite transition of $NdVO_4$. The transition is irreversible, as the HP phase could be recovered at ambient pressure, with a band gap of 3.26(2) eV. The obtained results have been compared with previous studies. We hope that these conclusions will contribute to deepening the understanding of the optical properties of $NdVO_4$ and its behavior in high-pressure conditions.

Author Contributions: All authors contributed equally to this work; being all involved in experiments, analysis, interpretation of results, and writing of the manuscript.

Funding: This work was supported by the Spanish Ministry of Science, Innovation and Universities, the Spanish Research Agency, and the European Fund for Regional Development under grant MAT2016-75586-C4-1-P and by Generalitat Valenciana under grant Prometeo/2018/123 (EFIMAT). E. Bandiello thanks the Generalitat Valenciana for his postdoctoral contract (VaII+D, APOSTD2017).

Acknowledgments: The authors thank Erica Viviani (Department of Biotechnology, University of Verona) for the technical assistance in the synthesis of $NdVO_4$ crystals.

Conflicts of Interest: The authors declare no conflict of interest.

References

1. Fujimoto, Y.; Yanagida, T.; Yokota, Y.; Chani, V.; Kochurikhin, V.V.; Yoshikawa, A. Comparative study of optical and scintillation properties of YVO_4, $(Lu_{0.5}Y_{0.5})VO_4$, and $LuVO_4$ single crystals. *Nucl. Instrum. Methods Phys. Res. A* **2011**, *635*, 53–56. [CrossRef]
2. Oshikiri, M.; Ye, J.; Boero, M. Inhomogeneous RVO_4 Photocatalyst Systems (R = Y, Ce, Pr, Nd, Sm, Eu, Gd, Tb, Dy, Ho, Er, Tm, Yb, Lu). *J. Phys. Chem. C* **2014**, *118*, 8331–8841. [CrossRef]
3. Wang, F.; Liu, X. Multicolor tuning of lanthanide-doped nanoparticles by single wavelength excitation. *Acc. Chem. Res.* **2014**, *47*, 1378–1385. [CrossRef]
4. Errandonea, D.; Garg, A.B. Recent progress on the characterization of the high-pressure behaviour of AVO_4 orthovanadates. *Prog. Mater. Sci.* **2018**, *97*, 123–169. [CrossRef]
5. Palacios, E.; Evangelisti, M.; Sáez-Puche, R.; Dos Santos-García, A.J.; Fernández-Martínez, F.; Cascales, C.; Castro, C.; Burriel, R.; Fabelo, O.; Rodríguez-Velamazán, J.A. Magnetic structures and magnetocaloric effect in RVO_4 (R = Gd, Nd). *Phys. Rev. B* **2018**, *97*, 214401. [CrossRef]
6. Monsef, R.; Ghiyasiyan-Arani, M.; Salavati-Niasari, M. Utilizing of neodymium vanadate nanoparticles as an efficient catalyst to boost the photocatalytic water purification. *J. Environ. Manag.* **2019**, *230*, 266–281. [CrossRef]
7. Xu, J.; Hu, C.; Liu, G.; Liu, H.; Du, G.; Zhang, Y. Synthesis and visible-light photocatalytic activity of $NdVO_4$ nanowires. *J. Alloy Compd.* **2011**, *509*, 7968–7972. [CrossRef]
8. Wu, X.; Tao, Y.; Dong, L.; Zhu, J.; Hu, Z. Preparation of Single-Crystalline $NdVO_4$ Nanorods, and Their Emissions in the Ultraviolet and Blue under Ultraviolet Excitation. *J. Phys. Chem. B* **2005**, *109*, 11544–11547. [CrossRef]
9. Dragomir, M.; Valant, M. Room-Temperature Synthesis and Optical Properties of $NdVO_4$ Nanoneedles. *Acta Chim. Slov.* **2018**, *65*, 679–686. [CrossRef]
10. Dragomir, M.; Arcon, I.; Gardonio, S.; Valant, M. Phase relations and optoelectronic characteristics in the $NdVO_4$-$BiVO_4$ system. *Acta Mater.* **2013**, *61*, 1126–1135. [CrossRef]
11. Panchal, V.; Errandonea, D.; Segura, A.; Rodriguez-Hernandez, P.; Muñoz, A.; Lopez-Moreno, S.; Bettinelli, M. The electronic structure of zircon-type orthovanadates: Effects of high-pressure and cation substitution. *J. Appl. Phys.* **2011**, *110*, 043723. [CrossRef]
12. Shur, J.W.; Kochurikhin, V.V.; Borisova, A.E.; Ivanov, M.A.; Yoon, D.H. Photoluminescence properties of Nd:YVO_4 single crystals by multi-die EFG method. *Opt. Mater.* **2004**, *26*, 347. [CrossRef]
13. Marqueño, T.; Monteseguro, V.; Cova, F.; Errandonea, D.; Santamaria-Perez, D.; Bandiello, E.; Bettinelli, M. High-pressure phase transformations in $NdVO_4$ under hydrostatic, conditions: A structural powder x-ray diffraction study. *J. Phys. Condens. Matter* **2019**, *31*, 235401. [CrossRef]
14. Panchal, V.; Errandonea, D.; Manjón, F.J.; Muñoz, A.; Rodríguez-Hernández, P.; Achary, S.N.; Tyagi, A.K. High-pressure lattice-dynamics of $NdVO_4$. *J. Phys. Chem. Solids* **2017**, *100*, 126–133. [CrossRef]

15. Panchal, V.; Errandonea, D.; Manjon, F.J.; Muñoz, A.; Rodriguez-Hernandez, P.; Bettinelli, M.; Achary, S.N.; Tyagi, A.K. High Pressure phase transitions in NdVO$_4$. *AIP Conf. Proc.* **2015**, *1665*, 030006.

16. Errandonea, D.; Popescu, C.; Achary, S.N.; Tyagi, A.K.; Bettinelli, M. In situ high-pressure synchrotron X-ray diffraction study of the structural stability in NdVO$_4$ and LaVO$_4$. *Mater. Res. B* **2014**, *50*, 279–284. [CrossRef]

17. Errandonea, D.; Tu, C.; Jia, G.; Martın, I.R.; Rodrıguez-Mendoza, U.R.; Lahoz, F.; Torres, M.E.; Lavın, V. Effect of pressure on the luminescence properties of Nd^{3+} doped SrWO$_4$ laser crystal. *J. Alloy Compd.* **2008**, *451*, 212–214. [CrossRef]

18. Muñoz-Santiuste, J.E.; Lavín, V.; Rodríguez-Mendoza, U.R.; Ferrer-Roca, Ch.; Errandonea, D.; Martínez-García, D.; Rodríguez-Hernández, P.; Muñoz, A.; Bettinelli, M. Experimental and theoretical study on the optical properties of LaVO$_4$ crystals under pressure. *Phys. Chem. Chem. Phys.* **2018**, *20*, 27314–27328. [CrossRef]

19. Feigelson, R. Flux Growth of Type RVO$_4$ Rare-Earth Vanadate Crystals. *J. Am. Ceram. Soc.* **1968**, *51*, 538–539. [CrossRef]

20. Garton, G.; Smith, S.; Wanklyn, B.M. Crystal Growth from the Flux Systems PbO-V$_2$O$_5$ and Bi$_2$O$_3$-V$_2$O$_5$. *J. Cryst. Growth* **1972**, *13–14*, 588–592. [CrossRef]

21. Wanklyn, B.M. Two New Flux Systems, PbO-V$_2$O$_5$ and Bi$_2$O$_3$-V$_2$O$_5$. Part I. *J. Cryst. Growth* **1970**, *7*, 368–370. [CrossRef]

22. Smith, S.H.; Wanklyn, B.M. Flux Growth of Rare Earth Vanadates and Phosphates. *J. Cryst. Growth* **1974**, *21*, 23–28. [CrossRef]

23. Segura, A.; Sanz, J.A.; Errandonea, D.; Martinez-Garcia, D.; Fages, V. High conductivity of Ga-doped rock-salt ZnO under pressure: Hint on deep-ultraviolet-transparent conducting oxides. *Appl. Phys. Lett.* **2006**, *88*, 011910. [CrossRef]

24. Lacomba-Perales, R.; Errandonea, D.; Segura, A.; Ruiz-Fuertes, J.; Rodriguez-Hernandez, P.; Radescu, S.; Lopez-Solano, J.; Mujica, A.; Munoz, A. A combined high-pressure experimental and theoretical study of the electronic band-structure of scheelite-type AWO$_4$ (A = Ca, Sr, Ba, Pb) compounds. *J. Appl. Phys.* **2011**, *110*, 043703. [CrossRef]

25. Klotz, S.; Chervin, J.-C.; Munsch, P.; Le Marchand, G. Hydrostatic limits of 11 pressure transmitting media. *J. Phys. D Appl. Phys.* **2009**, *42*, 075413. [CrossRef]

26. Errandonea, D.; Muñoz, A.; Gonzalez-Platas, J. Comment on "High-pressure x-ray diffraction study of YBO$_3$/Eu^{3+}, GdBO$_3$, and EuBO$_3$: Pressure-induced amorphization in GdBO$_3$". *J. Appl. Phys.* **2014**, *115*, 216101. [CrossRef]

27. Mao, H.K.; Xu, J.; Bell, P.M. Calibration of the ruby pressure gauge to 800 kbar under quasi-hydrostatic conditions. *J. Geophys. Res.* **1986**, *91*, 4673. [CrossRef]

28. Piermarini, G.J.; Block, S.; Barnett, J.D. Hydrostatic limits in liquids and solids to 100 kbar. *J. Appl. Phys.* **1973**, *44*, 5377. [CrossRef]

29. Chakoumakos, B.C.; Abraham, M.M.; Boatner, L.A. Crystal Structure Refinements of Zircon-Type MVO$_4$ (M = Sc, Y, Ce, Pr, Nd, Tb, Ho, Er, Tm, Yb, Lu). *J. Solid State Chem.* **1994**, *109*, 197–202. [CrossRef]

30. Santos, C.C.; Silva, E.N.; Ayala, A.P.; Guedes, I.; Pizani, P.S.; Loong, C.-K.; Boatner, L.A. Raman investigations of rare earth orthovanadates. *J. Appl. Phys.* **2007**, *101*, 053511. [CrossRef]

31. Nguyen, A.-D.; Murdoch, K.; Edelstein, M.; Boatner, L.A.; Abraham, M.M. Polarization dependence of phonon and electronic Raman intensities in PrVO$_4$ and NdVO$_4$. *Phys. Rev. B* **1997**, *56*, 7974–7987. [CrossRef]

32. Errandonea, D.; Martínez-García, D.; Lacomba-Perales, R.; Ruiz-Fuertes, J.; Segura, A. Effects of high pressure on the optical absorption spectrum of scintillating PbWO$_4$ crystals. *Appl. Phys. Lett.* **2006**, *89*, 091913. [CrossRef]

33. Botella, P.; Errandonea, D.; Garg, A.B.; Rodriguez-Hernandez, P.; Muñoz, A.; Achary, S.N.; Vomiero, A. High-pressure characterization of the optical and electronic properties of InVO$_4$, InNbO$_4$, and InTaO$_4$. *SN Appl. Sci.* **2019**, *1*, 389. [CrossRef]

34. Hakeem, M.A.; Jackson, D.E.; Hamlin, J.J.; Errandonea, D.; Proctor, J.E.; Bettinelli, M. High Pressure Raman, Optical Absorption, and Resistivity Study of SrCrO$_4$. *Inorg. Chem.* **2018**, *57*, 7550–7557. [CrossRef]

35. Carnall, W.T.; Crosswhite, H. *Energy Level Structure and Transition Probabilities in the Spectra of the Trivalent Lanthanides in LaF$_3$*; Technical Report ANL-78-XX-95; Argonne National Lab.: Argonne, IL, USA, 1978. [CrossRef]

36. Manjón, F.J.; Rodríguez-Hernández, P.; Muñoz, A.; Romero, A.H.; Errandonea, D.; Syassen, K. Lattice dynamics of YVO$_4$ at high pressures. *Phys. Rev. B* **2010**, *81*, 075202. [CrossRef]

37. Manjón, F.J.; Errandonea, D.; Segura, A.; Chervin, J.C.; Muñoz, V. Precursor effects of the Rhombohedral-to-Cubic Phase Transition in Indium Selenide. *High Press. Res.* **2002**, *22*, 261–266.

38. Manjón, F.J.; Jandl, S.; Riou, G.; Ferrand, B.; Syassen, K. Effect of pressure on crystal-field transitions of Nd-doped YVO$_4$. *Phys. Rev. B* **2004**, *69*, 165121. [CrossRef]

39. Mahlik, S.; Grinberg, M.; Cavalli, E.; Bettinelli, M.; Boutinaud, P. High pressure evolution of YVO$_4$:Pr^{3+} luminescence. *J. Phys. Condens. Matter* **2009**, *21*, 105401. [CrossRef]

40. Rivera-Lopez, F.; Lavin, V. Upconversion/back-transfer losses and emission dynamics in Nd^{3+}-Yb^{3+} co-doped phosphate glasses for multiple pump channel laser. *J. Non-Cryst. Solids* **2018**, *489*, 84–90. [CrossRef]

Review

A Practical Review of the Laser-Heated Diamond Anvil Cell for University Laboratories and Synchrotron Applications

Simone Anzellini [1,*] and **Silvia Boccato** [2,†]

1   Diamond Light Source Ltd., Harwell Science and Innovation Campus, Didcot OX11 0DE, UK
2   ESRF—European Synchrotron Radiation Facility, 38000 Grenoble, France; silvia.boccato@upmc.fr
*   Correspondence: simone.anzellini@diamond.ac.uk
†   Current address: Sorbonne Université, Muséum National d'Histoire Naturelle, UMR CNRS 7590, Institut de Minéralogie, de Physique des Matériaux, et de Cosmochimie (IMPMC), 75005 Paris, France.

Received: 24 April 2020; Accepted: 13 May 2020; Published: 1 June 2020

Abstract: In the past couple of decades, the laser-heated diamond anvil cell (combined with in situ techniques) has become an extensively used tool for studying pressure-temperature-induced evolution of various physical (and chemical) properties of materials. In this review, the general challenges associated with the use of the laser-heated diamond anvil cells are discussed together with the recent progress in the use of this tool combined with synchrotron X-ray diffraction and absorption spectroscopy.

Keywords: laser-heated diamond anvil cell; synchrotron radiation; extreme conditions

---

1. Introduction

Since the pioneering work of Ming and Basset in the early 1970s, Reference [1] the laser-heated diamond anvil cell (LH-DAC) has become a powerful and routinely used tool for studying materials under extreme conditions of pressures (*P*) and temperatures (*T*) [2–7]. Up to now, it represents the only static technique allowing continuous access to a *P-T* domain ranging from ambient condition up to 300 GPa and 6000 K, respectively (see Figure 1) [8].

This technique takes advantage from the mechanical (being the hardest material found in nature) and optical properties of diamonds. In particular, thanks to diamond's transparency in a wide range of wavelengths of the electromagnetic spectrum, it is possible to use the DAC together with: visible light (for sample visualization and spectroscopic methods), infrared light (IR; as heat source) and X and γ-rays (as probe for atomic, phononic, electronic and magnetic structure). Thanks to technological advancements, the range of in situ techniques now available to be used with LH-DAC has largely increased in recent years. In particular, LH-DAC Raman scattering, References [9–13] Nuclear Magnetic Resonance (NMR) [14] and Brillouin scattering [11,15,16] techniques are now available to be used within university laboratories. Several techniques are also available to be used combined with LH-DAC at third generation synchrotrons (both as permanent installation or using portable systems): X-ray diffraction (XRD), References [17–23] X-ray absorption (XAS), References [24–30] inelastic X-ray spectroscopy (IXS), References [31,32] X-ray fluorescence (XRF), Reference [33] Mössbauer spectroscopy (SMS) [34–36], X-ray transmission spectroscopy (XTM) [37] and nuclear forward and inelastic scattering (NFS and NIS) [35,38].

These techniques enable investigating in situ a wide variety of physical phenomena induced by the extreme *P-T* conditions: such as long and short order structural modifications, phase transitions, chemical reactions and thermal and electronic excitations. The use of fast synchrotron-based techniques, enables performing time-resolved analysis of the *P-T*-induced evolution of the sample, providing important insight on the actual dynamics of the observed transformations. Furthermore, thanks to technological advancements, it is now possible to combine multiple investigation techniques with LH-DAC. Therefore, several information can be obtained from the sample at the same time, providing a better picture of the observed evolution [11,12,36,39–41].

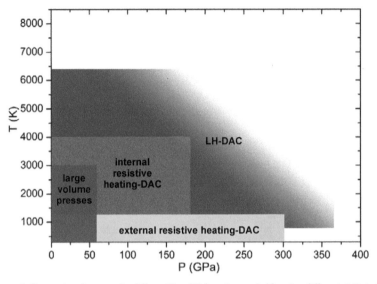

Figure 1. Comparison between the different *P* and *T* domains reachable using different static techniques. Large volume presses [42–46], including the Paris-Edinburgh cell, equipped with resistive heaters are a well-established static method to create high *P-T* environments. They allow large samples to be used (improving the signal/noise) but they are limited in the *P-T* range reachable. During external resistive heating in DAC, References [47–50] the entire DAC body is heated by Joule effect, providing a nice and uniform heating. However, such a configuration causes big thermal dispersion and mechanical instability, limiting the *P-T* range reachable with this technique. Internal resistive heating in DAC [51–58] provides advantages similar to the latter method, extending the reachable *P-T* range (as the heating is localized only to the high-pressure chamber of the DAC). Although the *P-T* limits of the internally heated DAC are constantly expanding, the laser-heating in DAC remains the only static technique allowing *P-T* in excess of 300 GPa and 6000 K [8,59].

For these reasons, the LH-DAC is extensively used in various scientific fields, ranging from solid-state physics to geological science, engineering and chemistry.

Thanks to its versatility, the LH-DAC technique has started attracting scientists from different experiences and scientific backgrounds, not necessarily expert of this technique. The aim of this review is therefore to provide a general knowledge of the LH-DAC technique and, to discuss the various challenges associated with its use (such as: sample preparation, pressure and temperature metrology, etc.) in both university laboratories and synchrotron radiation facilities. As different groups are adopting different solutions (often conflicting with each other), in this review the state-of-the-art is presented in the most objective way as possible, in order to provide the reader with all the information needed to make his (her) own mind.

Finally, the last part of the review gives an overview of the advancements obtained combining LH-DAC with synchrotron-based XRD and XAS techniques. Although several references are provided in this final part, they represent only a selection (as it would be impossible to cite the plethora of works done using these techniques), therefore the reader is advised to independently look for more works to read for further information.

2. Experimental Methods

In this section, we will focus our attention on the general challenges encountered while performing experiments using LH-DAC. First of all, we will provide insights on DAC and samples preparation. We will then give an overview of the lasers and on the system geometries generally adopted to perform these experiments. Finally, we will provide details on the *P* and *T* metrologies associated with LH-DAC.

2.1. DAC and Sample Preparation

A generalized design and working concept of the DAC is represented in Figure 2. A metallic gasket (generally made of Re, W or Stainless Steel; Be gaskets are also used when lateral optical access is needed [60]) is placed between two gem-quality diamond anvils. The space between the two culets of the anvils and the gasket's hole forms the cell high-pressure chamber, where the sample is loaded. The pressure on the sample is obtained by squeezing the two anvils toward each other. This is generally obtained using screws [61–63] or by inflating a metallic membrane [64]. As the pressure is defined by the force applied per surface area, the smaller the culet size of the diamonds, the higher the pressure that can be reached during the experiment for the same applied force.

Different DAC designs and diamond anvil types are used for different experiments. The choice, is mainly dictated by the experimental geometry needed by the adopted technique and the *P-T* range intended to reach. In particular, when performing LH-DAC experiments, the opening angle of the cell is generally maximized. Such a precaution eases the simultaneous access to the sample chamber of different optical paths (e.g., visible and IR lasers, X-rays, sample visualization, etc.), enabling several characterizations to be performed at the same time.

The geometry and nature of the diamonds also play an important role in DAC experiments. In particular, beveled diamonds are generally used to increase the pressure range accessible by a nominal culet size of the diamonds [65]. This is due to the increase in the deformation path the diamond can undergo before breaking while "cupping" under pressure [66]. Following a similar concept design, toroidal anvils have recently granted the access to the multi-megabar domain for static DAC experiments [67,68].

Despite the relatively high transparency of diamonds to light and X-rays, the radiation absorption of a pair of diamond anvils must be taken into account, as it may create a significant impediment to radiation transmission. For experiments requiring a wide cell opening and thinner diamond anvils (e.g., XAS or Pair Distribution Function (PDF) experiments), diamond and seat with an Boehler-Almax design are often chosen for their geometry [63]. Perforated diamonds can also be used to reduce the Compton scattering from the diamond [49,69,70] (although the presence of the perforation impedes the use of these anvils for laser-heating).

Finally, the crystalline nature of the diamonds also plays an important role for certain experimental techniques. In fact, the gem-quality diamonds generally used in standard DAC experiments are made from single crystals cut along certain crystallographic planes. Therefore, when these diamonds interact with an X-ray source, they produce intense single-crystal diffraction spots. For most synchrotron techniques combined with DAC, the presence of this spots does not cause any particular problem (aside from possible detector saturation). However, when using XAS techniques, these Bragg peaks cause the presence of "glitches" in the spectra, hindering the quality of the collected data and compromising any possible analysis [71]. Recently, new nano-polycrystalline diamonds (NPD) have been developed and become a major breakthrough for XAS techniques coupled with DAC [71,72]. In fact, the continuous and broad diffraction produced by these diamonds results into a monotonic and smooth background that does not affect the XAS spectrum. Furthermore, the NPD maintain a high degree of optical transparency in the visible and IR region of the electromagnetic spectrum and are harder than standard diamonds [72]. A pressure transmitting medium (PTM) is generally loaded together with the sample in the high-pressure chamber of the DAC. The combined action of PTM and pre-indented gasket is to transform the uniaxial compression exercised by the diamonds compression to a quasi-hydrostatic one, homogenizing the stresses and the strain acting upon the sample under compression. PTM are generally chosen among materials with low shear strength (fluids or soft solids) and are loaded in the high-pressure chamber of the DAC to completely embed the sample.

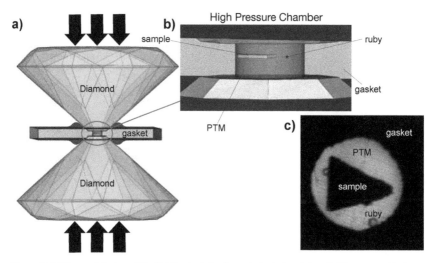

Figure 2. (**a**) Concept design of the DAC principle. A pre-indented metal gasket is squeezed between two diamond anvils along the direction indicated by the black arrows. (**b**) Zoom of the resulting high-pressure chamber containing the sample, the PTM and a pressure gauge (ruby). (**c**) Real image obtained from the top view of the high-pressure chamber of a loaded DAC.

Generally, for measurements performed at ambient T, aiming to characterize the mechanical properties or the compression curve of a material, it is important to reproduce the most hydrostatic conditions as possible. For this purpose, gases represent the ideal pressure media [73]. Among the available gases, He is considered to be the best PTM, as it presents the highest freezing pressure (11.6 GPa at 300 K) [74] and, even in its solid phase, it releases stress through re-crystallization [75].

The choice of the right PTM becomes particularly critical for LH-DAC experiments. This is mainly due to three reasons: (i) the extremely high thermal conductivity of the diamonds causes an important heat loss (it is practically impossible heating the sample up if it touches the diamonds); (ii) increased probability of T-induced chemical reactions between the PTM and the sample or caused by the carbon diffusion due to the Soret effect [76]; (iii) sample confinement (especially for melting curve and liquid characterization experiments).

Pressure media with high melting points and low thermal conductivity (such as KCl, NaCl, MgO, LiF, etc.) are generally used to insulate the sample from the diamond anvils. In particular, it is important to choose a PTM whose insulating conditions will not be affected by the P-T conditions reached during the experiment (no big re-crystallization and no extensive melting) [77].

Undesired chemical reactions present one of the most frequent issues encountered in LH-DAC experiments and have been often the cause of data misinterpretation [29,78,79]. It is, therefore, vital to choose PTM that remains inert under such extreme P-T conditions. It is also important to reduce any possible water content in the PTM that can cause oxidation at high T. This is generally done by keeping the PTM in an oven (or vacuum oven) before and after the actual loading (once loaded in the cell) [6,72].

A good sample confinement is ideal to obtain good LH-DAC data [80]. In fact, it has been shown to be the perfect recipe to achieve high thermal and chemical insulation of the sample. Furthermore, it reduces the risk of diffusion of the molten samples, facilitating experiments aimed to characterize melting temperatures and liquid structures at extreme P-T conditions [59,78,81,82]. For this reason, several groups are now dedicating lots of time optimizing the sample loading technique and geometry, taking advantage from the new advances in technology. PTM and samples can now be cut into specific shapes using FIB (Focused Ion Beam) [59,82] or femto-laser cutting techniques [29,83,84] and, the obtained parts can be loaded in the DAC using special robotic micro-manipulators [85]. Special protective layers (coating the diamond's culets and/or the gasket inner walls) have also been used to reduce any chemical reactions between the DAC high-pressure chamber and the sample (PTM) [78,86,87].

PTM can also be especially chosen for the possible chemical reactions they can have with the sample. Typical examples are experiments performed to create new metal-gases compounds e.g., Fe_xH_y, [88] Fe_xN_y [89,90] or to study the incorporation of noble gases in solids such as Kr in $(Mg_{1-x}Fe_x)O$ [91]. During these experiments, the sample is loaded in the DAC with an excess of the gas it will be intercalated with. Once the sample is heated for several minutes at a certain P, a portion of the gas (also acting as PTM) will react with the sample, forming the new compound.

2.2. Laser-Heating System: Concept Design and Associated Issues

Generally, as discussed in Refs. [18,26], the design of a laser-heating system for DAC experiments is defined by three main components: (i) the laser-beam delivery optics (to heat the sample), (ii) the sample imaging and T measurement optics and (iii) the coupling of the setup with the desired experimental analysis technique. As the aim of this review is to talk about the challenges associated with the use of the LH-DAC and therefore common to all the possible experimental analysis techniques, we will first focus our attention on the first two components. The final component will be discussed in the last paragraphs applied to the specific cases of synchrotron XRD and XAS techniques.

2.2.1. Lasers

To achieve temperatures of the order of thousands of Kelvins, high power (50–200 W) lasers need to be focused on the sample surface. Two different types of laser systems are generally used combined with the DAC. Nominally: (i) near-IR lasers such as Nd:YAG (Nd^{3+} doped yttrium aluminum garnet), Nd:YLF (Nd^{3+} yttrium lithium fluoride) or orthovanadate crystals and (ii) CO_2 lasers. The choice between the two different types of lasers depends on the nature of the sample needing characterization.

In particular, near-IR lasers (wavelength = 1.064 µm) are generally used to heat metallic and semiconducting materials [18,92,93], where the laser absorption of the material is defined by the interaction between the photons of the laser and the mainly free (or bounded) electrons of the metal (semiconductor). The excited electrons move from the conduction band to higher energy states and, the collisions between those excited electrons and the lattice phonons creates the thermal energy [7,94]. When used in combination with DAC to produce extreme T, these lasers need to be tightly focused, as they are generally weakly absorbed by the materials. The obtained focused beam has a minimal penetration depth into the material, leading to radial and axial temperature gradients of the order of hundreds of kelvins per µm [95–97]. The axial thermal gradient is generally minimized by heating the sample from both sides (double-sided heating) [5,92,98,99], but it can also be minimized by using multimode lasers [7]. The radial thermal gradient can be minimized by slightly defocusing the lasers (towards the sample position: to avoid drilling the diamond base) or, by using special optics converting the standard Gaussian profile of the laser (TEM_{00}) into a flat-top one [100]. In this way, it is possible to obtain a relatively large and uniform hotspot on the sample surface.

The CO_2 lasers (wavelength = 10.6 µm) are specially used to heat optical transparent minerals, organic materials and oxides [3,7]. In fact, the phononic frequencies of those materials are of the same order of the one emitted by the CO_2 lasers. The emitted light is therefore directly absorbed by the lattice and the resulting vibration develops thermal energy. Due to the nature of the CO_2 laser emission, special optics, in particular, special transmissive (e.g., ZnSe lenses) and reflective (e.g., Cu total reflectors) collimators must be used to focus and reflect the laser, respectively [3,15]. Furthermore, the penetration depth of the CO_2 laser into the material is large (often larger than the sample itself). Therefore, axial thermal gradients are basically nonexistent and only one laser is needed to heat the sample.

Focusing and positioning of the lasers on the sample surface can be achieved either via independent optics (off-axis geometry) or by sharing a common lens with the imaging part of the system (on-axis) geometry (see Figure 3). Both geometries are equally used by different groups around the world and both bring their advantages and disadvantages.

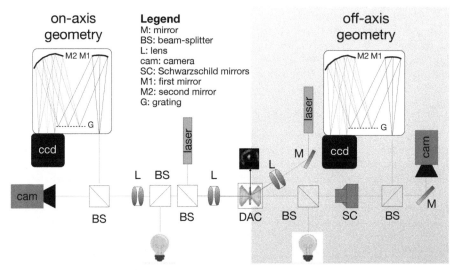

Figure 3. Concept design of a LH system showing both the on-axis geometry with lenses and the off-axis geometry, here combined with the Schwarzschild mirrors. Different color legends are used for different optical paths: red for the lasers, yellow for the illumination and green for the imaging part and temperature measurement. An actual image of a heated sample is also presented above the DAC.

The on-axis geometry is generally used in both university laboratories and synchrotron facilities [12,101–104] for the simplified geometry (making the system more compact) and user-friendliness. In fact, in this geometry, once the laser beams are aligned with the sample image, they do not need further alignments. However, by using the same lenses for both imaging collection and laser delivery, the lenses undergo a thermal expansion that compromises the reliability of the temperature measurements (see temperature measurement section of this review). Therefore, some groups use a cooling system for the lenses to keep this thermal expansion to the minimum [105,106]. In this case, a large and uniform hotspot on the sample surface can be obtained only adding additional optics to the laser path, such as beam-expanders and π-shapers [19].

Compared to the on-axis, the off-axis geometry is more flexible, as the lasers can be moved independently from the imaging optics. However, the increased complexity of the system (now requiring additional optical and mechanical components), might result in a laser alignment more complicated to non-expert users. Despite these technical difficulties, the off-axis geometry is used by several groups and beamlines [17,18,24,26] as it allows reflective optics or iris-equipped refractive optics to be used to collect the temperature with minimal aberration effects (*vide infra*) [29,107].

During LH experiments, the lasers can be used in two different operational modes: continuous (CW) and pulsed (PL). Historically, the CW was the first to be used for LH-DAC experiments and it is still used by several groups [17,18,26]. However, keeping the laser for several seconds on the sample, promotes mobilization and diffusion of carbon from the anvils into the sample chamber [76,108]. Such a diffusion can lead to undesired chemical reactions with the sample [6,109]. The PL mode is thought to minimize the risk of chemical reactions by reducing the heating time [95,110,111]. For this reason, several experiments have been reported using this mode (also referred as "flash heating" [111]) However, a recent work by Aprilis et al. [40] performed a comparison between the CW and PL modes, finding the same carbon diffusion in both cases. Confirming that, the best way to reduce chemical reactions, remains a perfect sample confinement in an inert and dry PTM.

2.2.2. Sample Imaging and T Measurement

The coupling of the imaging and the T reading setup is an extremely important component of every LH system. In fact, the same optics are used to visualize the sample and to collect the thermally radiated signal from the sample surfaces. Furthermore, as the typical dimension of the sample used in LH-DAC can vary from few to hundreds of microns, it is important to use optics allowing high image quality and magnification.

The most common way used to create a magnified image of the sample for LH-DAC experiments is by using a pair of refractive lenses. This solution has been adopted by several groups around the world owing its affordability, easy and compact design and image quality [18,20,21,23,26,41,100]. In this setup, the first lens is used to collect the signal from the sample inside the DAC, whereas the second one is used to focus the signal simultaneously into a CCD camera and into a spectrometer. The camera is used to visualize the actual sample, whereas the spectrometer is used to collect the radiated thermal signal from the laser-heated portion of the sample surface. Typically, the spectrometers used for LH-DAC experiments belong to the Czerny-Turner family. Therefore, the signal focused on the spectrometer slits (placed at the spectrometer working distance), is collimated by the spectrometer first concave mirror, energy-dispersed by the spectrometer grating (typical value of 150/300 lines per mm) and focused on a CCD by a second concave mirror (see Figure 3). The intensities collected on the spectrometer CCD are energy-dispersed and once integrated, reproduce the black body radiation coming from the sample. Therefore, a Planck fit of the obtained signal provides the T at the sample surface. In these experiments the black body radiation is collected in a wavelength range between 400 nm and 1000 nm, therefore the measurable temperatures start from a minimum of 1100 K. Typically, band-pass or notch filters are also added before the spectrometer to avoid any parasitic signal from the laser that might saturate the CCD of the spectrometer, whereas neutral density filters are used to attenuate the thermal emitted light for the highest temperatures (>3500 K). Temperature measurements between 500 K and 1700 K can be performed using an InGaAs OMA camera with sensitivity for near-IR (900 nm–1800 nm) [112]. However, resistively heated DAC are generally preferred to LH-DAC for working at this temperature range, as they provide a more stable and homogeneous heating during the experiment [48–51,113–118].

As the focal distance of refractive lenses depends on their wavelength-dependent refractive index, different wavelengths of the collected light are focused at different distances (see Figure 4a) [119]. In particular, wavelengths in the green region of the electromagnetic spectrum are focused at the nominal focal distance of the lens, whereas wavelengths in the blue and in the red regions are focused at distances shorter and longer than the nominal one, respectively. Therefore, when the light is collected (and energy-dispersed) by the spectrometer, this chromatic dispersion (known as aberration) introduces random chromatic displacements, where intensities belonging to different wavelengths will end on the wrong region of the CCD of the spectrometer. This results in erroneous T readings, with errors of several hundreds of kelvins. In order to reduce these aberration effects, special apochromatic lenses are generally used (see Figure 4b) [18,120]. As an alternative to the refractive lenses, reflective objectives with a Schwarzschild design can be used [17,77]. Those objectives are made of two spherical mirrors and are therefore purely achromatic in nature. Such a solution leads to an aberration free signal at the cost of a worse image quality.

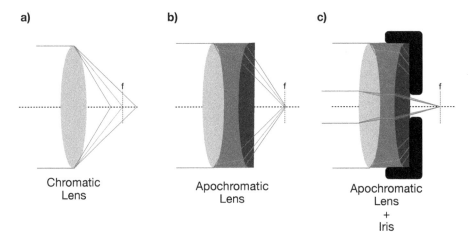

Figure 4. Comparison between the wavelength-dependent focusing obtained with (**a**) chromatic, (**b**) apochromatic and (**c**) apochromatic combined with iris lenses. The use of apochromatic lenses massively reduces the chromatic aberration observed with standard chromatic lenses. However, the remaining aberration caused by the lenses geometry can only be minimized by reducing the numerical aperture, and therefore, by collecting the signal only using the part of the lens closer to the optical axis (less affected by aberration).

To simplify the experimental setup, some groups use additional optical fibers to collect the signal radiated from the sample surfaces and bring it to the entrance of the spectrometer [37,94,104]. Such a solution makes the system more compact and reduces the time spent to perfectly align the spectrometer with the rest of the system (especially when using portable systems). However, the possible aberration effects that can be introduced using the optical fiber are still unclear.

For several years, a long standing debate was carried on in the scientific community about the effects that chromatic aberration could have on the reliability of the temperature measurements. In particular, it has often been argued how the use of reflective or refractive lenses might (or might not) be the cause (or at least one of the causes) for the discrepancies observed in the results obtained by different groups while studying materials under such extreme conditions [5,6,29,121–124]. However, a recent study by Giampaoli et al. [107] has empirically shown how the two optical systems can provide similar results if some precautions are put in place. In particular, it is important to reduce the numerical aperture of the refractive lenses using an iris in front of the collecting lens. In this way, the signal collected by the spectrometer is mainly composed by rays that are paraxial to the lens (see Figure 4c), therefore less affected by aberration [24,119]. Such a precaution worsens the image quality (as the number of photons reaching the CCD is notably reduced) but increases the depth of field, improving the reliability of the temperature measurement.

Finally, the sample illumination is generally achieved using ultra-bright white LEDs. The white light can be brought to the optical path by either mounting the LEDs on pneumatic mount (resulting in sample imaged using transmitted light only) or by sending the LED-generated light into beamsplitters mounted (directly or on pneumatic mounts) on the main optical path.

2.3. LH-DAC Metrology

Despite all the available geometries and designs (more or less complicated), the LH-DAC is simply a tool used to produce extreme P and T conditions. If the actual values of these conditions are not known or are erroneously known, any result experimentally observed becomes unreproducible. An accurate P and T metrology is therefore extremely important and will be discussed in detail in the following paragraphs.

2.3.1. Temperature Measurement

Accurate *T* measurements are extremely important when performing LH-DAC experiments. In particular, they are doubly important as their values also affects the *P* measurements. In fact, when laser-heating a nearly isochoric environment, such as the sample chamber of a DAC, the induced thermal expansion of the sample, causes the development of a thermal pressure. Therefore, the resulting *P* applied on the sample is *T*-dependent.

As introduced in the previous paragraph, temperature measurement in LH-DAC is generally performed via spectral radiometry, i.e., by fitting the radiated thermal signal *I(λ,ε,T)* from the sample surface with the Planck's law:

$$I(\lambda, \epsilon, T) = \epsilon \frac{2\pi hc^2}{\lambda^5} \frac{1}{exp(hc/\lambda K_B T) - 1} \tag{1}$$

Here *h*, K_B, *c* and *T* are the Planck constant, the Boltzmann constant, the speed of light and the temperature corresponding to the collected radiation, respectively; *λ* is the wavelength of the measured signal. *ε* is the emissivity, it is a function of *P*, *T* and *λ* and it is generally not known. However, for the wavelength range used in LH-DAC experiments (400–1000 nm), the emissivity variation with wavelength can be considered negligible and the gray-body approximation (*ε* < 1) is assumed. The emissivity is therefore used as fitting parameter together with the *T*. Such an assumption is not always valid due to several aspects such as: aberration effects and changes in emissivity due to phase transitions. A critical evaluation of the temperature measurement is therefore extremely important.

Different groups around the world use different methods to identify the reliability of the temperature measurement. One of the most common one is to fit the *I(λ,ε,T)* with a Wien function [83,105,125,126]. In fact, for *T* where exp(hc/λ K_BT) » 1, the Planck function can be approximated with the Wien law:

$$I_{Wien}(\lambda, T) = \epsilon \frac{2\pi hc^2}{\lambda^5} e^{-\frac{hc}{\lambda K_B T}} \tag{2}$$

This equation can be linearized:

$$Wien = \frac{K_B}{hc} ln(I_{Wien} \frac{\lambda^5}{2\pi hc^2}) = \frac{K_B}{hc} ln\epsilon - \frac{1}{\lambda T} \tag{3}$$

Equation (3) can therefore be used to check the reliability of the *T* measurement, as any deviation from the linear behavior can be easily detected. The error introduced by using the Wien function (instead of the Planck one) is of the order of 1 % for *T* < 5000 K. [127]

The use of Equation (3) also allows a more accurate determination of the error in the *T* measurement to be performed. In fact, thanks to the linearity of Equation (3), only two wavelengths are needed to measure the *T*. Therefore, if *ε* wavelength dependence can be neglected, temperatures calculated from the ratio of a pairs of intensities separated by a fixed spectral difference should give the same result [119,125]. If we define $\omega_i = -\frac{hc}{K_B \lambda_i}$, the equation to obtain the temperature from two wavelengths (two-colors pyrometry) can be written as:

$$T_{two-colors} = \frac{\omega_2 - \omega_1}{Wien(\omega_2) - Wien(\omega_1)} \tag{4}$$

If Equation (4) is applied to each collected wavelength at constant Δω (sliding two-colors pyrometry), it is possible to obtain the wavelength-dependent temperature distribution of the collected radiation. The Δω must be chosen to be small enough to allow the observation of possible wavelength-dependent variation in the $T_{two-colors}$, but large enough so that the corresponding intensity difference is detectable (see Figure 2 in [125]). Regular data show substantial noise, but the gray-body radiation must yield a constant temperature across the wavelength range investigated.

However, several factors such as: wavelength dependence of the emissivity, temperature gradients and aberration effects (from lenses and diamonds), will introduce a distinct wavelength dependence in the temperature versus wavelength plot [125]. For these reasons, some groups use as the actual error in the T measurement, the full width at half maximum (FWHM) of the histogram obtained from the sliding-two-colors pyrometry, instead of the error obtained from the fit with the Planck (or Wien) law [18,29,30,81,82,126,128]. An example of a complete set of spectral radiometry measurements is reported in Figure 5.

Regardless of the adopted method, in order to obtain a reliable T measurement during a LH-DAC experiment, the spectrometer needs to be calibrated. In fact, in addition to the standard wavelengths calibration with a NIST (National Institute of Standards and Technology) lamp (e.g., Ne or Ar), it is necessary to find the optical response of the LH system $S(\lambda)$. This is generally done by collecting the radiated intensity from a calibrated tungsten (W) lamp at a specific temperature $I(\lambda, T_0)$. The W lamp is placed at the sample position and the signal is collected with the spectrometer via the LH optical system (this procedure must be repeated for each adopted configuration *i.e.*, with or without filters or iris). The collected intensity can be then written as:

$$I(\lambda, T_0) = \epsilon S(\lambda)\frac{2\pi hc^2}{\lambda^5}\frac{1}{exp(hc/\lambda K_B T_0) - 1} \tag{5}$$

The system response can be obtained by dividing the measured intensity by the ideal Planck intensity at the nominal temperature T_0. Therefore, during a LH experiment, the Planck function is obtained by normalizing the collected radiated intensity to the obtained system response $S(\lambda)$.

Due to the presence of large thermal gradients and changes in the insulation conditions, in-homogeneous distributions of the T on the sample surfaces are often observed during LH-DAC experiments [129,130]. The actual determination of both the peak temperature and the temperature field can then become very useful. For this purpose, new methods providing a 2D temperature maps of the heated sample in LH-DAC have been developed and are used by different groups for providing information about the dynamic changes occurring during these experiments [131]. Among them, the peak-scaling method [130,132–134], adopts a pseudo-Planck curve obtained using a spectrometer with a wide entrance slit and averaging the light collected on the entire region of interest (ROI) of the spectrometer CCD. The peak temperature obtained from correcting the obtained pseudo-Planck curve is used to scale the monochromatic image of the sample collected at the same time on a CCD camera. This results in a temperature map. This relatively simple method is insensitive to small optical misalignments, but depends on the model adopted to correct the pseudo-Planck curve [133]. An alternative is represented by the multispectral imaging radiometry (MIR) [95,135,136]. Developed by Campbell in 2008, MIR does not require the use of a spectrometer. In fact, the original sample image is decomposed in four near-monochromatic images (therefore also referred to as four colors pyrometry) using a combination of beamsplitters, mirrors and narrow band-pass filters. The decomposed images, each at a different wavelength, are independently focused and are simultaneously acquired on a single CCD camera. The images are then spatially correlated to provide four intensity-wavelength data at each pixel that can be fitted to provide maps of both temperature and emissivity (see Figure 6). Like for the spectral radiometry case, the system needs to be calibrated with a known radiation source, following the same procedure described above. Compared to the peak-scaling method, MIR has the advantage that the temperature of each pixel is determined directly, without recourse to any model dependent parameters. Furthermore, as each image is focused independently, any chromatic aberration is largely reduced.

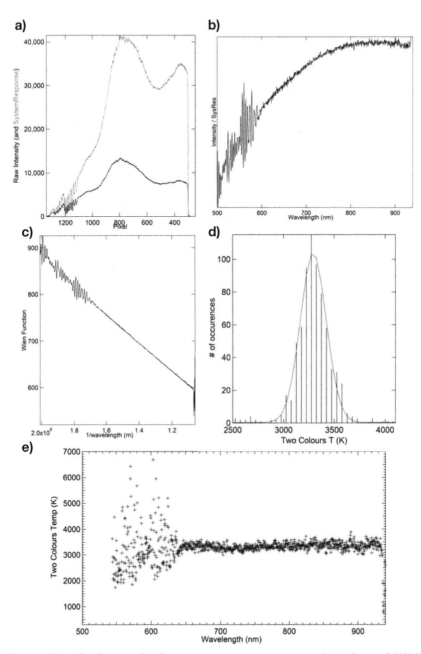

Figure 5. Example of a spectral radiometry temperature measurement obtained around 3300 K. In particular: (**a**) shows the calibrated system response (green) and the raw measurement (black); (**b**) shows the obtained Planck and (**c**) Wien functions and (**d**) the histogram of the obtained (**e**) two-colors signal. In the graphs, the red lines indicate the region used for the fit.

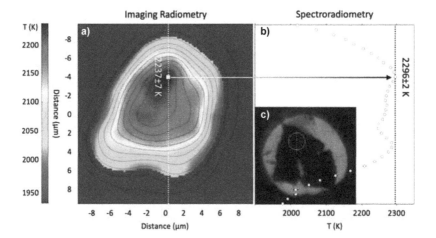

Figure 6. Comparison between 1-D spectroradiometry and 2-D imaging radiometry for temperature measurement on a sample of FeS$_2$ (pyrite) at 54 GPa in the LH-DAC. (**a**) Imaging radiometry temperature map generated from a four-color system and processed using the MIRRORS software [129]. Contours indicate light intensity in 10% intervals relative to the brightest pixel. (**b**) Corresponding spectroradiometric temperature cross-section collected within 1 sec of the temperature map at the same nominal location (the white dotted line in (**a**)). Note the similarity in shape and peak temperature, denoted by the white square in (**a**) and the black dotted line in (**b**). (**c**) A photomicrograph of the sample before heating, with the 25-micron diameter heated spot indicated by the red circle. The red dotted line indicates the location of the spectroradiometric measurement presented in (**b**).

2.3.2. Pressure Measurements

The P measurement in a DAC is one of the major contribution of uncertainty in high-pressure experiments. In fact, differently from other static apparatus (such as piston-cylinder or multi-anvil devices), it is impossible to find a direct correspondence between the applied force and the resulting P. This is due to non-reproducible loss in the given load caused by internal frictions, diamonds and gasket deformation [66,127,137,138]. Therefore, internal standard materials (pressure gauges), whose changes in physical properties have been calibrated as a function of P, are generally loaded (together with the sample) in the high-pressure chamber of the DAC.

The most widely used method (in both synchrotrons and university laboratories) to determine the pressure inside a DAC is by measuring the pressure-induced shifts in the fluorescence signal (see Figure 7a), obtained using visible lasers on optical pressure gauges such as ruby (Cr:Al$_2$O$_3$) or Sm doped Y$_3$Al$_5$O$_{12}$. Such materials are secondary standards, since they are calibrated using reduced shockwave, ultrasonic and static data obtained from a series of metals [139–145]. If the standards are used for measuring P outside the calibrated region, the uncertanties can be quite large. For this reason, over the years new calibrations have been performed so that the advances in the DAC technologies and the P metrology can move with a similar pace (refer to Ref. [146] for a more detailed review).

The Raman signal obtained from the center of the diamond culet is also used for P measurement. In fact, under high-pressure, the obtained spectra reflect the stress gradient of the anvil and they are therefore characterized by steep edges at high and low frequencies [144,147–149]. The calibration was performed on the high frequency edge of the plateau, where the strong first-order Raman peak shifts to higher frequency with increasing pressure (see Figure 7c).

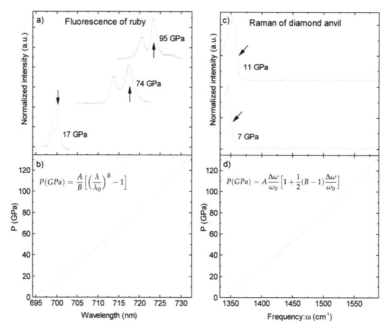

Figure 7. Left panel: (**a**) Ruby fluorescence signals at different pressures. The arrows indicate the position of the more intense peak, where the calibration is taken. (**b**) Pressure vs. wavelength relation according to the most recent calibration by Liu et al. [139]. The relation is obtained through the equation here represented, where λ is taken as indicated in panel (**a**) and λ_0 = 694.24 nm. According to the calibration A = 1904 GPa and B = 9.827 [139]. Right panel: (**c**) Raman spectra of diamond anvils at different pressures. The arrows indicate where the calibration is taken. (**d**) Pressure vs. frequency relation according to the most recent calibration by Akahama and Kawamura [147]. The relation is obtained following the equation represented, where $\Delta\omega = \omega - \omega_0$, ω is the wavelength taken at the edge of the signal as indicated by the arrows in panel (**c**) and ω_0 = 1334 cm^{-1}. In the chosen calibration A = 547 GPa and B = 3.75 [147].

Compared to the other luminescence methods (Ruby and Sm:Y$_3$Al$_5$O$_{12}$), the measured Raman shift is less sensitive to pressure-modification for $P < 100$ GPa but can still be used for pressures higher than 200 GPa (when the ruby signal is broader and weaker) [149]. In addition, it has been calibrated up to 400 GPa [147], therefore it can still be used for measuring pressure while using toroidal-DACs [67,68].

Unfortunately, NPD do not present any Raman signal, therefore the ruby fluorescence (or Sm:Y$_3$Al$_5$O$_{12}$) remains the only method for measuring P in DAC experiments coupled with XAS [71,72].

Due to the temperature effects on the fluorescence signal and to the geometry of the loaded DAC, P in the LH-DAC can be measured with spectrometric methods only before and after the actual heating. Such a constraint removes the possibility of directly measure the thermal pressure experienced by the sample. However, as the resulting thermal pressure is intrinsically linked to the PTM used, this value can be estimated empirically from reported XRD data of the used PTM at extreme P-T conditions. An example of this is presented in Lord et al. [105], where they find for KCl the empirical relation $P_m = 0.03 \cdot P_{300K}$, where P_m is the thermal pressure at the melting temperature of the sample and P_{300K} is the pressure measured on the ruby at room temperature after heating. If $T_0 = 300$ K and T_m is the melting temperature, the thermal pressure can be evaluated as:

$$P_{th}(T) = \frac{0.03 \cdot P_{300K}}{T_m - T_0}(T - T_0). \tag{6}$$

The value of the pressure is then given by $P(T) = P_{300K} + P_{th}(T)$ [105].

When performing DAC experiments combined with XRD, it is possible to evaluate the *P-T*-induced evolution of the lattice parameters of a standard material whose equation of state (EoS) *P(V,T)* is known. Therefore, if the material is placed close to the sample (few μm apart) in the high-pressure chamber of a DAC, it is possible to measure the pressure on the sample by assuming the two materials are experiencing the same stress.

At ambient temperature, P(V,300 K) EoS reduced from shockwave measurements [150,151] and ultrasonic measurements [152,153] are generally used to calibrate the X-ray gauges [146]. The accuracy of these EoS are cross-checked by compressing several standards in the DAC and comparing the corresponding pressure obtained from the volumes measured by XRD [140,143].

During LH-DAC experiments, the adopted PTM is also used as X-ray gauge. Such a trick, used for reducing the amount of elements contained in the high-pressure chamber, has also the advantage of collecting *P* and sample information from the same region of the DAC (minimizing any radial *P* gradient). According to Ref. [154], the general thermal EoS can be expressed as:

$$P(V,T) = P(V,300K) + (P_{Th}(V,T) - P_{Th}(V,300K)) \tag{7}$$

where $P(V, 300 \text{ K})$ represents the ambient temperature part of the total EoS and P_{Th} is the thermal pressure essentially created by the thermal motion of the atoms in the lattice. This contribution is generally obtained from thermodynamic models or ab initio simulations [50,155,156] and it is provided with the thermal EoS.

Considering Equation (7), the accuracy of the measured *P* can be determined from:

$$\Delta P = \left(\frac{\partial P}{\partial V}\right)_T \Delta V + \left(\frac{\partial P}{\partial T}\right)_V \Delta T \tag{8}$$

where ΔV and ΔT are respectively the uncertainties of the measured *V* and *T*. From thermodynamic relations the last equation can be expressed as:

$$\Delta P = K_T \frac{\Delta V}{V} + \alpha K_T \Delta T \tag{9}$$

where K_T is the isothermal bulk modulus and α represents the volumetric thermal expansion coefficient.

During an actual LH-DAC experiment, the insulating PTM (also used as pressure gauge) is placed between the heated sample and the diamond. As the diamond thermal conductivity is extremely high, the PTM is under a big temperature gradient between the temperature of the sample ($T_{measured}$) and the one of the diamond (T_{anvil}, generally considered at ambient temperature). Given that the temperature gradient is not linear along the heating axis, according to Campbell et al. [157] the temperature to be used in Equation (7) for the PTM can be approximated as follows:

$$T_{eff} = \frac{(3 * T_{measured} + T_{anvil})}{4} \pm \frac{(T_{measured} - T_{anvil})}{2}, \tag{10}$$

Therefore, from the measured volume of the insulating material and the *T* on the sample surface, it is possible to estimate the *P* (and the corresponding error) experienced by the laser-heated sample. This can be obtained from Equations (7) and (8), if the values of K_T, α and $P_{Th}(V,T)$ are provided.

3. LH-DAC at Synchrotron Radiation Facilities

The last few decades have seen a rise in the LH-DAC systems combined with in situ synchrotron techniques. In fact, the extremely high flux and collimation of the synchrotron radiation, combined with the use of fast detectors with a low signal/noise [109,158,159], opens the possibility to perform time-resolved characterizations of samples contained in the micrometric high-pressure chambers of the DAC. Therefore, it allows an in-depth analysis of the actual transition rate (kinetics) of the transformations undergone by the sample to be performed [39]. Furthermore, it can also be used for mapping eventual thermal and pressure gradients developed on the sample at this extreme conditions. Several LH-systems are now available to be used at synchrotron radiation facilities as permanent installations [17–23,26,94] (or portable systems [24,36,101,112,160]). However, there are several factors that need to be taken into account when combining a LH-DAC system with a beamline setup.

First of all, as discussed in the previous paragraphs, major thermal gradients can develop during a LH-DAC experiment, with typical regions at homogeneous T with a diameter of 10–40 μm. It is, therefore, important to use well collimated and clean X-rays beam, with dimensions smaller than the region of the sample at homogeneous T. Furthermore, as the size of the sample contained in the DAC is relatively small, in order to maximize the signal it is important to use high flux X-rays. This is why LH-DAC systems are generally mounted on undulator beamlines [17,19,20,22,23,26,94] (with the only exception of Refs. [18,21]). In fact, undulator sources produce a high flux beam with an extremely low divergence. The beam is generally focused further down to a diameter of a few μm by using a pairs of Kirkpatrick–Baez (KB) mirrors or compound refractive lenses (CRL) [161,162].

The alignment of the X-ray beam with the hotspot and the T measurement must be precise within a μm or less, otherwise the obtained information will be collected from different regions of the sample at different T conditions. A possible solution to this problem was proposed by Boehler et al. [24] and it is now used by several groups [17,18,20,26,94]. The idea is to place a polished mirror, with a pinhole drilled on it, at the entrance of the spectrometer. A high sensitivity camera focused on the mirror will then see the superposition of the sample image with the drilled pinhole. Therefore, the pinhole will effectively act as the spectrometer pupil (the spectrometer CCD will need to be adjusted to the new focal position) and as reference for the temperature measurement. The high sensitivity camera can be used to observe the X-ray-induced fluorescence of the PTM (or the gasket material). The fluorescence maximum intensity is observed at the X-rays position and can be aligned with the spectrometer pupil (by moving the sample image). Finally, if also the laser-induced hotspot is aligned with the spectrometer pupil, all the information will be collected from the same region of the sample. While laser-heating the sample, a large amount of the heat is transferred to the DAC body (through the diamonds) and the surrounding optics. This leads to a progressive misalignment of the image with the spectrometer pupil. For this reason, the DAC is placed in a water-cooled holder, and the relative alignment of the X-rays, T measurement and hotspot must be regularly checked.

Finally, as the combination of LH-DAC with any synchrotron-based technique is done by adding the LH system to the X-ray path, it is important to consider the impact that the new optics and components will have on the collected signal from the X-rays. As each technique shows specific challenges and (as usual) different groups use different solutions, we will talk about this more in detail for XRD and XAS in the next paragraphs.

3.1. LH-DAC and XRD

Historically, XRD was the first synchrotron technique to be ever combined with LH-DAC. Since the pioneering work of Shen et al. [92], this technique evolved into a productive and routinely used experimental method at synchrotron beamlines, leading to numerous major scientific advances across the disciplines of chemistry, material science, physics and geoscience [163–166].

In fact, the nature of the XRD (combined with LH) technique, allows the characterization of the structural, textural and chemical evolution of the sample (and the PTM) to be performed in situ as a function of P and T. Furthermore, thanks to the use of fast detectors with an improved signal/noise, it is now possible to probe (in situ) intermediate phases, revealing important crystallographic information.

In recent years this technique has been successfully used to (first) synthesize and (successively) characterize new materials, both for geological interest or in the rush for finding new ultra-hard materials [88,89,167,168]. The use of LH-DAC with XRD (and the help of computational methods such as ab initio simulation or thermodynamic modeling) allows detailed characterizations of phase diagrams and thermal equations of state (EoS) of materials at extreme *P-T* conditions (see Figure 8 for an example) [81,169–173].

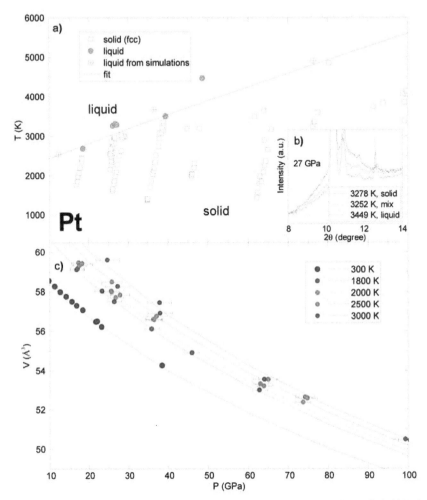

Figure 8. Example results of a study on Pt performed using LH-DAC combined with XRD [81]. (a) High *P*-high *T* phase diagram of Pt. (b) Example of a solid/liquid transition (characterized by the appearance of a diffuse signal in the diffraction pattern) observed in Pt at around 3300 K and 27 GPa. (c) Thermal equation of state of Pt determined along four isotherms.

Special experimental set-ups, allowing single-crystal XRD experiments using LH-DAC, are opening the possibility of characterizing the chemical compositions of complex multiphase and/or multigrain samples and the unambiguous solution of structures as well as the determination of *P-T*-induced structural distortions [104,137,174].

The XRD signals obtained from liquid (and amorphous) and crystalline samples are very different (see Figure 8b). In particular, while the latter is characterized by the presence of Bragg peaks, the first (and second) is characterized by a diffuse scattering. Such a difference in the two signals, has made this technique particularly attractive for both experiments aimed to characterize the melting curve of elements under extreme conditions of *P* and *T* and [47,59,81,82,84,128,131,175], for those aimed to characterize their liquid structures [176–179].

For these reasons, several XRD beamlines worldwide are now equipped with permanent (or portable) LH-systems for DACs [17,18,20–24,36]. However, when combining LH-DAC with a synchrotron beamline, the interaction of the X-rays with the LH collecting optics can hinder the collected XRD data. In particular, the scattering from the collecting optics, creates additional contributions (such as shadows and diffraction signal) to the signal from the high-pressure chamber of the DAC. In order to avoid this situation, some beamlines only perform *T* measurements from the upstream side. In this way, the diffraction from the optical components is blocked by the DAC body, but it is not possible to measure the axial thermal gradients. Other beamlines use X-ray transparent mirrors (generally made of glassy carbon). This allows the *T* to be measured from both sides simultaneously, adding a minor diffuse contribution to the XRD signal from the amorphous mirrors. The temperature can also be measured simultaneously from both sides using a perforated mirror (few mm holes) at each side of the cell. In this way, the micrometric direct X-ray beam passes through both mirrors without causing any scattering (or losing any flux).

3.2. LH-DAC and XAS

XAS is an element selective technique, suitable for probing the local environment around a given element in a molecule or compound. When an X-ray photon impinges on the sample, a photoelectron is generated, which propagates as a wave inside the sample and then scatters against the first neighbors. For photoelectrons generated with a kinetic energy of a few tenths of eV, the interference between the outgoing and the incoming wave modulates the oscillations in the EXAFS (Extended X-ray Absorption Fine Structure) region of the spectrum. This part of the signal is thus sensitive to the local structure around the absorbing atom. However, the XANES (X-ray Absorption Near Edge Structure) part of the signal corresponds to photoelectrons generated with a lower kinetic energy and a longer mean free path. Therefore, this signal reflects the multiple scattering of the photoelectron with the atoms of the sample. As the probability of absorption in the near edge region is strongly influenced by the electronic density of states just above the Fermi level, the XANES part of the absorption signal contains information on the electronic structure.

These characteristics make XAS a very informative technique, able to provide (in situ) both structural and chemical information of the sample at high *P* and *T*. The technique can be used to probe local, electronic and magnetic structure at high *P* and *T*, obtaining information about phase diagrams, compressibility for solids and liquids and chemical partitioning with implications for fundamental and planetary science.

Due to technical challenges, the coupling between LH-DAC and XAS has been finalized only recently [26]. One of the fundamental aspects that was considered during the development, was the need for fast acquisition time (as previously discussed). For this reason, the system was built on an energy-dispersive beamline [180], where the high flux "pink beam" from the beamline undulators is both focused and energy-dispersed by a polychromator, allowing the acquisition of a complete XAS spectrum to be performed in a few ms using a fast position sensitive detector, without beam stability problems. This system was built in the off-axis geometry, where the lenses face the diamonds, the lasers and the X-rays reach the sample with two different directions. More details on the system's geometry are provided in Kantor et al. [26]

Pioneering high *P* and high *T* experiments were performed with portable systems and consisted of the detection of melting of iron [24] and study of the chemistry of mantle minerals [160]. The past years, have seen a rise in LH-DAC experiments coupled with XAS. In particular, since the first pioneering work by Boehler et al. [24], a lot of efforts have been dedicated to the XANES determination of phase diagrams and in particular the melting curves of late 3d transition metals such as Fe [29,124], Ni [28], Co [181], and late 3d transition metal-bearing binary systems such as Fe-C, Fe-O, Fe-S and Fe-Si [182] as well as Fe-Ni [183]. The example of Ni melting curve detected with XANES is represented in Figure 9, where solid and liquid spectra are represented. In Boccato et al. [182] it has also been proposed to use XANES to track the variation of light elements due to the non-congruent melting.

The potential in the EXAFS determination of the compressibility of liquid nickel and cobalt has also been proven [25,30], therefore preparing the ground for the study of the high *P-T* behavior of more complex liquids. Furthermore, XAS can be exploited to study the chemical behavior of geochemically interesting systems at high pressure and temperature [91,184–187] and to study the disorder in laser annealed systems under high pressure [188]. The sensitivity of XAS to the chemical properties also allowed the comparison between PL and CW heating [40].

Finally, Kantor et al. [93] showed that by exploiting the fast integration time of the FReLoN detector (at maximum performance down to ~250 μs—full frame rate of 4230 frames/s), it is possible to use LH and XAS to characterize in situ the kinetics of reactions happening at high *T*.

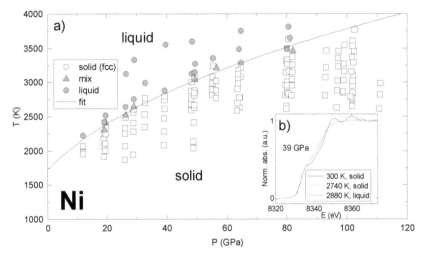

Figure 9. (**a**) Example results of a study on Ni performed using LH-DAC combined with XAS [28]. In the inset (**b**), the three XANES collected at 39 GPa show: solid nickel at ambient temperature, hot solid at 2740 K and a liquid at 2880 K.

Author Contributions: Both authors contributed equally to this work. All authors have read and agreed to the published version of the manuscript.

Acknowledgments: We want to thank C.M. Beavers and D. Errandonea for their precious comments on the manuscript. O.T. Lord for the help with Figure 6 and N.C. Siersch for the Raman data. We also want to thank our three anonymous reviewers for their constructive remarks and comments that have improved the quality of this work.

Conflicts of Interest: The authors declare no conflict of interest.

Abbreviations

The following abbreviations are used in this manuscript:

CRL	Compounds Refractive Lenses
CW	Continuous Wave, laser in continuous mode
DAC	Diamond Anvil Cell
EoS	Equation of State
ESRF	European Synchrotron Radiation Facility
FIB	Focused Ion Beam
FWHM	Full Width Half Maximum
EXAFS	Extended X-ray Absorption Fine Structure
IR	Infrared

KB	Kirkpatrick–Baez mirrors
LH-DAC	Laser-Heated Diamond Anvil Cell
NIST	National Institute of Standards and Technology
NPD	Nano-Polycrystalline Diamond
P	Pressure
PDF	Pair Distribution Function
PL	Pulsed Laser
PTM	Pressure Transmitting Medium
T	Temperature
XANES	X-ray Absorption Near Edge Structure
XAS	X-ray Absorption Spectroscopy
XRD	X-ray Diffraction

References

1. Ming, L.; Bassett, W. Laser heating in the diamond anvil press up to 2000 C sustained and 3000 C pulsed at pressures up to 260 kilobars. *Rev. Sci. Instrum.* **1974**, *45*, 1115. [CrossRef]
2. Saxena, S.; Dubrovinsky, L.; Haggkvist, P.; Cerenius, Y.; Shen, G.; Mao, H. Synchrotron X-ray study of iron at high pressure and temperature. *Science* **1995**, *269*, 1703. [CrossRef] [PubMed]
3. Zerr, A.; Diegeler, A.; Boehler, R. Solidus of Earth's Deep Mantle. *Science* **1998**, *281*, 243. [CrossRef] [PubMed]
4. Goncharov, A.; Beck, P.; Struzhkin, V.; Hemley, R.; Crowhurst, J. Laser-heating diamond anvil cell studies of simple molecular systems at high pressures and ttemperature. *J. Phys. Chem. Solids* **2008**, *69*, 2217. [CrossRef]
5. Boehler, R. High-pressure experiments and the phase diagram of lover mantle and core materials. *Rev. Geophys.* **2000**, *38*, 221–245. [CrossRef]
6. Anzellini, S.; Dewaele, A.; Mezouar, M.; Loubeyre, P.; Morard, G. Melting of iron at Earth's inner core boundary based on fast X-ray diffraction. *Science* **2013**, *340*, 464–466. [CrossRef]
7. Salamat, A.; Fischer, R.; Briggs, R.; McMahon, M.; Petitgirard, S. In situ synchrotron X-ray diffraction in the laser-heated diamond anvil cell: Melting phenomena and synthesis of new materials. *Coord. Chem. Rev.* **2014**, *277–278*, 15–30. [CrossRef]
8. Tateno, S.; Hirose, K.; Ohishi, Y.; Tatsumi, Y. The structure of iron in Earth's Inner core. *Science* **2010**, *330*, 359. [CrossRef]
9. Zhou, Q.; Ma, Y.; Cui, Q.; Cui, T.; Zhang, J.; Xie, Y.; Yang, K.; Zou, G. Raman scattering system for a laser heated diamond anvil cell. *Rev. Sci. Instrum.* **2004**, *75*, 2432. [CrossRef]
10. Goncharov, A.; Crowhurst, J. Pulsed laser Raman spectroscopy in the laser-heated diamond anvil cell. *Rev. Sci. Instrum.* **2005**, *76*, 063905. [CrossRef]
11. Zhang, J.; Bass, J.; Zhou, G. Single-crystal Brillouin spectroscopy with CO_2 laser heating and variable q. *Rev. Sci. Instrum.* **2015**, *86*, 063905. [CrossRef] [PubMed]
12. Zinin, P.; Prakapenka, V.; Burgess, K.; Odake, S.; Chigarev, N.; Sharma, S. Combined laser ultrasonic, laser heating and Raman scattering in diamond anvil cell system. *Rev. Sci. Instrum.* **2016**, *87*, 123908. [CrossRef] [PubMed]
13. Lin, J.; Santoro, M.; Struzhkin, V.; Mao, H.; Hemley, R. In situ high pressure-temperature Raman spectroscopy technique with laser-heated diamond anvil cells. *Rev. Sci. Instrum.* **2004**, *75*, 3302. [CrossRef]
14. Meier, T.; Dwivedi, A.; Khandarkhaeva, S.; Fedotenko, T.; Dobroviskaia, N.; Dubrovinsky, L. Table-top nuclear magnetic resonance system for high-pressure studies with in situ laser-heating. *Rev. Sci. Instrum.* **2019**, *90*, 123901. [CrossRef] [PubMed]
15. Kurnosov, A.; Marquardt, H.; Dubrovinsky, L.; Potapkin, V. A waveguide-based flexible CO_2-laser heating system for diamond-anvil cell applications. *C. R. Geosci.* **2019**, *351*, 280–285. [CrossRef]
16. Li, F.; Cui, Q.; He, Z.; Cui, T.; Gao, C.; Zhou, Q.; Zhou, G. Brillouin scattering spectroscopy for a laser heated diamond anvil cell. *Appl. Phys. Lett.* **2006**, *88*, 203507. [CrossRef]
17. Petitgirard, S.; Salamat, A.; Beck, P.; Weck, G.; Bouvier, P. Strategies for in situ laser heating in the diamond anvil cell at an X-ray diffraction beamline. *J. Synchrotron Radiat.* **2014**, *21*, 89–96. [CrossRef]
18. Anzellini, S.; Kleppe, A.; Daisenberger, D.; Wharmby, M.; Giampaoli, R.; Boccato, S.; Baron, M.; Miozzi, F.; Keeble, D.; Ross, A.; et al. Laser-heating system for high-pressure X-ray diffraction at the extreme condition beamline I15 at Diamond Light Source. *J. Synchrotron Radiat.* **2018**, *25*, 1860–1868. [CrossRef]

19.   Prakapenka, V.; Kubo, A.; Kuznetsov, A.; Laskin, A.; Shkurikhin, O.; Dera, P.; Rivers, M.; Sutton, S. Advanced flat top laser heating system for high pressure research at GSECARS: Application to the melting behaviour of germanium. *High Press. Res.* **2008**, *28*, 225. [CrossRef]

20.   Meng, Y.; Hrubiak, R.; Rod, E.; Boehler, R.; Shen, G. New development in laser-heated diamond anvil cell with in situ synchrotron X-ray diffraction at high pressure collaborative access team. *Rev. Sci. Instrum.* **2015**, *86*, 072201. [CrossRef]

21.   Stan, C.; Beavers, C.; Kunz, M.; Tamura, N. X-ray diffraction under extreme conditions at the advanced light source. *Quantum Beam Sci.* **2018**, *2*, 4. [CrossRef]

22.   Liermann, H.; Konopkova, Z.; Morgenroth, W.; Glazyrin, K.; Bednarcik, J.; McBride, E.; Petitgirard, S.; Delitz, J.; Wendt, M.; Bican, Y.; et al. The extreme condition beamline P02.2 and the Extreme Conditions Science Infrastructure at PETRA III. *J. Synchrotron Radiat.* **2015**, *22*, 908. [CrossRef] [PubMed]

23.   Watanuki, T.; Shimomura, O. Construction of laser-heated diamond anvil cell system for in situ X-ray diffraction study at Spring-8. *Rev. Sci. Instrum.* **2001**, *72*, 1289. [CrossRef]

24.   Boehler, R.; Musshoff, H.G.; Ditz, R.; Aquilanti, G.; Trapananti, A.; Boehler, R.; Musshoff, H.G.; Ditz, R.; Aquilanti, G.; Trapananti, A. Portable laser-heating stand for synchrotron applications. *Rev. Sci. Instrum.* **2009**, *80*, 045103. [CrossRef] [PubMed]

25.   Torchio, R.; Boccato, S.; Cerantola, V.; Morard, G.; Irifune, T.; Kantor, I. Probing the local, electronic and magnetic structure of matter under extreme conditions of temperature and pressure. *High Press. Res.* **2016**, *36*, 293–302. [CrossRef]

26.   Kantor, I.; Marini, C.; Mathon, O.; Pascarelli, S. A laser heating facility for energy-dispersive X-ray absorption spectroscopy. *Rev. Sci. Instrum.* **2018**, *89*, 1–13. [CrossRef]

27.   Boccato, S.; Sanson, A.; Kantor, I.; Mathon, O.; Dyadkin, V.; Chernyshov, D.; Carnera, A.; Pascarelli, S. Thermal and magnetic anomalies of α -iron : An exploration by extended X-ray absorption fine structure spectroscopy and synchrotron X-ray diffraction. *J. Phys. Condens. Matter* **2016**, *28*, 355401. [CrossRef]

28.   Boccato, S.; Torchio, R.; Kantor, I.; Morard, G.; Anzellini, S.; Giampaoli, R.; Briggs, R.; Smareglia, A.; Irifune, T.; Pascarelli, S. The Melting Curve of Nickel Up to 100 GPa Explored by XAS. *J. Geophys. Res. Solid Earth* **2017**, *122*, 1–10. [CrossRef]

29.   Morard, G.; Boccato, S.; Rosa, A.; Anzellini, S.; Miozzi, F.; Henry, L.; Garbarino, G.; Mezouar, M.; Harmand, M.; Guyot, F.; et al. Solving Controversies on the Iron Phase Diagram Under High Pressure. *Geophys. Res. Lett.* **2018**, *45*, 11074 – 11082. [CrossRef]

30.   Boccato, S.; Torchio, R.; D'Angelo, P.; Trapananti, A.; Kantor, I.; Recoules, V.; Anzellini, S.; Morard, G.; Irifune, T.; Pascarelli, S. Compression of liquid Ni and Co under extreme conditions explored by X-ray absorption spectroscopy. *Phys. Rev. B* **2019**, *100*, 180101. [CrossRef]

31.   Sakamaki, T.; Ohtani, E.; Fukui, H.; Kamada, S.; Takahashi, S.; Sakairi, T.; Takahata, A.; Sakai, T.; Tsusui, S.; Ishikawa, D.; et al. Constraints on Earth's inner core composition inferred from measurements of the sound velocity of hcp-iron in extreme conditions. *Sci. Adv.* **2016**, *2*, e1500802. [CrossRef] [PubMed]

32.   Kawaguchi, S.I.; Nakajima, Y.; Hirose, K.; Komabayashi, T.; Ozawa, H.; Tateno, S.; Kuwayama, Y.; Tsutsui, S.; Baron, A.Q.R. Sound velocity of liquid Fe-Ni-S at high pressure. *J. Geophys. Res. Solid Earth* **2017**, *122*, 3624–3634. [CrossRef]

33.   Andrault, D.; Petitgirard, S.; Nigro, G.L.; Devidal, J.; Veronesi, G.; Garbarino, G.; Mezouar, M. Solid-liquid iron partitioning in Earth's deep mantle. *Nature* **2012**, *487*, 354. [CrossRef] [PubMed]

34.   Potapkin, V.; McCammon, C.; Glazyrin, K.; Kantor, A.; Kupenko, I.; Prescher, C.; Sinmyo, R.; Smirnov, G.; Chumakov, A.; Ruffer, R.; et al. Effect of iron oxidation state on the electrical conductivity of the Earth's lowe mantle. *Nat. Commun.* **2013**, *4*, 1427. [CrossRef]

35.   McCammon, C.; Dubrovinsky, L.; Narygina, O.; Kantor, I.; Wu, X.; Glazyrin, K.; Sergueev, I.; Chumakov, A. Low-spin Fe^{2+} in silicate perovskite and a possible layer at the base of the lower mantle. *Phys. Earth Planet. Inter.* **2010**, *180*, 215. [CrossRef]

36.   Kupenko, I.; Dubrovinsky, L.; Dubrovinskaia, N.; McCammon, C.; Glazyrin, K.; Bykova, E.; Ballaran, T.B.; Sinmyo, R.; Chumakov, A.; Potapkin, V.; et al. Portable double-sided laser-heating system for Mössbaur spectroscopy and X-ray diffraction experiments at synchrotron facilities with diamond anvil cells. *Rev. Sci. Instrum.* **2012**, *83*, 124501. [CrossRef]

37. Fedotenko, T.; Dubrovinsky, L.; Aprilis, G.; Koemets, E.; Snigirev, A.; Snigireva, I.; Barannikov, A.; Ershov, P.; Cova, F.; Dubrovinskaia, N. Laser heating setup for diamond anvil cells for in situ synchrotron and in house high and ultra-high pressure studies. *Rev. Sci. Instrum.* **2019**, *90*, 104501. [CrossRef]

38. Fukui, H.; Sakai, T.; Sakamaki, T.; Kamada, S.; Takahashi, S.; Ohtani, E.; Baron, A. A compact system for generating extreme pressures and temperatures: an application of laser-heated diamond anvil cell to inelastic X-ray scattering. *Rev. Sci. Instrum.* **2013**, *84*, 113902. [CrossRef]

39. Petitgirard, S.; Borchet, M.; Andrault, D.; Appel, K.; Mezouar, M.; Liermann, H. An in situ approach to study trace element partitioning in the laser heated diamond anvil cell. *Rev. Sci. Instrum.* **2012**, *83*, 013904. [CrossRef]

40. Aprilis, G.; Kantor, I.; Kupenko, I.; Cerantola, V.; Pakhomova, A.; Collings, I.E.; Torchio, R.; Fedotenko, T.; Chariton, S.; Bykov, M.; et al. Comparative study of the influence of pulsed and continuous wave laser heating on the mobilization of carbon and its chemical reaction with iron in a diamond anvil cell. *J. Appl. Phys.* **2019**, *125*, 095901. [CrossRef]

41. Murakami, M.; Asahara, Y.; Ohishi, Y.; Hirao, N.; Hirose, K. Development of in situ Brillouin spectroscopy at high pressure and high temperature with synchrotron radiation and infrared laser heating system: Application to the Earth's deep interior. *Phys. Earth Planet. Inter.* **2009**, *174*, 282. [CrossRef]

42. Funamori, N.; Yamamoto, S.; Yagi, T.; Kikegawa, T. Exploratory studies of silicate melt structure at high pressures and temperatures by in situ X-ray diffraction. *J. Geophys. Res. Solid Earth* **2004**, *109*, B03203. [CrossRef]

43. Yamada, A.; Wang, Y.; Inoue, T.; Yang, W.; Park, C.; Yu, T.; Shen, G. High-pressure X-ray diffraction studies on the structure of liquid silicate using a Paris–Edinburgh type large volume press. *Rev. Sci. Instrum.* **2011**, *82*, 015103. [CrossRef] [PubMed]

44. Sanloup, C.; Drewitt, J.; Crepisson, C.; Kono, Y.; Park, C.; McCammon, C.; Hennet, L.; Brassamin, S.; Bytchkov, A. Structure and density of molten fayalite at high pressure. *Geochim. Cosmochim. Acta* **2013**, *118*, 118. [CrossRef]

45. Wang, Y.; Sakamaki, T.; Skinner, L.; Jing, Z.; Yu, T.; Kono, Y.; Park, C.; Shen, G.; Rivers, M.; Sutton, S. Atomistic insight into viscosity and density of silicate melts under pressure. *Nat. Commun.* **2014**, *5*, 3241. [CrossRef] [PubMed]

46. Kono, Y.; Park, C.; Kenney-Benson, C.; Shen, G.; Wang, Y. Toward comprehensive studies of liquids at high pressures and high temperatures: Combined structure, elastic wave velocity, and viscosity measurements in the Paris–Edinburgh cell. *Phys. Earth Planet. Inter.* **2014**, *228*, 269. [CrossRef]

47. Stinton, G.; MacLeod, S.; Cynn, H.; Errandonea, D.; Evans, W.; Proctor, J.; Meng, Y.; McMahon, M. Equation of state and high-pressure/high-temperature phase diagram of magnesium. *Phys. Rev. B* **2014**, *90*, 134105. [CrossRef]

48. Errandonea, D.; MacLeod, S.; Ruiz-Fuertes, J.; Burakovsky, L.; McMahon, M.; Wilson, C.; Ibanez, J.; Daisenberger, D.; Popescu, C. High-pressure/high-temperature phase diagram of zinc. *J. Phys. Condens. Matter* **2018**, *30*, 295402. [CrossRef]

49. Louvel, M.; Drewitt, J.; Ross, A.; Thwaites, R.; Heinen, B.; Keeble, D.; Beavers, C.; Walter, M.; Anzellini, S. The HXD95: A modified Bassett-type hydrothermal diamond-anvil cell for in situ XRD experiments up to 5 GPa and 1300 K. *J. Synchrotron Radiat.* **2020**, *27*, 529. [CrossRef]

50. Anzellini, S.; Errandonea, D.; Cazorla, C.; MacLeod, S.; Monteseguro, V.; Boccato, S.; Bandiello, E.; Anichtchenko, D.; Popescu, C.; Beavers, C. Thermal equation of state of ruthenium characterized by resistively heated diamond anvil cell. *Sci. Rep.* **2020**, *10*, 7092. [CrossRef]

51. Ozawa, H.; Tateno, S.; Xie, L.; Nakajima, Y.; Sakamoto, N.; Kawaguchi, S.; Yoneda, A.; Hirao, N. Boron-doped diamond as a new heating element for internal-resistive heated diamond-anvil cell. *High Press. Res.* **2018**, *38*, 120. [CrossRef]

52. Sinmyo, R.; Hirose, K.; Ohishi, Y. Melting curve of iron to 290 GPa determined in a resistance-heated diamond-anvil cell. *Earth Planet. Sci. Lett.* **2019**, *510*, 45. [CrossRef]

53. Dubrovinsky, L.; Dubrovinskaia, N.; Narygina, O.; Kantor, I.; Kuznetzov, A.; Prakapenka, V.; Vitos, L.; Johansson, B.; Mikhaylushkin, A.; Simak, S.; et al. Body-Centered Cubic Iron-Nickel Alloy in Earth's Core. *Science* **2007**, *316*, 1880. [CrossRef] [PubMed]

54. Boehler, R.; Nicol, M.; Johnson, M. *Internally-Heated Diamond-Anvil Cell: Phase Diagram and P-V-T of Iron*; Terra Scientific Publishing Company (TERRAPUB): Tokyo, Japan; American Geophysical Union: Washington, DC, USA, 1987; p. 173.

55. Antonangeli, D.; Komabayashi, T.; Occelli, F.; Borissenko, E.; Walters, A.; Fiquet, G.; Fei, Y. Simultaneous sound velocity and density measurements of hcp iron up to 93 GPa and 1100 K: An experimental test of the Birch's law at high temperature. *Earth Planet. Sci. Lett.* **2012**, *331–332*, 210. [CrossRef]

56. Zha, C.; Bassett, W. Internal resistive heating in diamond anvil cell for in situ X-ray diffraction and Raman scattering. *Rev. Sci. Instrum.* **2003**, *74*, 1255. [CrossRef]

57. Jenei, Z.; Cynn, H.; Visbeck, K.; Evans, W. High-temperature experiments using a resistively heated high-pressure membrane diamond anvil cell. *Rev. Sci. Instrum.* **2013**, *84*, 095114. [CrossRef] [PubMed]

58. Komabayashi, T.; Hirose, K.; Ohishi, Y. In situ X-ray diffraction measurements of the fcc–hcp phase transition boundary of an Fe–Ni alloy in an internally heated diamond anvil cell. *Phys. Chem. Miner.* **2012**, *39*, 329. [CrossRef]

59. Hrubiak, R.; Meng, Y.; Shen, G. Microstructures define melting of molybdenum at high pressure. *Nat. Commun.* **2016**, *8*, 14562. [CrossRef] [PubMed]

60. Lin, J.; Sturhahn, W.; Zhao, J.; Shen, G.; Mao, H.; Hemley, R. Nuclear resonant inelastic X-ray scattering and synchrotron Mossbauer spectroscopy with laser-heated diamond anvil cells. In *Advances in High-Pressure Technology for Geophysical Applications*; Elsevier: Amsterdam, The Netherlands, 2005; p. 397.

61. Kantor, I.; Prakapenka, V.; Kantor, A.; Dera, P.; Kurnosov, A.; Sinogeikin, S.; Dubrovinskaia, N.; Dubrovinsky, L. BX90: A new diamond anvil cell design for X-ray diffraction and optical measurements. *Rev. Sci. Instrum.* **2012**, *83*, 125102. [CrossRef]

62. Merrill, L.; Bassett, W.A. Miniature diamond anvil pressure cell for single crystal X-ray diffraction studies. *Rev. Sci. Instrum.* **1974**, *45*, 290. [CrossRef]

63. Boehler, R. New diamond cell for single-crystal X-ray diffraction. *Rev. Sci. Instrum.* **2006**, *77*, 115103. [CrossRef]

64. Letoullec, R.; Pinceaux, J.; Loubeyre, P. The membrane diamond anvil cell: A new device for generating continuous pressure and temperature variations. *High Press. Res.* **1988**, *1*, 77. [CrossRef]

65. Mao, H. High-Pressure Physics: Sustained Static Generation of 1.36 to 1.72 Megabars. *Science* **1978**, *200*, 1145. [CrossRef]

66. Hemley, R.; Mao, H.; Shen, G.; Badro, J.; Gillet, P.; Hanfland, M.; Hausermann, D. X-ray Imaging of Stress and Strain of Diamond, Iron, and Tungsten at Megabar Pressures. *Science* **1997**, *276*, 1242. [CrossRef]

67. Dewaele, A.; Loubeyre, P.; Occelli, F.; Marie, O.; Mezouar, M. Toroidal diamond anvil cell for detailed measurements under extreme static pressures. *Nat. Commun.* **2018**, *9*, 1–9. [CrossRef]

68. Jenei, Z.; O'Bannon, E.; Weir, S.; Cynn, H.; Lipp, M.; Evans, W. Single crystal toroidal diamond anvils for high pressure experiments beyond 5 megabar. *Nat. Commun.* **2018**, *9*, 1–6. [CrossRef]

69. Soignard, E.; Benmore, C.; Yarger, J. A perforated diamond anvil cell for high-energy X-ray diffraction of liquids and amorphous solids at high pressure. *Rev. Sci. Instrum.* **2010**, *81*, 035110. [CrossRef]

70. Chapman, K.W.; Chupas, P.J.; Halder, G.J.; Hriljac, J.A.; Kurtz, C.; Greve, B.K.; Ruschman, C.J.; Wilkinson, A.P. Optimizing high-pressure pair distribution function measurements in diamond anvil cells. *J. Appl. Crystallogr.* **2010**, *43*, 297. [CrossRef]

71. Ishimatsu, N.; Matsumoto, K.; Maruyama, H.; Kawamura, N.; Mizumaki, M.; Sumiya, H.; Irifune, T. Glitch-free X-ray absorption spectrum under high pressure obtained using nano-polycrystalline diamond anvils. *J. Synchrotron Radiat.* **2012**, *19*, 768–772. [CrossRef]

72. Rosa, A.D.; Mathon, O.; Torchio, R.; Jacobs, J.; Pasternak, S.; Irifune, T.; Pascarelli, S. Nano-polycrystalline diamond anvils: Key devices for XAS at extreme conditions: their use, scientific impact, present status and future needs. *High Press. Res.* **2019**, *40*, 65–81. [CrossRef]

73. Klotz, S.; Chervin, J.C.; Munsch, P.; Le Marchand, G. Hydrostatic limits of 11 pressure. *J. Phys. D Appl. Phys.* **2009**, *42*, 075413. [CrossRef]

74. Besson, J.; Pinceaux, J. Melting of Helium at Room Temperature and High Pressure REPORTS Melting of Helium at Room Temperature and High Pressure. *Science* **1979**, *206*, 1073. [CrossRef] [PubMed]

75. Mao, H.K.; Hemley, R.J.; Wu, Y.; Jephcoat, A.P.; Finger, L.W.; Zha, C.S.; Bassett, W.A. High-Pressure Phase Diagram and Equation of State of Solid Helium from Single-Crystal X-ray Diffraction to 23.3 GPa. *Phys. Rev. Lett.* **1988**, *60*, 2649. [CrossRef] [PubMed]

76. Sinmyo, R.; Hirose, K. The Soret diffusion in laser-heated diamond-anvil cell. *Phys. Earth Planet. Inter.* **2010**, *180*, 172–178. [CrossRef]

77. Mezouar, M.; Giampaoli, R.; Garbarino, G.; Kantor, I.; Dewaele, A.; Weck, G.; Boccato, S.; Svitlyk, V.; Rosa, A.; Torchio, R.; et al. Methodology for in situ synchrotron X-ray studies in the laser-heated diamond anvil cell. *High Press. Res.* **2017**, *37*, 170–180. [CrossRef]

78. Dewaele, A.; Mezouar, M.; Guignot, N.; Loubeyre, P. High melting points of tantalum in a laser-heated diamond anvil cell. *Phys. Rev. Lett.* **2010**, *104*, 29–31. [CrossRef]

79. Karandikar, A.; Boehler, R. Flash melting of tantalum in a diamond cell to 85 GPa. *Phys. Rev. B* **2016**, *93*, 1–6. [CrossRef]

80. Hrubiak, R.; Sinogeikin, S.; Rod, E.; Shen, G. The laser micro-machining system for diamond anvil cell experiments and general precision machining applications at the High Pressure Collaborative Access Team. *Rev. Sci. Instrum.* **2015**, *86*, 072202. [CrossRef]

81. Anzellini, S.; Monteseguro, V.; Bandiello, E.; Dewaele, A.; Burakovsky, L.; Errandonea, D. In situ characterization of the high pressure- high temperature meltig curve of platinum. *Sci. Rep.* **2019**, *9*, 13034. [CrossRef]

82. Weck, G.; Recoules, V.; Queyroux, J.; Datch, F.; Bouchet, J.; Ninet, S.; Garbarino, G.; Mezouar, M.; Loubeyre, P. Determination of the melting curve of gold up to 110 GPa. *Phys. Rev. B* **2020**, *101*, 014106. [CrossRef]

83. Drewitt, J.; Walter, M.; Zhang, H.; MacMahon, S.; Edwards, D.; Heinen, B.; Lord, O.; Anzellini, S.; Kleppe, A. The fate of carbonate in oceanic crust subducted into Earth's lower mantle. *Earth Planet. Sci. Lett.* **2019**, *511*, 213. [CrossRef]

84. Baron, M.; Lord, O.; Myhill, R.; Thompson, A.; Wang, W.; Tronnes, R.; Walter, M. Experimental constraints on melting temperatures in the MgO–SiO2 system at lower mantle pressures. *Earth Planet. Sci. Lett.* **2017**, *472*, 186. [CrossRef]

85. Dubrovinsky, L.; Dubrovinskaia, N.; Prakapenka, V.; Abakumov, A. Article Open Access Published: 23 October 2012 Implementation of micro-ball nanodiamond anvils for high-pressure studies above 6 Mbar. *Nat. Commun.* **2012**, *3*, 1163. [CrossRef] [PubMed]

86. Zha, H.; Krasnicki, S.; Meng, Y.; Yan, C.; Lai, J.; Liang, Q.; Mao, H.; Hemley, R. Composite chemical vapor deposition diamond anvils for high-pressure/high-temperature experiments. *High Press. Res.* **2009**, *29*, 317. [CrossRef]

87. Pepin, C.; Dewaele, A.; Geneste, G.; Loubeyre, P. New Iron Hydrides under High Pressure. *Phys. Rev. Lett.* **2014**, *113*, 265504. [CrossRef]

88. Pepin, C.; Geneste, G.; Dewaele, A.; Mezouar, M.; Loubeyre, P. Synthesis of FeH5: A layered structure with atomic hydrogen slabs. *Science* **2017**, *357*, 382. [CrossRef]

89. Laniel, D.; Dewaele, A.; Anzellini, S.; Guignot, N. Study of the iron nitride FeN into the megabar regime. *J. Alloys Compd.* **2018**, *733*, 53. [CrossRef]

90. Breton, H.; Komabayashi, T.; Thompson, S.; Potts, N.; McGuire, C.; Suehiro, S.; Anzellini, S.; Ohishi, Y. Static compression of Fe4N to 77 GPa and its implications for nitrogen storage in the deep Earth. *Am. Mineral.* **2019**, *104*, 1781. [CrossRef]

91. Rosa, A.; Bouhifd, M.; Morard, G.; Briggs, R.; Garbarino, G.; Irifune, T.; Mathon, O.; Pascarelli, S. Krypton storage capacity of the Earth's lower mantle. *Earth Planet. Sci. Lett.* **2020**, *532*, 116032. [CrossRef]

92. Shen, G.; Prakapenka, V.B.; Rivers, M.L.; Sutton, S.R. Structure of liquid iron at pressures up to 58 GPa. *Phys. Rev. Lett.* **2004**, *92*, 185701. [CrossRef]

93. Kantor, I.; Labiche, J.C.; Collet, E.; Siron, L.; Thevenin, J.J.; Ponchut, C.; Borrel, J.; Mairs, T.; Marini, C.; Strohm, C.; et al. A new detector for sub-millisecond EXAFS spectroscopy at the European Synchrotron Radiation Facility. *J. Synchrotron Radiat.* **2014**, *21*, 1240–1246. [CrossRef] [PubMed]

94. Smith, D.; Smith, J.; Childs, C.; Rod, E.; Hrubiak, R.; Shen, G.; Salamat, A. A CO2 laser heating system fo in situ high pressure-temperature experiments at HPCAT. *Rev. Sci. Instrum.* **2018**, *89*, 083901. [CrossRef] [PubMed]

95. Du, Z.; Amulele, G.; Benedetti, L.; Lee, K. Mapping temperatures and temperatures gradients during flash heating in a diamond anvil cell. *Rev. Sci. Instrum.* **2013**, *84*, 075111. [CrossRef]

96. Geballe, Z.M.; Jeanloz, R. Origin of temperature plateaus in laser-heated diamond anvil cell experiments. *J. Appl. Phys.* **2012**, *111*, 123518. [CrossRef]

97. Geballe, Z.; Collins, G.; Jeanloz, R. Modulation calorimetry in diamond anvil cells. I. Heat flow models. *J. Appl. Phys.* **2017**, *121*, 145902. [CrossRef]

98. Boehler, R. Melting of the FeFeO and the FeFeS systems at high pressure: Constraints on core temperatures. *Earth Planet. Sci. Lett.* **1992**, *111*, 217–227. [CrossRef]

99. Errandonea, D. Phase behavior of metals at very high P-T conditions: A review of recent experimental studies. *J. Phys. Chem. Solids* **2006**, *67*, 2018. [CrossRef]

100. Shen, G.; Rivers, M.; Wang, Y.; Sutton, S. Laser heated diamond anvil cell system at the Advanced Photon Source for in situ X-ray measurements at high pressure and temperature. *Rev. Sci. Instrum.* **2001**, *72*, 1273. [CrossRef]

101. Spiekermann, G.; Kupenko, I.; Petitgirard, S.; Harder, M.; Nyrow, A.; Weis, C.; Albers, C.; Biedermann, N.; Libon, L.; Sahle, C.; et al. A portable on-axis laser-heating system for near-90 X-ray spectroscopy: Application to ferropericlase and iron silicide. *J. Synchrotron Radiat.* **2019**, *27*, 414. [CrossRef]

102. Hasegawa, A.; Yagi, T.; Ohta, K. Combination of pulsed light heating thermoreflectance and laser-heated diamond anvil cell for in-situ high pressure-temperature thermal diffusivity measurements. *Rev. Sci. Instrum.* **2019**, *90*, 074901. [CrossRef]

103. Hirao, N.; Kawaguchi, S.; Hirose, K.; Shimuzu, K.; Ohtani, E.; Ohishi, Y. New developments in high-pressure X-ray diffraction beamline for diamond anvil cell at Spring-8. *Matter Radiat. Extrem.* **2020**, *5*, 018403. [CrossRef]

104. Bykova, E.; Aprilis, G.; Bykov, M.; Glazyrin, K.; Wendt, M.; Wenz, S.; Liermann, H.; Roeh, J.; Ehnes, A.; Dubrovinskaia, N.; et al. Single-crystal diffractometer coupled with double-sided laser heating system at the Extreme Conditions Beamline P02.2 at PETRAIII. *Rev. Sci. Instrum.* **2019**, *90*, 073907. [CrossRef] [PubMed]

105. Lord, O.; Wood, I.; Dobson, D.; Vocadlo, L.; Wang, W.; Thomson, A.; Wann, E.; Morard, G.; Mezouar, M.; Walter, M. The melting curve of Ni to 1 Mbar. *Earth Planet. Sci. Lett.* **2014**, *408*, 226–236. [CrossRef]

106. Lord, O.T.; Wann, E.T.H.; Hunt, S.A.; Walker, A.M.; Santangeli, J.; Walter, M.J.; Dobson, D.P.; Wood, I.G.; Vocadlo, L.; Morard, G.; et al. The NiSi melting curve to 70 GPa. *Phys. Earth Planet. Inter.* **2014**, *233*, 13–23. [CrossRef]

107. Giampaoli, R.; Kantor, I.; Mezouar, M.; Boccato, S.; Rosa, A.D.; Torchio, R.; Garbarino, G.; Mathon, O.; Pascarelli, S. Measurement of temperature in the laser heated diamond anvil cell: comparison between reflective and refractive optics. *High Press. Res.* **2018**, *38*, 250–269. [CrossRef]

108. Prakapenka, V.; Shen, G.; Dubrovinsky, L. Carbon transport in diamond anvil cell. *High Temp. High Press.* **2003**, *35/36*, 237. [CrossRef]

109. Dewaele, A.; Mezouar, M.; Guignot, N.; Loubeyre, P. Melting of lead under high pressure studied using second-scale time-resolved X-ray diffraction. *Phys. Rev. B Condens. Matter Mater. Phys.* **2007**, *76*, 1–5. [CrossRef]

110. Goncharov, A.; Prakapenka, V.; Struzhkin, V.; Kantor, I.; Rivers, M.; Dalton, D. X-ray diffraction in the pulsed laser heated diamond anvil cell. *Rev. Sci. Instrum.* **2010**, *81*, 113902. [CrossRef]

111. Yang, L.; Karandikar, A.; Boehler, R. Flash heating in the diamond cell: Melting curve of rhenium. *Rev. Sci. Instrum.* **2012**, *83*, 063905. [CrossRef]

112. Shen, G.; Wang, L.; Ferry, R.; Mao, H.; Hemley, R. A portable laser heating microscope for high pressure research. *J. Phys. Conf. Ser.* **2010**, *215*, 012191. [CrossRef]

113. Errandonea, D. The melting curve of ten metals up to 12 GPa and 1600 K. *J. Appl. Phys.* **2010**, *108*, 033517. [CrossRef]

114. Anzellini, S.; Errandonea, D.; MacLeod, S.; Botella, P.; Daisenberger, D.; DeAth, J.; Gonzalez-Platas, J.; Ibanez, J.; McMahon, M.; Munro, K.; et al. Phase diagram of calcium at high pressure and high temperature. *Phys. Rev. Mater.* **2018**, *2*, 083608. [CrossRef]

115. Sanloup, C.; de Grouchy, C. *X-ray Diffraction Structure Measurements*; Elsevier: Amsterdam, The Netherlands, 2018; Chapter 5, p. 137.

116. Rosa, A.; Merkulova, M.; Garbarino, G.; Svitlyk, V.; Jacobs, J.; Sahle, C.; Mathon, O.; Munoz, M.; Merkel, S. Amorphous boron composite gaskets for in situ high-pressure and high-temperature studies. *High Press. Res.* **2016**, *36*, 564. [CrossRef]

117. Mijiti, Y.; Trapananti, A.; Minicucci, M.; Ciambezi, M.; Coquet, J.; Nataf, L.; Baudelet, F.; Cicco, A.D. Development of a high temperature diamond anvil cell for x ray absorption experiments under extreme conditions. *Radiat. Phys. Chem.* **2018**, *108106*. [CrossRef]

118. Zhang, F.; Wu, Y.; Lou, H.; Zeng, Z.; Prakapenka, V.; Greenberg, E.; Ren, Y.; Yan, J.; Okasinski, J.; Liu, X.; et al. Polymorphism in a high-entropy alloy. *Nat. Commun.* **2017**, *8*, 15687. [CrossRef]

119. Walter, M.J.; Koga, K.T. The effects of chromatic dispersion on temperature measurement in the laser-heated diamond anvil cell. *Phys. Earth Planet. Inter.* **2004**, *143–144*, 541–558. [CrossRef]

120. Aprilis, G.; Strohm, C.; Kupenko, I.; Linhardt, S.; Laskin, A.; Vasiukov, D.; Cerantola, V.; Koemets, E.; McCammon, C.; Kurnosov, A.; et al. Portable double-sided pulsed laser heating system for time-resolved geoscience and materials science applications. *Rev. Sci. Instrum.* **2017**, *88*, 084501. [CrossRef]

121. Jeanloz, R.; Kavner, A. Melting criteria and imaging spectroradiometry in laser-heated diamond-cell experiments. *Philos. Trans. R. Soc. A* **1996**, *354*, 1279.

122. Williams, Q.; Knittle, E.; Jeanloz, R. The high-pressure melting curve of iron: A technical discussion. *J. Geophys. Res.* **1991**, *96*, 2171. [CrossRef]

123. Jackson, J.M.; Sturhahn, W.; Lerche, M.; Zhao, J.; Toellner, T.S.; Alp, E.E.; Sinogeikin, S.V.; Bass, J.D.; Murphy, C.A.; Wicks, J.K. Melting of compressed iron by monitoring atomic dynamics. *Earth Planet. Sci. Lett.* **2013**, *362*, 143–150. [CrossRef]

124. Aquilanti, G.; Trapananti, A.; Karandikar, A.; Kantor, I.; Marini, C.; Mathon, O.; Pascarelli, S.; Boehler, R. Melting of iron determined by X-ray absorption spectroscopy to 100 GPa. *Proc. Natl. Acad. Sci. USA* **2015**, *112*, 12042–12045. [CrossRef] [PubMed]

125. Benedetti, L.; Loubeyre, P. Temperature gradients, wavelength-dependent emissivity, and accuracy of high and very-high temperatures measured in the laser-heated diamond cell. *High Press. Res.* **2004**, *24*, 423–445. [CrossRef]

126. Queyroux, J.; Ninet, S.; Weck, G.; Garbarino, G.; Plisson, T.; Mezouar, M.; Datchi, F. Melting curve and chemical stability of ammonia at high pressure: Combined X-ray diffraction and Raman study. *Phys. Rev. B* **2019**, *99*, 134107. [CrossRef]

127. Anzellini, S. Phase Diagram of Iron under Extreme Conditions Measured with Time-Resolved Methods. Ph.D. Thesis, Université Pierre et Marie Curie, Paris, France, 2014.

128. Parisiades, P.; Cova, F.; Garbarino, G. Melting curve of elemental zirconium. *Phys. Rev. B* **2019**, *100*, 054102. [CrossRef]

129. Lord, O.; Wang, W. Mirrors: A Matlab(R) GUI for temperature measurment by multispectral imaging radiometry. *Rev. Sci. Instrum.* **2018**, *89*, 104903. [CrossRef]

130. Kavner, A.; Panero, W.R. Temperature gradients and evaluation of thermoelastic properties in the synchrotron-based laser-heated diamond cell. *Phys. Earth Planet. Inter.* **2004**, *143–144*, 527–539. [CrossRef]

131. Briggs, R.; Daisenberger, D.; Lord, O.T.; Salamat, A.; Bailey, E.; Walter, M.J.; McMillan, P.F. High-pressure melting behavior of tin up to 105 GPa. *Phys. Rev. B* **2017**, *95*, 054102. [CrossRef]

132. Kavner, A.; Nugent, C. Precise measurements of radial temperature gradients in the laser-heated diamond anvil cell. *Rev. Sci. Instrum.* **2008**, *79*, 024902. [CrossRef]

133. Rainey, E.; Kavner, A. Peak scaling method to measure temperatures in the laser-heated diamond anvil cell and application to the thermal conductivity of MgO. *J. Geophys. Res. Solid Earth* **2014**, *119*, 8154. [CrossRef]

134. Kunz, M.; Yan, J.; Cornell, E.; Domning, E.; Yen, C.; Doran, A.; Beavers, C.; Treger, A.; Williams, Q.; MacDowell, A. Implementation and application of the peak scaling method for temperature measurement in the laser heated diamond anvil cell. *Rev. Sci. Instrum.* **2018**, *89*, 083903. [CrossRef]

135. Campbell, A. Measurement of temperature distributions across laser heated samples by multispectral imaging radiometry. *Rev. Sci. Instrum.* **2008**, *79*, 015108. [CrossRef] [PubMed]

136. Walter, M.; Thompson, A.; Wang, W.; Lord, O.; Ross, J.; McMahon, S.C.; Baron, M.A.; Melekhova, E.; Kleppe, A.; Kohn, S. The stability of hydrous silicates in Earth's lower mantle: Experimental constraints from the systems MgO–SiO$_2$–H$_2$O and MgO–Al$_2$O$_3$–SiO$_2$–H$_2$O. *Chem. Geol.* **2015**, *418*, 16. [CrossRef]

137. Ballaran, T.; Kurnosov, A.; Trots, D. Single-crystal X-ray diffraction at extreme conditions: A review. *High Press. Res.* **2013**, *33*, 453. [CrossRef]

138. Merkel, S.; Hemley, R.; Mao, H. Finite-element modeling of diamond deformation at multimegabar pressures. *Appl. Phys. Lett.* **1999**, *74*, 656. [CrossRef]

139. Liu, L.; Bi, Y.; Xu, J.A. Ruby fluorescence pressure scale: Revisited. *Chin. Phys. B* **2013**, *22*, 056201. [CrossRef]

140. Dewaele, A.; Loubeyre, P.; Mezouar, M. Equations of state of six metals above 94 GPa. *Phys. Rev. B* **2004**, *094112*, 1–8.

141. Chijioke, A.D.; Nellis, W.J.; Soldatov, A.; Silvera, I.F. The ruby pressure standard to 150GPa. *J. Appl. Phys.* **2005**, *98*, 114905. [CrossRef]
142. Silvera, I.F.; Chijioke, A.D.; Nellis, W.J.; Soldatov, A.; Tempere, J. Calibration of the ruby pressure scale to 150 GPa. *Phys. Status Solidi* **2007**, *244*, 460–467. [CrossRef]
143. Dorogokupets, P.I.; Oganov, A.R. Ruby, metals, and MgO as alternative pressure scales: A semiempirical description of shock-wave, ultrasonic, X-ray, and thermochemical data at high temperatures and pressures. *Phys. Rev. B* **2007**, *75*, 024115. [CrossRef]
144. Datchi, F.; Dewaele, A.; Loubeyre, P.; Letoullec, R.; Le Godec, Y.; Canny, B. Optical pressure sensors for high-pressure-high-temperature studies in a diamond anvil cell. *High Press. Res.* **2007**, *27*, 447. [CrossRef]
145. Dewaele, A.; Torrent, M.; Loubeyre, P.; Mezouar, M. Compression curves of transition metals in the Mbar range: Experiments and projector augmented-wave calculations. *Phys. Rev. B Condens. Matter Mater. Phys.* **2008**, *78*, 1–13. [CrossRef]
146. Syassen, K. Ruby under pressure. *High Press. Res.* **2008**, *28*, 75. [CrossRef]
147. Akahama, Y.; Kawamura, H. Pressure calibration of diamond anvil Raman gauge to 410 GPa. *J. Phys. Conf. Ser.* **2010**, *215*, 012195. [CrossRef]
148. Dubrovinskaia, N.; Dubrovinsky, L.; Caracas, R.; Hanfland, M. Diamond as a high pressure gauge up to 2.7 Mbar. *Appl. Phys. Lett.* **2010**, *97*, 251903. [CrossRef]
149. Mao, H.K.; Chen, X.J.; Ding, Y.; Li, B.; Wang, L. Solids, liquids, and gases under high pressure. *Rev. Mod. Phys.* **2018**, *90*, 015007. [CrossRef]
150. Holmes, N.C.; Moriarty, J.A.; Gathers, G.R.; Nellis, W.J. The equation of state of platinum to 660 GPa (6.6 Mbar). *J. Appl. Phys.* **1989**, *66*, 2962–2967. [CrossRef]
151. Hixson, R.S.; Fritz, J.N. Shock compression of tungsten and molybdenum. *J. Appl. Phys.* **1992**, *71*, 1721. [CrossRef]
152. Bassett, W.; Reichmann, H.J.; Angel, R.; Spetzler, H.; Smyth, J. New diamond anvil cells for gigahertz ultrasonic interferometry and X-ray diffraction. *Am. Mineral.* **2000**, *85*, 283. [CrossRef]
153. Li, B.; Kung, J. Pressure calibration to 20GPa by simultaneous use of ultrasonic and X-ray techniques. *J. Appl. Phys.* **2005**, *98*, 013521. [CrossRef]
154. Anderson, O.L. *Equations of State of Solids for Geophysics and Ceramic Science*; Oxford University Press: Oxford, UK, 1995.
155. Jacobs, M.; Schmid-Fetzer, R.; van den Berg, A. Phase diagrams, thermodynamic properties and sound velocities derived from a multiple Einstein method using vibrational densities of states: An application to MgO–SiO$_2$. *Phys Chem Miner.* **2016**, *44*, 43. [CrossRef]
156. Belmonte, D. First Principles Thermodynamics of Minerals at HP–HT Conditions: MgO as a Prototypical Material. *Minerals* **2017**, *7*, 183. [CrossRef]
157. Campbell, A.J.; Danielson, L.; Righter, K.; Seagle, C.T.; Wang, Y.; Prakapenka, V.B. High pressure effects on the iron-iron oxide and nickel-nickel oxide oxygen fugacity buffers. *Earth Planet. Sci. Lett.* **2009**, *286*, 556–564. [CrossRef]
158. Labiche, J.C.; Mathon, O.; Pascarelli, S.; Newton, M.; Ferre, G.G.; Curfs, C.; Vaughan, G.; Homs, A. Invited article: The fast readout low noise camera as a versatile X-ray detector for time resolved dispersive extended X-ray absorption fine structure and diffraction studies of dynamic problems in materials science, chemistry, and catalysis. *Rev. Sci. Instrum.* **2007**, *78*, 091301. [CrossRef] [PubMed]
159. Toyokawa, H.; Furukawa, Y.; Hirono, T.; Ikeda, H.; Kajiwara, K.; Kawase, M.; Ohata, T.; Sato, G.; Sato, M.; Takahashi, T.; et al. Si and CdTe pixel detector developments at SPring-8. *Nucl. Instrum. Methods Phys. Res. Sect. A Accel. Spectrometers Detect. Assoc. Equip.* **2011**, *636*, S218. [CrossRef]
160. Dubrovinsky, L.; Glazyrin, K.; McCammon, C.; Narygina, O.; Greenberg, E.; Übelhack, S.; Chumakov, A.I.; Pascarelli, S.; Prakapenka, V.; Bock, J.; et al. Portable laser-heating system for diamond anvil cells. *J. Synchrotron Radiat.* **2009**, *16*, 737–741. [CrossRef]
161. Kirkpatrick, P.; Baez, A. Formation of Optical Images by X-rays. *J. Opt. Soc. Am.* **1948**, *38*, 766. [CrossRef]
162. Drakopoulos, M.; Snigirev, A.; Snigireva, I.; Schilling, J. X-ray high-resolution diffraction using refractive lenses. *Appl. Phys. Lett.* **2005**, *86*, 014102. [CrossRef]
163. Shim, S.; Duffy, T.; Jeanloz, R.; Shen, G. Stability and crystal structure of MgSiO$_3$ perovskite to the core-mantle boundary. *Geophys. Res. Lett.* **2004**, *31*, L10603. [CrossRef]

164. Lavina, B.; Dera, P.; Kim, E.; Meng, Y.; Downs, R.; Weck, P.; Sutton, S.; Zhao, Y. Discovery of the recoverable high-pressure iron oxide Fe_4O_5. *Proc. Natl. Acad. Sci. USA* **2011**, *108*, 17281. [CrossRef]

165. Zhang, L.; Meng, Y.; Yang, W.; Wang, L.; Mao, W.; Zeng, Q.S.; Jeong, J.; Wagner, A.; Mkhoyan, K.; Liu, W.; et al. Disproportionation of (Mg,Fe)SiO3 perovskite in Earth's deep lower mantle. *Science* **2014**, *344*, 877. [CrossRef]

166. Mao, W.; Meng, Y.; Shen, G.; Prakapenka, V.; Campbell, A.; Heinz, D.; Shu, J.; Caracas, R.; Cohen, R.; Fei, Y.; et al. Iron-rich silicates in the Earth's D" layer. *Proc. Natl. Acad. Sci. USA* **2005**, *102*, 9751. [CrossRef] [PubMed]

167. McMillan, P. New nitrides: From high pressure–high temperature synthesis to layered nanomaterials and energy applications. *Phylosophical Trans. R. Soc. A* **2019**, *377*, 20180244. [CrossRef] [PubMed]

168. Laniel, D.; Weck, G.; Gaiffe, G.; Garbarino, G.; Loubeyre, P. High-Pressure Synthesized Lithium Pentazolate Compound Metastable under Ambient Conditions. *J. Phys. Chem. Lett.* **2018**, *9*, 1600. [CrossRef] [PubMed]

169. Fischer, R.; Campbell, A.; Caracas, R.; Reaman, D.M.; Heinz, D.; Dera, P.; Prakapenka, V.B. Equations of state in the Fe-FeSi system at high pressures and temperatures. *J. Geophys. Res. Solid Earth* **2014**, *119*, 2810. [CrossRef]

170. McGuire, C.; Santamaria-Perez, D.; Makhluf, A.; Kavner, A. Isothermal equation of state and phase stability of Fe5Si3 up to 96 GPa and 3000 K. *J. Geophys. Res. Solid Earth* **2017**, *122*, 4328. [CrossRef]

171. Miozzi, F.; Morard, G.; Antonangeli, D.; Clark, A.N.; Mezouar, M.; Dorn, C.; Rozel, A.; Fiquet, G. Equation of State of SiC at Extreme Conditions: New Insight Into the Interior of Carbon-Rich Exoplanets. *J. Geophys. Res. Planets* **2018**, *123*, 2295–2309. [CrossRef]

172. Lazicki, A.; Dewaele, A.; Loubeyre, P.; Mezouar, M. High-pressure–temperature phase diagram and the equation of state of beryllium. *Phys. Rev. B* **2012**, *86*, 174118. [CrossRef]

173. Komabayashi, T.; Fei, Y. Internally consistent thermodynamic database for iron to the Earth's core conditions. *J. Geophys. Res. Solid Earth* **2010**, *115*, B03202. [CrossRef]

174. Dubrovinsky, L.; Ballaran, T.B.; Glazyrin, K.; Kurnosov, A.; Frost, D.; Merlini, M.; Hanfland, M.; Prakapenka, V.; Schouwink, P.; Pippinger, T.; et al. Single-crystal X-ray diffraction at megabar pressures and temperatures of thousands of degree. *High Press. Res.* **2010**, *30*, 620. [CrossRef]

175. Errandonea, D.; MacLeod, S.G.; Burakovsky, L.; Santamaria-Perez, D.; Proctor, J.E.; Cynn, H.; Mezouar, M. Melting curve and phase diagram of vanadium under high-pressure and high-temperature conditions. *Phys. Rev. B* **2019**, *100*, 094111. [CrossRef]

176. Morard, G.; Garbarino, G.; Antonangeli, D.; Andrault, D.; Guignot, N.; Siebert, J.; Roberge, M.; Boulard, E.; Lincot, A.; Denoeud, A.; et al. Density measurements and structural properties of liquid and amorphous metals under high pressure. *High Press. Res.* **2014**, *34*, 9. [CrossRef]

177. Andrault, D.; Morard, G.; Garbarino, G.; Mezouar, M.; Bouhifd, M.; Kawamoto, T. Melting behavior of SiO_2 up to 120 GPa. *Phys. Chem. Miner.* **2020**, *47*, 10. [CrossRef]

178. Sanloup, C.; Drewitt, J.; Konopkova, Z.; Dalladay-Simpson, P.; Morton, D.; Rai, N.; Westrenen, W.V.; Morgenroth, W. Structural change in molten basalt at deep mantle conditions. *Nature* **2013**, *503*, 104. [CrossRef] [PubMed]

179. Drewitt, J.; Jahn, S.; Sanloup, C.; de Grouchy, C.; Garbarino, G.; Hennet, L. Development of chemical and topological structure in aluminosilicate liquids and glasses at high pressure. *J. Phys. Condens. Matter* **2015**, *27*, 105103. [CrossRef]

180. Pascarelli, S.; Mathon, O.; Mairs, T.; Kantor, I.; Agostini, G.; Strohm, C.; Pasternak, S.; Perrin, F.; Berruyer, G.; Chappelet, P.; et al. The Time-resolved and Extreme-conditions XAS (TEXAS) facility at the European Synchrotron Radiation Facility: The energy-dispersive X-ray absorption spectroscopy beamline ID24. *J. Synchrotron Radiat.* **2016**, *23*, 353–368. [CrossRef]

181. Boccato, S. Local Structure of Liquid 3d Metals under Extreme Conditions of Pressure and Temperature. Ph.D. Thesis, Université Grenoble Alpes, Grenoble, France, 2017.

182. Boccato, S.; Torchio, R.; Anzellini, S.; Boulard, E.; Irifune, T.; Harmand, M.; Kantor, I.; Miozzi, F.; Parisiades, P.; Rosa, A.D.; et al. Melting properties by X-ray absorption spectroscopy : Common signatures in binary Fe-C, Fe-O, Fe-S and Fe-Si systems. *Sci. Rep.* **2020**, under revision.

183. Torchio, R.; Boccato, S.; Miozzi, F.; Rosa, A.D.; Ishimatsu, N.; Kantor, I.; Briggs, R.; Irifune, T.; Morard, G. Melting curve and phase relations of Fe-Ni alloys at high pressure and high temperature: Implications for the Earth's core composition. *Geophys. Res. Lett.* **2020**, under revision.

184. Dewaele, A.; Worth, N.; Pickard, C.J.; Needs, R.J.; Pascarelli, S.; Mathon, O.; Mezouar, M.; Irifune, T. Synthesis and stability of xenon oxides Xe_2O_5 and Xe_3O_2 under pressure. *Nat. Chem.* **2016**, *8*, 784–790. [CrossRef]
185. Andrault, D.; Muñoz, M.; Pesce, G.; Cerantola, V.; Chumakov, A.; Kantor, I.; Pascarelli, S.; Rüffer, R.; Hennet, L. Large oxygen excess in the primitive mantle could be the source of the Great Oxygenation Event. *Geochem. Perspect. Lett.* **2018**, *6*, 5–10. [CrossRef]
186. Boulard, E.; Harmand, M.; Guyot, F.; Lelong, G.; Morard, G.; Cabaret, D.; Boccato, S.; Rosa, A.D.; Briggs, R.; Pascarelli, S.; et al. Ferrous Iron Under Oxygen - Rich Conditions in the Deep Mantle Geophysical Research Letters. *Geophys. Res. Lett.* **2019**, *46*, 1348–1356. [CrossRef]
187. Cerantola, V.; Wilke, M.; Kantor, I.; Ismailova, L.; Kupenko, I.; McCammon, C.; Pascarelli, S.; Dubrovinsky, L.S. Experimental investigation of $FeCO_3$ (siderite) stability in Earth's lower mantle using XANES spectroscopy. *Am. Mineral.* **2019**, *104*, 1083–1091. [CrossRef]
188. Sneed, D.; Kearney, J.S.; Smith, D.; Smith, J.S.; Park, C.; Salamat, A. Probing disorder in high-pressure cubic tin (IV) oxide: A combined X-ray diffraction and absorption study. *J. Synchrotron Radiat.* **2019**, *26*, 1245. [CrossRef] [PubMed]

Review

A Short Review of Current Computational Concepts for High-Pressure Phase Transition Studies in Molecular Crystals

Denis A. Rychkov [1,2]

[1] Laboratory of Physicochemical Fundamentals of Pharmaceutic Materials, Novosibirsk State University, 2 Pirogova st., 630090 Novosibirsk, Russia; rychkov.dennis@gmail.com; Tel.: +7-905-957-7447

[2] Institute of Solid State Chemistry and Mechanochemistry SB RAS, 18 Kutateladze st., 630128 Novosibirsk, Russia

Received: 1 December 2019; Accepted: 29 January 2020; Published: 31 January 2020

Abstract: High-pressure chemistry of organic compounds is a hot topic of modern chemistry. In this work, basic computational concepts for high-pressure phase transition studies in molecular crystals are described, showing their advantages and disadvantages. The interconnection of experimental and computational methods is highlighted, showing the importance of energy calculations in this field. Based on our deep understanding of methods' limitations, we suggested the most convenient scheme for the computational study of high-pressure crystal structure changes. Finally, challenges and possible ways for progress in high-pressure phase transitions research of organic compounds are briefly discussed.

Keywords: high-pressure phase transitions; molecular crystals; computational methods; DFT and Force Field methods; energy calculations; intermolecular interactions

1. Introduction

High-pressure chemistry and particularly crystallography is developing fast in recent decades [1–3]. High-pressure effects are known to be studied originally and mainly by physicists and geologists [4,5]. Such significant interest in various minerals at high pressure and extreme temperatures is caused by questions that usually arise from geology—how substances act in Earth's crust and mantle and how do they transform further when temperature and pressure decreases. To describe mineral behavior, many theoretical and experimental works have been done, resulting in a deep understanding of the formation of our and other planets [6]. Among more practical results, one should point out equations of state (EoS) many of which were originally developed for minerals and combined in special software [7–10] but now used frequently in different fields [11–13].

Construction and functional materials are also known for both precise and vast investigations [14–16]. One can easily understand that many of the materials we use every day are exposed to relatively high pressures and sometimes high temperatures and, for sure, should be studied for possible undesirable phase transitions. Another application of high pressures is discovering new forms of different elements and substances, which may exhibit new valuable properties [17–22]. Stabilizing new forms obtained at high pressures at ambient conditions is an ambitious aim for applied science and industry—graphite transforming to diamonds is the most popular example of this possible process. The importance of finding and stabilizing new inorganic and metal-organic phases that may arise at high pressure is evident. These materials are almost in every part of different devices and constructions surrounding us.

Nevertheless, new forms of organic substances are also in demand by industry [23–25]. Production of different forms of active pharmaceutical ingredients is a hot topic of crystallography of organic substances and can be used in practice [25–28]. Among others, one can find the application of high-pressure being a non-trivial way for obtaining new polymorphs of desired substances [29–32]. Several groups are doing extensive work looking over many organic substances for new phases at high pressure. Nowadays, there are more than 1000 structures at high pressure according to the CCDC database, and its number is growing steadily [33].

Thus, one can understand that a high-pressure phase transition study in molecular crystals is an important field of modern chemistry and as any advanced direction should be studied both experimentally and computationally.

2. The Main Research Directions

High-pressure phase transitions can be studied mainly using two different concepts:

1. Experimental techniques usually provided in diamond anvil cells (DAC) by X-Ray diffraction and Raman spectroscopy [1] but not limited to these methods [34,35]. Significant progress in engineering made these experiments possible and even routine in some sense, but still, very time consuming and complicated. These experiments give atomic coordinates and unit cell parameters of molecular crystal and rarely some information on atom/functional groups motion, which can be interpreted in terms of energetic characteristics.
2. Computational methods, which can be divided into DFT and Force Field (FF) groups, provide direct energetic characteristics of studied materials.

    a. FF methods are implemented in PIXEL [36,37] and CrystalExplorer17 [38–40] software for convenient work with molecular crystals. Very accurate parametrization [39] makes these methods being reliable and very fast in the estimation of lattice energies of organic crystals.
    b. DFT methods have many implementations for periodic systems, most popular are VASP [41–43], QuantumEspresso [44,45], CPMD [46], CASTEP [47] and CRYSTAL [48] codes, but researcher's choice is not limited to abovementioned software. Reasonable choice of level of theory (LOT), which combines many parameters, gives very accurate energies of investigated systems but requires much more resources and time in comparison to FF methods.

Here can be also mentioned many specific methods that can shed some light on different aspects of structures' nature or phase transitions, including electron/charge density analysis [49,50]. Nevertheless, these methods are relatively rarely used for molecular crystals at high-pressure conditions and have a specific discussion in literature [51,52], thus would not be described further in this work. We also do not describe molecular dynamics (MD) methods which are very useful [53–55], having not much experience in these methods. Modern machine learning and big data approaches are also not specified here, showing significant progress in the crystal structure and properties prediction, but not being applied to high-pressure phase transitions in molecular crystals [56–60].

Important to understand that both computational DFT and FF methods can barely be used without experiments being done at all because atom coordinates are needed for calculations [13,61]. Nevertheless, computational methods should be used when possible to understand and explain the nature of phase transitions of molecular crystals at high pressure because crystallographic and spectroscopic data rarely gives unambiguous reasons for phase transitions and can be interpreted differently by different scientific groups [61–67].

2.1. Force Field Methods

Force Field methods represent the functional form and parameter sets used to calculate the potential energy of a system of atoms or coarse-grained particles in molecular mechanics and molecular dynamics simulations. They are known widely due to their implementation in molecular dynamics for different systems, including biological ones [68,69]. Much less known but still frequently used "static"

FF methods can evaluate lattice energies and some properties of molecular crystals [70–74]. The first successful attempt (accepted by the scientific community) was made by A. Gavezzotti, showing AA-CLP method for very fast evaluation of lattice energy and enthalpy of molecular crystals, based on atomic charges [36]. In this work [36], it has been postulated that *"Structural papers with discussions of crystal packing involving user-selected atom–atom distances or geometries only should no longer be allowed in the literature"* and gave the right direction of further research on molecular crystals and their forms. Later this method was substituted by more precise and sophisticated PIXEL [37], which used electron density for energy evaluation and Gaussian software [75] as a backend for this type of calculations. The energy of pair-wise interactions was described as a sum of different terms ($E_{tot} = E_{ele} + E_{pol} + E_{dis} + E_{rep}$), which are given in analytical form [37]. Summation of these pair-wise interactions over a cluster of molecules in the crystal structure of radius equal to 20–50 A^3 (depending on the system) gives total lattice energy (E_{tot}). This leads to a very fast estimation of energies and, what is more important, to the "chemical" partition of different energy terms. Definitely, this very clear energy concept together with DFT was applied to high-pressure research to find the reasons for phase transition at extreme conditions [67,76,77]. One of the pioneer works in this direction was done by Wood et al. [67] who combined experimental study with extensive computational work, where both DFT and FF methods were used. Relatively good convergence with DFT methods has been shown, taking into account some limitations applied to DFT optimization procedure. Moreover, energies of structures of L-serine polymorphs were reasonable and close for both DFT and PIXEL calculations. The most valuable feature is that the energy of intermolecular interactions can be monitored for experimental structures at different pressures, and it is possible to divide into chemically sensible terms and visualize using new CE17 implementation [40]. Another important feature is the possibility to monitor the energy of any pair-wise interaction in the crystal structure and thus follow phase transitions at high pressure at a molecular level. If one needs to understand deeply what the reason for a phase transition is and why some interactions or H-bonds are broken and new appeared, energy calculations are mandatory and should be done if atom coordinates are known for different crystal phases at high pressures.

Nevertheless, one should understand that despite FF methods are very fast and relatively easy to use, they should be checked carefully for specific tasks, one of which is high-pressure research. Energy is written as a summation of different terms with coefficients $E_{tot} = k_{ele}E_{ele} + k_{pol}E_{pol} + k_{disp}E_{disp} + k_{rep}E_{rep}$, where k_i is parametrized on DFT energies or experimental enthalpies (depending on implementation) for molecular crystals at *ambient* conditions. H-bond length is also parametrized for ambient condition calculations, and cannot be switched off due to the very low precision of X-ray experimental measurements in defining hydrogen bond lengths. Summing up, an extensive benchmark on high-pressure data is needed to use FF methods unambiguously for high-pressure research in molecular crystals or should be critically checked by DFT calculations.

2.2. DFT Methods

In contrast to FF methods, DFT calculations for systems with periodic boundary conditions prove to give very accurate energies if multiple parameters are used correctly [78–82]. Discussing any phase transition, we assume that Gibbs energy change should be negative for spontaneous phase transition, which can be estimated from enthalpy and entropy terms. Entropy term is frequently neglected (especially if space group preserves after phase transition), supposing to have a small impact on Gibbs energy, while enthalpy can be calculated relatively easy [61,67,80,81,83]. Nevertheless, a lack of entropy calculations can lead to significant mistakes in the prediction of phase stability, which was shown for many inorganic materials [84,85]. The introduction of thermal effects in phase stability (DFT-QHA) increases computational costs drastically, but phonon effects are crucial to define accurate thermodynamic properties and Gibbs energies [86]. Calculation of all terms of Gibbs energy is strongly recommended to correctly predict stability and (subtle) phase transitions at high-pressure and temperature conditions [85,87,88]. Thus, the enthalpy diagram should be used as the first step of an investigation. Enthalpy is a sum of internal energy and PV term, where internal energy can be described as lattice energy—a sum of intermolecular and intramolecular (conformational) interactions. The thing that can be calculated using FF methods is intermolecular interactions only, while conformational (relaxation) energies should be calculated using DFT methods [80,81]. Constant-pressure (or, more precisely, fixed-stress) geometry optimization has to be carried out in DFT energy calculations to accurately define static pressure. Then, P-V-T EoS of the substance must be known via phonon dispersion calculations to define total pressure as a sum of static, zero-point and thermal contributions. Correct assessment of PV term at high-pressure and temperature conditions could be rather complex and expensive from a computational point of view and could strongly affect P-T location of phase transition boundaries [88]. Despite the abovementioned statements were supported mostly by examples from inorganic materials, we do not see any significant difference when these calculations applied to molecular crystals. Following Gibbs energy contributions can lead to a new understanding of reasons for phase transitions for any crystalline material. In some cases, it is possible to claim volume change (PV), energy (E_{inter} or E_{intra}), enthalpy (H) or entropy (S) as a driving force for phase transition [61,67,81,89]

DFT methods can be applied to experimental structures, obtained at some pressure, and all the above-mentioned parameters can be calculated. If one needs thermodynamic parameters for crystal structure at a pressure where no experimental data available, there are two main possibilities. The first is to calculate the equation of state for the phase and estimate the volume of the system, which will be fixed during ionic relaxation (structure optimization) to obtain energies. It is important to note that all calculations, in this case, should be performed for fixed unit cell volume. This technique is dependent on the accuracy of the equation of state, which needs at least 5–7 pressure-energy points within a pressure interval of a couple of GPa [90]. In this case, the quality of experimental data is very important because it influences the whole procedure of calculations. More convenient and independent from the number and quality of experimental data is the procedure with programmed pressure, e.g., as a starting structure can be taken one at ambient conditions and optimized to programmed pressure of, e.g., 3 GPa (e.g. PSTRESS option at VASP package). It was shown before that this kind of optimization to programmed pressure works accurately [61,91]. In this case, the volume is not fixed. In the case of DFT calculations enthalpy, the sum of inter- and intramolecular interactions ($U_{crystal}$), entropy and PV term are obtained, which may give unambiguous answers with reasons for a pressure-induced phase transition. It is recommended to verify computational parameters (functional, dispersion correction scheme, k-points, and E_{cutoff}, etc.) on the structure, which is used as a guess structure for further calculations, usually ambient-pressure experimental data. One of the most important parameters is dispersion correction, which is crucial for accurate simulation of molecular crystals at ambient and non-ambient conditions. Nowadays, highly sophisticated density functionals are available to account for dispersive interactions in DFT (e.g., DFT-D3, B3LYP-D*, HF-3c, DFT-TS, M06, etc.) [92–99].

In relation to FF methods no "chemical" (electrostatic, polarization, dispersion, repulsion) terms are obtained in case of DFT calculations, but in common case gives more reliable energies. The advantages and disadvantages of DFT and FF methods applied to high-pressure researches are summarized in Table 1.

Table 1. Advantages and disadvantages of DFT and FF methods applied for high-pressure research of organic substances.

Method	High-Pressure Experimental Data is Mandatory	Time and Resource-Consuming	Energy Estimation Accuracy	"Chemical" Energy Terms	Intermolecular Energies (E_{inter})	Intramolecular Energies (E_{intra})
FF	yes	no	lower	yes	yes	no[2]
DFT	yes/no[1]	yes	higher	no	yes	yes

[1] not mandatory if programmed pressure is used during optimization procedure but preferable for selection of parameters of optimization [61]. [2] Not implemented in PIXEL and CE17 software.

2.3. Combined Techniques

Recently we suggested the scheme where different phases are simulated not only in the intervals of their stability according to experimental data but also *out of these intervals* [61], and used it later for another system [91]. To the best of our knowledge, these are the first examples of such calculations for organic systems, studying intrinsic vs. extrinsic phase stability of polymorphs for molecular crystals. Nevertheless, it is widely used for inorganic systems, e.g., [85,100]. This helps to understand what would happen to the polymorph structure if no phase transition occurs. Such simulation provides structural and energy data to direct comparison of all phases at the same conditions, even if some phases are not found at these pressures experimentally (Figure 1).

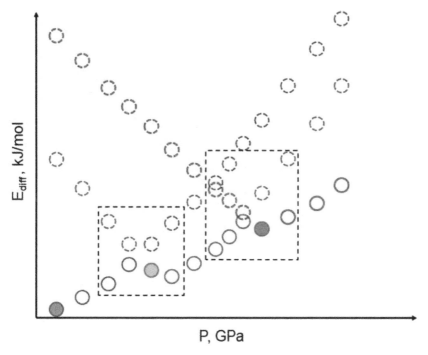

Figure 1. Schematic representation of calculated energy parameters ($\Delta U_{crystal}$, ΔH, etc.) for different phases in the whole pressure range, even at pressures where a specific phase is not found experimentally. Solid dots filled with color, starting experimental structure for DFT optimization; solid empty dots, DFT optimized structure at fixed pressure as observed experimentally; dashed dots, DFT optimized structure at fixed pressure out of experimental pressure range. Blue, phase I; green, phase II; red, phase III. Dashed squares show pressure-energy conditions close to high-pressure phase transitions.

In previous work, we also simulated pair-wise interactions using DFT gas-phase calculations with fixed geometries after solid state optimization. This simulation shows energy changes when distance decreases at pressure, building energy well for each H-bond [61]. This kind of energy well simulation is definitely not perfect due to the absence of molecular conformation change because of the molecular surrounding. This possibly could be also simulated using FF methods as described before and was done in other works [67,80,81,101].

Summing up, we suggest that the following scheme should be applied for high-pressure phase transition research of molecular crystals (Scheme 1).

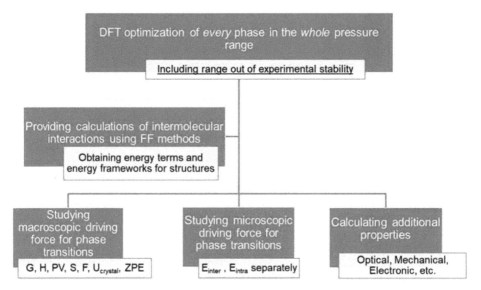

Scheme 1. A suggested scheme for the computational study of high-pressure phase transitions of molecular crystals. All steps are calculated for structures both in and out of structural stability regions.

Finally, we would like to point out that scrupulous checks for the validity of all parameters and methods should be done for every particular system. If this is done correctly, one can assume *this computational scheme is a numerical experiment* aimed at understanding the reasons and mechanisms of phase transitions of molecular crystals at high pressure.

In this review, we put aside many questions related to high-pressure polymorphism of organic compounds: kinetic barriers for phase transitions which play an important role [102–105], temperature corrections for DFT methods [106–108], prediction of new phases (which in future can significantly decrease amount of experimental work) [109–111], etc. Nevertheless, as it is shown above, current computational methods can bring out many more answers than pure experimental results.

3. Prospects

To the best of our knowledge, there is no specific method or software for computationally studying high-pressure phase transitions in molecular crystals. Still, wise usage of already developed methods and their implementations with appropriate validation of different parameters can unveil reasons and even some mechanisms for phase transitions. We hope that in the future all experimental works be complemented with computational techniques. This may bring out true reasons and a deep understanding of already observed phase transitions and the prediction of new polymorphs. Finally, taking into account progress in crystal structure prediction [109,112], one can expect the prediction of new structures of molecular crystals at extreme conditions. Nevertheless, we are sure that the concept shown in this work gives the most convenient and all-round scheme for high-pressure research of organic compounds using computational methods and can be improved by obtaining different more specific properties: mechanical, electronic, optical, etc. Finally, we do not see any issues that can prevent usage of the proposed scheme for metal-organic and inorganic materials if proper parameters of the simulation are chosen.

Funding: Funding for this research was provided by Russian Science Foundation grant No. 18-73-00154.

Acknowledgments: The Siberian Branch of the Russian Academy of Sciences (SB RAS) Siberian Supercomputer Center is gratefully acknowledged for providing supercomputer facilities (http://www.sscc. icmmg.nsc.ru). The author also acknowledges the Supercomputing Center of the Novosibirsk State University (http:// nusc.nsu.ru) for provided computational resources. ISSCM SB RAS is acknowledged for free access to the literature used for this review.

Conflicts of Interest: The author declares no conflict of interest.

References

1. Katrusiak, A. High-pressure crystallography. *Acta Crystallogr. Sect. A Found. Crystallogr.* **2008**, *64*, 135–148. [CrossRef] [PubMed]
2. Mao, H.-K.; Chen, B.; Chen, J.; Li, K.; Lin, J.-F.; Yang, W.; Zheng, H. Recent advances in high-pressure science and technology. *Matter Radiat. Extrem.* **2016**, *1*, 59–75. [CrossRef]
3. Tse, J.S. A Chemical Perspective on High Pressure Crystal Structures and Properties. *Natl. Sci. Rev.* **2019**, *1*, 53. [CrossRef]
4. Chen, J.; Wang, Y.; Duffy, T.S.; Shen, G.; Dobrzhinetskaya, L.F. (Eds.) *Advances in High-Pressure Technology Geophysical Application*; Elsevier: Amsterdam, The Netherlands, 2005; ISBN 9780444519795.
5. Gonzalez-Platas, J.; Rodriguez-Hernandez, P.; Muñoz, A.; Rodríguez-Mendoza, U.R.; Nénert, G.; Errandonea, D. A High-Pressure Investigation of the Synthetic Analogue of Chalcomenite, $CuSeO_3 \cdot 2H_2O$. *Crystals* **2019**, *9*, 643. [CrossRef]
6. Oganov, A.R.; Price, G.D.; Scandolo, S. Ab initio theory of planetary materials. *Zeitschrift für Krist. Cryst. Mater.* **2005**, *220*, 531–548. [CrossRef]
7. Angel, R.J. Equations of State. *Rev. Mineral. Geochem.* **2000**, *41*, 35–59. [CrossRef]
8. Angel, R.J. High-Pressure Structural Phase Transitions. *Rev. Mineral. Geochem.* **2000**, *39*, 85–104. [CrossRef]
9. Angel, R.J.; Alvaro, M.; Gonzalez-Platas, J. EosFit7c and a Fortran module (library) for equation of state calculations. *Zeitschrift für Krist.Cryst. Mater.* **2014**, *229*, 405–419. [CrossRef]
10. Gonzalez-Platas, J.; Alvaro, M.; Nestola, F.; Angel, R. EosFit7-GUI: A new graphical user interface for equation of state calculations, analyses and teaching. *J. Appl. Crystallogr.* **2016**, *49*, 1377–1382. [CrossRef]
11. Bull, C.L.; Playford, H.Y.; Knight, K.S.; Marshall, W.G.; Stenning, G.B.G.; Smith, R.I.; Hart, Z. New insights into the phase diagram of a magnetic perovskite, LaCo 1/3 Mn 2/3 O 3. *J. Phys. Condens. Matter* **2015**, *27*, 165401. [CrossRef]
12. Errandonea, D.; Muñoz, A.; Gonzalez-Platas, J. Comment on "High-pressure x-ray diffraction study of YBO 3 /Eu 3+, GdBO 3, and EuBO 3: Pressure-induced amorphization in GdBO 3". *J. Appl. Phys.* **2014**, *115*, 216101. [CrossRef]
13. Hunter, S.; Sutinen, T.; Parker, S.F.; Morrison, C.A.; Williamson, D.M.; Thompson, S.; Gould, P.J.; Pulham, C.R. Experimental and DFT-D Studies of the Molecular Organic Energetic Material RDX. *J. Phys. Chem. C* **2013**, *117*, 8062–8071. [CrossRef]
14. Sharma, S.M.; Garg, N. Material Studies at High Pressure. In *Materials Under Extreme Conditions*; Elsevier: Amsterdam, The Netherlands, 2017; pp. 1–47. ISBN 9780128014424.
15. Millar, D.I.A. *Energetic Materials at Extreme Conditions*; Springer Berlin Heidelberg: Berlin, Heidelberg, 2012; ISBN 978-3-642-23131-5.
16. Dubrovinsky, L.; Dubrovinskaia, N.; Bykova, E.; Bykov, M.; Prakapenka, V.; Prescher, C.; Glazyrin, K.; Liermann, H.-P.; Hanfland, M.; Ekholm, M.; et al. The most incompressible metal osmium at static pressures above 750 gigapascals. *Nature* **2015**, *525*, 226–229. [CrossRef] [PubMed]
17. Gao, G.; Oganov, A.R.; Ma, Y.; Wang, H.; Li, P.; Li, Y.; Iitaka, T.; Zou, G. Dissociation of methane under high pressure. *J. Chem. Phys.* **2010**, *133*, 144508. [CrossRef]
18. Zhang, W.; Oganov, A.R.; Goncharov, A.F.; Zhu, Q.; Boulfelfel, S.E.; Lyakhov, A.O.; Stavrou, E.; Somayazulu, M.; Prakapenka, V.B.; Konopkova, Z. Unexpected Stable Stoichiometries of Sodium Chlorides. *Science* **2013**, *342*, 1502–1505. [CrossRef] [PubMed]
19. Zentkova, M.; Mihalik, M. The Effect of Pressure on Magnetic Properties of Prussian Blue Analogues. *Crystals* **2019**, *9*, 112. [CrossRef]
20. Manjón, F.J.; Sans, J.A.; Ibáñez, J.; Pereira, A.L.J. Pressure-Induced Phase Transitions in Sesquioxides. *Crystals* **2019**, *9*, 630. [CrossRef]

21. Yang, X.; Wang, X.; Wang, Y.; Li, K.; Zheng, H. From Molecules to Carbon Materials—High Pressure Induced Polymerization and Bonding Mechanisms of Unsaturated Compounds. *Crystals* **2019**, *9*, 490. [CrossRef]

22. Cazorla, C.; MacLeod, S.G.; Errandonea, D.; Munro, K.A.; McMahon, M.I.; Popescu, C. Thallium under extreme compression. *J. Phys. Condens. Matter* **2016**, *28*, 445401. [CrossRef]

23. Barrio, M.; Maccaroni, E.; Rietveld, I.B.; Malpezzi, L.; Masciocchi, N.; Céolin, R.; Tamarit, J.-L. Pressure-temperature state diagram for the phase relationships between benfluorex hydrochloride forms I and II: A case of enantiotropic behavior. *J. Pharm. Sci.* **2012**, *101*, 1073–1078. [CrossRef]

24. Neumann, M.A.; van de Streek, J.; Fabbiani, F.P.A.; Hidber, P.; Grassmann, O. Combined crystal structure prediction and high-pressure crystallization in rational pharmaceutical polymorph screening. *Nat. Commun.* **2015**, *6*, 7793. [CrossRef]

25. Fabbiani, F.P.A.; Pulham, C.R. High-pressure studies of pharmaceutical compounds and energetic materials. *Chem. Soc. Rev.* **2006**, *35*, 932. [CrossRef] [PubMed]

26. Bernstein, J. Polymorphism—A Perspective. *Cryst. Growth Des.* **2011**, *11*, 632–650. [CrossRef]

27. Bučar, D.-K.; Lancaster, R.W.; Bernstein, J. Disappearing Polymorphs Revisited. *Angew. Chemie Int. Ed.* **2015**, *54*, 6972–6993. [CrossRef]

28. Konieczny, K.; Ciesielski, A.; Bąkowicz, J.; Galica, T.; Turowska-Tyrk, I. Structural Transformations in Crystals Induced by Radiation and Pressure. Part 7. Molecular and Crystal Geometries as Factors Deciding about Photochemical Reactivity under Ambient and High Pressures. *Crystals* **2018**, *8*, 299. [CrossRef]

29. Bernstein, J. *Polymorphism in Molecular Crystals*; Oxford University Press: New York, NY, USA, 2002; Volume 14, ISBN 978-0-19-923656-5.

30. Boldyreva, E.V.; Shakhtshneider, T.P.; Ahsbahs, H.; Sowa, H.; Uchtmann, H. Effect of high pressure on the polymorphs of paracetamol. *J. Therm. Anal. Calori.* **2002**, *68*, 437–452.

31. Seryotkin, Y.V.; Drebushchak, T.N.; Boldyreva, E.V. A high-pressure polymorph of chlorpropamide formed on hydrostatic compression of the α-form in saturated ethanol solution. *Acta Crystallogr. Sect. B Struct. Sci. Cryst. Eng. Mater.* **2013**, *69*, 77–85. [CrossRef]

32. Görbitz, C.H. Crystal structures of amino acids: from bond lengths in glycine to metal complexes and high-pressure polymorphs. *Crystallogr. Rev.* **2015**, *21*, 160–212. [CrossRef]

33. Groom, C.R.; Bruno, I.J.; Lightfoot, M.P.; Ward, S.C. The Cambridge Structural Database. *Acta Crystallogr. Sect. B Struct. Sci. Cryst. Eng. Mater.* **2016**, *72*, 171–179. [CrossRef]

34. Anzellini, S.; Errandonea, D.; Cazorla, C.; MacLeod, S.; Monteseguro, V.; Boccato, S.; Bandiello, E.; Anichtchenko, D.D.; Popescu, C.; Beavers, C.M. Thermal equation of state of ruthenium characterized by resistively heated diamond anvil cell. *Sci. Rep.* **2019**, *9*, 14459. [CrossRef]

35. Knudson, M.D.; Desjarlais, M.P.; Dolan, D.H. Shock-Wave Exploration of the High-Pressure Phases of Carbon. *Science* **2008**, *322*, 1822–1825. [CrossRef] [PubMed]

36. Gavezzotti, A. Efficient computer modeling of organic materials. The atom–atom, Coulomb–London–Pauli (AA-CLP) model for intermolecular electrostatic-polarization, dispersion and repulsion energies. *New J. Chem.* **2011**, *35*, 1360. [CrossRef]

37. Gavezzotti, A. Calculation of lattice energies of organic crystals: the PIXEL integration method in comparison with more traditional methods. *Zeitschrift für Krist. Cryst. Mater.* **2005**, *220*, 499–510. [CrossRef]

38. Edwards, A.J.; Mackenzie, C.F.; Spackman, P.R.; Jayatilaka, D.; Spackman, M.A. Intermolecular interactions in molecular crystals: What's in a name? *Faraday Discuss.* **2017**, *203*, 93–112. [CrossRef]

39. Thomas, S.P.; Spackman, P.R.; Jayatilaka, D.; Spackman, M.A. Accurate Lattice Energies for Molecular Crystals from Experimental Crystal Structures. *J. Chem. Theory Comput.* **2018**, *14*, 1614–1623. [CrossRef]

40. Mackenzie, C.F.; Spackman, P.R.; Jayatilaka, D.; Spackman, M.A. CrystalExplorer model energies and energy frameworks: extension to metal coordination compounds, organic salts, solvates and open-shell systems. *IUCrJ* **2017**, *4*, 575–587. [CrossRef]

41. Kresse, G.; Furthmüller, J. Efficient iterative schemes for ab initio total-energy calculations using a plane-wave basis set. *Phys. Rev. B* **1996**, *54*, 11169–11186. [CrossRef]

42. Kresse, G.; Furthmüller, J. Efficiency of ab-initio total energy calculations for metals and semiconductors using a plane-wave basis set. *Comput. Mater. Sci.* **1996**, *6*, 15–50. [CrossRef]

43. Kresse, G.; Joubert, D. From ultrasoft pseudopotentials to the projector augmented-wave method. *Phys. Rev. B* **1999**, *59*, 1758–1775. [CrossRef]

44. Giannozzi, P.; Baroni, S.; Bonini, N.; Calandra, M.; Car, R.; Cavazzoni, C.; Ceresoli, D.; Chiarotti, G.L.; Cococcioni, M.; Dabo, I.; et al. QUANTUM ESPRESSO: A modular and open-source software project for quantum simulations of materials. *J. Phys. Condens. Matter* **2009**, *21*, 395502. [CrossRef]
45. Giannozzi, P.; Andreussi, O.; Brumme, T.; Bunau, O.; Buongiorno Nardelli, M.; Calandra, M.; Car, R.; Cavazzoni, C.; Ceresoli, D.; Cococcioni, M.; et al. Advanced capabilities for materials modelling with Quantum ESPRESSO. *J. Phys. Condens. Matter* **2017**, *29*, 465901. [CrossRef] [PubMed]
46. CPMD.org. Available online: http://www.cpmd.org/ (accessed on 7 October 2015).
47. Clark, S.J.; Segall, M.D.; Pickard, C.J.; Hasnip, P.J.; Probert, M.I.J.; Refson, K.; Payne, M.C. First principles methods using CASTEP. *Zeitschrift für Krist.* **2005**, *220*, 567–570. [CrossRef]
48. Dovesi, R.; Erba, A.; Orlando, R.; Zicovich-Wilson, C.M.; Civalleri, B.; Maschio, L.; Rérat, M.; Casassa, S.; Baima, J.; Salustro, S.; et al. Quantum-mechanical condensed matter simulations with CRYSTAL. *Wiley Interdiscip. Rev. Comput. Mol. Sci.* **2018**, *8*, e1360. [CrossRef]
49. Macchi, P.; Casati, N.; Kleppe, A.; Jephcoat, A. Electron density of molecular crystals at high pressure from synchrotron data. *Acta Crystallogr. Sect. A Found. Adv.* **2014**, *70*, C1340. [CrossRef]
50. Casati, N.; Kleppe, A.; Jephcoat, A.P.; Macchi, P. Putting pressure on aromaticity along with in situ experimental electron density of a molecular crystal. *Nat. Commun.* **2016**, *7*, 10901. [CrossRef] [PubMed]
51. Spackman, M.A. How Reliable Are Intermolecular Interaction Energies Estimated from Topological Analysis of Experimental Electron Densities? *Cryst. Growth Des.* **2015**, *15*, 5624–5628. [CrossRef]
52. Kumar, P.; Cabaj, M.K.; Dominiak, P.M. Intermolecular Interactions in Ionic Crystals of Nucleobase Chlorides—Combining Topological Analysis of Electron Densities with Energies of Electrostatic Interactions. *Crystals* **2019**, *9*, 668. [CrossRef]
53. Mahesta, R.; Mochizuki, K. Stepwise Homogeneous Melting of Benzene Phase I at High Pressure. *Crystals* **2019**, *9*, 279. [CrossRef]
54. Nemkevich, A.; Bürgi, H.-B.; Spackman, M.A.; Corry, B. Molecular dynamics simulations of structure and dynamics of organic molecular crystals. *Phys. Chem. Chem. Phys.* **2010**, *12*, 14916. [CrossRef]
55. Iyer, S.; Slagg, N. *Energetic Materials*; Boddu, V., Redner, P., Eds.; CRC Press: Boca Raton, FL, USA, 2010; Volume 2, ISBN 978-1-4398-3513-5.
56. Musil, F.; De, S.; Yang, J.; Campbell, J.E.; Day, G.M.; Ceriotti, M. Machine learning for the structure-energy-property landscapes of molecular crystals. *Chem. Sci.* **2018**, *9*, 1289–1300. [CrossRef]
57. Wicker, J.G.P.; Cooper, R.I. Will it crystallise? Predicting crystallinity of molecular materials. *CrystEngComm* **2015**, *17*, 1927–1934. [CrossRef]
58. Wicker, J.G.P.; Crowley, L.M.; Robshaw, O.; Little, E.J.; Stokes, S.P.; Cooper, R.I.; Lawrence, S.E. Will they co-crystallize? *CrystEngComm* **2017**, *19*, 5336–5340. [CrossRef]
59. Podryabinkin, E.V.; Tikhonov, E.V.; Shapeev, A.V.; Oganov, A.R. Accelerating crystal structure prediction by machine-learning interatomic potentials with active learning. *Phys. Rev. B* **2019**, *99*, 064114. [CrossRef]
60. Evans, J.D.; Coudert, F.-X. Predicting the Mechanical Properties of Zeolite Frameworks by Machine Learning. *Chem. Mater.* **2017**, *29*, 7833–7839. [CrossRef]
61. Rychkov, D.A.; Stare, J.; Boldyreva, E.V. Pressure-driven phase transition mechanisms revealed by quantum chemistry: l-serine polymorphs. *Phys. Chem. Chem. Phys.* **2017**, *19*, 6671–6676. [CrossRef]
62. Kolesnik, E.N.; Goryainov, S.V.; Boldyreva, E.V. Different behavior of L- and DL-serine crystals at high pressures: Phase transitions in L-serine and stability of the DL-serine structure. *Dokl. Phys. Chem.* **2005**, *404*, 169–172. [CrossRef]
63. Moggach, S.A.; Allan, D.R.; Morrison, C.A.; Parsons, S.; Sawyer, L. Effect of pressure on the crystal structure of L-serine-I and the crystal structure of L-serine-II at 5.4 GPa. *Acta Crystallogr. Sect. B Struct. Sci.* **2005**, *61*, 58–68. [CrossRef]
64. Drebushchak, T.N.; Sowa, H.; Seryotkin, Y.V.; Boldyreva, E.V.; Ahsbahs, H. L-Serine III at 8.0 GPa. *Acta Crystallogr. Sect. E Struct. Rep. Online* **2006**, *62*, 4052–4054. [CrossRef]
65. Boldyreva, E.V.; Sowa, H.; Seryotkin, Y.V.; Drebushchak, T.N.; Ahsbahs, H.; Chernyshev, V.; Dmitriev, V. Pressure-induced phase transitions in crystalline l-serine studied by single-crystal and high-resolution powder X-ray diffraction. *Chem. Phys. Lett.* **2006**, *429*, 474–478. [CrossRef]
66. Moggach, S.A.; Marshall, W.G.; Parsons, S. High-pressure neutron diffraction study of L-serine-I and L-serine-II, and the structure of L-serine-III at 8.1 GPa. *Acta Crystallogr. Sect. B Struct. Sci.* **2006**, *62*, 815–825. [CrossRef]

67. Wood, P.A.; Francis, D.; Marshall, W.G.; Moggach, S.A.; Parsons, S.; Pidcock, E.; Rohl, A.L. A study of the high-pressure polymorphs of L-serine using ab initio structures and PIXEL calculations. *CrystEngComm* **2008**, *10*, 1154. [CrossRef]

68. Antila, H.S.; Salonen, E. *Biomolecular Simulations*; Monticelli, L., Salonen, E., Eds.; Methods in Molecular Biology; Humana Press: Totowa, NJ, USA, 2013; Volume 924, ISBN 978-1-62703-016-8.

69. Kukol, A. (Ed.) *Molecular Modeling of Proteins*; Methods in Molecular Biology; Springer New York: New York, NY, USA, 2015; Volume 1215, ISBN 978-1-4939-1464-7.

70. Jensen, F. *Introduction to Computational Chemistry*, 2nd ed.; Wiley: Hoboken, NJ, USA, 2007; Volume 2, ISBN 9780470058046.

71. Turner, M.J.; Thomas, S.P.; Shi, M.W.; Jayatilaka, D.; Spackman, M.A. Energy frameworks: Insights into interaction anisotropy and the mechanical properties of molecular crystals. *Chem. Commun.* **2015**, *51*, 3735–3738. [CrossRef]

72. Rychkov, D.; Arkhipov, S.; Boldyreva, E. Structure-forming units of amino acid maleates. Case study of L-valinium hydrogen maleate. *Acta Crystallogr. Sect. B Struct. Sci. Cryst. Eng. Mater.* **2016**, *72*, 160–163. [CrossRef]

73. Binns, J.; Parsons, S.; McIntyre, G.J. Accurate hydrogen parameters for the amino acid l -leucine. Acta Crystallogr. *Sect. B Struct. Sci. Cryst. Eng. Mater.* **2016**, *72*, 885–892. [CrossRef]

74. Moggach, S.A.; Marshall, W.G.; Rogers, D.M.; Parsons, S. How focussing on hydrogen bonding interactions in amino acids can miss the bigger picture: A high-pressure neutron powder diffraction study of ε-glycine. *CrystEngComm* **2015**, *17*, 5315–5328. [CrossRef]

75. Frisch, M.J.; Trucks, G.W.; Schlegel, H.B.; Scuseria, G.E.; Robb, M.A.; Cheeseman, J.R.; Scalmani, G.; Barone, V.; Mennucci, B. Gaussian 09. Expanding the limits of computational chemistry. 2009. Available online: https://gaussian.com/g09citation/ (accessed on 30 January 2020).

76. Johnstone, R.D.L.; Lennie, A.R.; Parker, S.F.; Parsons, S.; Pidcock, E.; Richardson, P.R.; Warren, J.E.; Wood, P.A. High-pressure polymorphism in salicylamide. *CrystEngComm* **2010**, *12*, 1065. [CrossRef]

77. Munday, L.B.; Chung, P.W.; Rice, B.M.; Solares, S.D. Simulations of High-Pressure Phases in RDX. *J. Phys. Chem. B* **2011**, *115*, 4378–4386. [CrossRef]

78. Dolgonos, G.A.; Hoja, J.; Boese, A.D. Revised values for the X23 benchmark set of molecular crystals. *Phys. Chem. Chem. Phys.* **2019**, *21*, 24333–24344. [CrossRef]

79. Colmenero, F. Mechanical properties of anhydrous oxalic acid and oxalic acid dihydrate. *Phys. Chem. Chem. Phys.* **2019**, *21*, 2673–2690. [CrossRef]

80. Giordano, N.; Beavers, C.M.; Kamenev, K.V.; Marshall, W.G.; Moggach, S.A.; Patterson, S.D.; Teat, S.J.; Warren, J.E.; Wood, P.A.; Parsons, S. High-pressure polymorphism in l-threonine between ambient pressure and 22 GPa. *CrystEngComm* **2019**, *21*, 4444–4456. [CrossRef]

81. Giordano, N.; Afanasjevs, S.; Beavers, C.M.; Hobday, C.L.; Kamenev, K.V.; O'Bannon, E.F.; Ruiz-Fuertes, J.; Teat, S.J.; Valiente, R.; Parsons, S. The Effect of Pressure on Halogen Bonding in 4-Iodobenzonitrile. *Molecules* **2019**, *24*, 2018. [CrossRef]

82. Abraham, B.M.; Vaitheeswaran, G. First principles study of pressure induced polymorphic phase transition in trimethylamine. In Proceedings of the AIP Conference Proceedings. *AIP Conf. Proc.* **2018**, *1942*, 030006.

83. Wen, X.D.; Hoffmann, R.; Ashcroft, N.W. Benzene under high pressure: A story of molecular crystals transforming to saturated networks, with a possible intermediate metallic phase. *J. Am. Chem. Soc.* **2011**, *133*, 9023–9035. [CrossRef]

84. Zucchini, A.; Prencipe, M.; Belmonte, D.; Comodi, P. Ab initio study of the dolomite to dolomite-II high-pressure phase transition. *Eur. J. Mineral.* **2017**, *29*, 227–238. [CrossRef]

85. Belmonte, D.; Ottonello, G.; Zuccolini, M.V. Ab initio-assisted assessment of the CaO-SiO$_2$ system under pressure. *Calphad* **2017**, *59*, 12–30. [CrossRef]

86. Belmonte, D.; Gatti, C.; Ottonello, G.; Richet, P.; Zuccolini, M.V. Ab initio thermodynamic and thermophysical properties of sodium metasilicate, Na$_2$SiO$_3$, and their electron-density and electron-pair-density counterparts. *J. Phys. Chem. A* **2016**, *120*, 8881–8895. [CrossRef] [PubMed]

87. Wdowik, U.D. Structural stability and thermal properties of BeO from the quasiharmonic approximation. *J. Phys. Condens. Matter* **2010**, *22*, 045404. [CrossRef] [PubMed]

88. Belmonte, D. First Principles Thermodynamics of Minerals at HP–HT Conditions: MgO as a Prototypical Material. *Minerals* **2017**, *7*, 183. [CrossRef]

89. Moses Abraham, B.; Adivaiah, B.; Vaitheeswaran, G. Microscopic origin of pressure-induced phase-transitions in urea: A detailed investigation through first principles calculations. *Phys. Chem. Chem. Phys.* **2019**, *21*, 884–900. [CrossRef]

90. Katrusiak, A.; McMillan, P.F. *High-Pressure Crystallography*; Katrusiak, A., McMillan, P., Eds.; Springer: Dordrecht, The Netherlands, 2004.

91. Fedorov, A.Y.; Rychkov, D.A.; Losev, E.A.; Zakharov, B.A.; Stare, J.; Boldyreva, E.V. Effect of pressure on two polymorphs of tolazamide: Why no interconversion? *CrystEngComm* **2017**, *19*, 2243–2252. [CrossRef]

92. Grimme, S.; Hansen, A.; Brandenburg, J.G.; Bannwarth, C. Dispersion-Corrected Mean-Field Electronic Structure Methods. *Chem. Rev.* **2016**, *116*, 5105–5154. [CrossRef] [PubMed]

93. Brandenburg, J.G.; Grimme, S. Dispersion Corrected Hartree–Fock and Density Functional Theory for Organic Crystal Structure Prediction. In *Peptide-Based Materials*; Springer: Dordrecht, The Netherlands, 2013; Volume 310, pp. 1–23.

94. Klimeš, J.; Michaelides, A. Perspective: Advances and challenges in treating van der Waals dispersion forces in density functional theory. *J. Chem. Phys.* **2012**, *137*, 120901. [CrossRef]

95. Zhao, Y.; Truhlar, D.G. The M06 suite of density functionals for main group thermochemistry, thermochemical kinetics, noncovalent interactions, excited states, and transition elements: Two new functionals and systematic testing of four M06-class functionals and 12 other function. *Theor. Chem. Acc.* **2008**, *120*, 215–241. [CrossRef]

96. Grimme, S.; Ehrlich, S.; Goerigk, L. Effect of the damping function in dispersion corrected density functional theory. *J. Comput. Chem.* **2011**, *32*, 1456–1465. [CrossRef] [PubMed]

97. Kronik, L.; Tkatchenko, A. Understanding Molecular Crystals with Dispersion-Inclusive Density Functional Theory: Pairwise Corrections and Beyond. *Acc. Chem. Res.* **2014**, *47*, 3208–3216. [CrossRef] [PubMed]

98. Brandenburg, J.G.; Grimme, S. Accurate Modeling of Organic Molecular Crystals by Dispersion-Corrected Density Functional Tight Binding (DFTB). *J. Phys. Chem. Lett.* **2014**, *5*, 1785–1789. [CrossRef]

99. Marom, N.; DiStasio, R.A.; Atalla, V.; Levchenko, S.; Reilly, A.M.; Chelikowsky, J.R.; Leiserowitz, L.; Tkatchenko, A. Many-Body Dispersion Interactions in Molecular Crystal Polymorphism. *Angew. Chemie Int. Ed.* **2013**, *52*, 6629–6632. [CrossRef]

100. Jahn, S. High-pressure phase transitions in $MgSiO_3$ orthoenstatite studied by atomistic computer simulation. *Am. Mineral.* **2008**, *93*, 528–532. [CrossRef]

101. Giordano, N.; Beavers, C.M.; Campbell, B.J.; Eigner, V.; Gregoryanz, E.; Marshall, W.G.; Peña-Álvarez, M.; Teat, S.J.; Vennari, C.E.; Parsons, S. High-pressure polymorphism in pyridine. *IUCrJ* **2020**, *7*, 58–70. [CrossRef]

102. Boldyreva, E. High-Pressure Polymorphs of Molecular Solids: When Are They Formed, and When Are They Not? *Some Examples of the Role of Kinetic Control. Cryst. Growth Des.* **2007**, *7*, 1662–1668.

103. Fisch, M.; Lanza, A.; Boldyreva, E.; Macchi, P.; Casati, N. Kinetic Control of High-Pressure Solid-State Phase Transitions: A Case Study on l-Serine. *J. Phys. Chem. C* **2015**, *119*, 18611–18617. [CrossRef]

104. Bhardwaj, P.; Singh, S. Pressure induced structural phase transitions—A review. *Open Chem.* **2012**, *10*, 1391–1422. [CrossRef]

105. Han, Y.; Liu, J.; Huang, L.; He, X.; Li, J. Predicting the phase diagram of solid carbon dioxide at high pressure from first principles. *npj Quantum Mater.* **2019**, *4*, 10. [CrossRef]

106. Smith, J.C.; Sagredo, F.; Burke, K. Warming Up Density Functional Theory. In *Frontiers of Quantum Chemistry*; Springer: Singapore, 2018.

107. Majewski, A.R.; Billman, C.R.; Cheng, H.-P.; Sullivan, N.S. DFT Calculations of Temperature-Dependent NQR Parameters in alpha-paradichlorobenzene and beta-HMX. *arXiv* **2019**, arXiv:1903.10097.

108. Hickel, T.; Grabowski, B.; Körmann, F.; Neugebauer, J. Advancing density functional theory to finite temperatures: Methods and applications in steel design. *J. Phys. Condens. Matter* **2012**, *24*, 053202. [CrossRef]

109. Reilly, A.M.; Cooper, R.I.; Adjiman, C.S.; Bhattacharya, S.; Boese, A.D.; Brandenburg, J.G.; Bygrave, P.J.; Bylsma, R.; Campbell, J.E.; Car, R.; et al. Report on the sixth blind test of organic crystal structure prediction methods. Acta Crystallogr. Sect. B Struct. *Sci. Cryst. Eng. Mater.* **2016**, *72*, 439–459.

110. Iuzzolino, L.; McCabe, P.; Price, S.L.; Brandenburg, J.G. Crystal structure prediction of flexible pharmaceutical-like molecules: Density functional tight-binding as an intermediate optimisation method and for free energy estimation. *Faraday Discuss.* **2018**, *211*, 275–296. [CrossRef]

111. Price, S.L.; Reutzel-Edens, S.M. The potential of computed crystal energy landscapes to aid solid-form development. Drug Discov. *Today* **2016**, *21*, 912–923.

112. Zurek, E.; Grochala, W. Predicting crystal structures and properties of matter under extreme conditions via quantum mechanics: The pressure is on. Phys. *Chem. Chem. Phys.* **2015**, *17*, 2917–2934. [CrossRef]

Review

Pressure-Tuned Interactions in Frustrated Magnets: Pathway to Quantum Spin Liquids?

Tobias Biesner * and Ece Uykur

1. Physikalisches Institut, Universität Stuttgart, Pfaffenwaldring 57, 70550 Stuttgart, Germany;
ece.uykur@pi1.physik.uni-stuttgart.de
* Correspondence: tobias.biesner@pi1.physik.uni-stuttgart.de

Received: 30 October 2019; Accepted: 21 November 2019; Published: 18 December 2019

Abstract: Quantum spin liquids are prime examples of strongly entangled phases of matter with unconventional exotic excitations. Here, strong quantum fluctuations prohibit the freezing of the spin system. On the other hand, frustrated magnets, the proper platforms to search for the quantum spin liquid candidates, still show a magnetic ground state in most of the cases. Pressure is an effective tuning parameter of structural properties and electronic correlations. Nevertheless, the ability to influence the magnetic phases should not be forgotten. We review experimental progress in the field of pressure-tuned magnetic interactions in candidate systems. Elaborating on the possibility of tuned quantum phase transitions, we further show that chemical or external pressure is a suitable parameter in these exotic states of matter.

Keywords: quantum spin liquids; frustrated magnets; quantum phase transitions; high-pressure measurements

1. Introduction

Quantum spin liquids (QSLs) possess nontrivial ground states, where a local order parameter does not exist. Moreover, it is not possible to observe spontaneous symmetry breaking even at very low temperatures. It is often thought that the QSLs are associated with topological phase transitions [1,2]. This make these systems a point of interest, and experimental evidences of this state are one of the central topics in the condensed matter physics.

QSLs are discussed in the framework of strongly correlated electron systems, while they are Mott insulators with half-filled electronic bands, and the electron–electron correlations play an important role. Possessing rich physics and properties, QSLs are subject to extensive experimental and theoretical efforts. This also bring the search for candidate materials, especially in 2D or 3D. The geometrically frustrated materials, where the resonating valance bond (RVB) model [3] is applicable, and the Kitaev QSL candidates, where the Kitaev physics [4] is relevant, are two groups of materials of which the candidates are searched for.

Technically speaking, it is difficult to identify the QSL state, as one needs to reach absolute zero temperature that is not achievable. Therefore, within the experimentally reachable limits, temperatures far below (2–3 orders of magnitude) the temperature that identify the magnetic exchange coupling (preferably antiferromagnetic (AFM) spin interactions) are assumed to show properties at the zero-temperature limits. The first step is to deduce the magnetic exchange coupling constant from the high-temperature behavior of the material via magnetic susceptibility measurements. To identify a QSL state, it is crucial to verify that there is no magnetic ordering or spin freezing down to very low temperatures. Magnetic susceptibility measurements are used initially to check the condition, where the absence (or existence) of a magnetic ordering can be identified. Absence of a sharp λ-type peak in specific heat vs. temperature curves is another indication of the absence of a magnetic ordering, albeit exceptions exist in the case of topological phase transitions [5]. Specific heat is a useful probe;

while it can give insight to the absence of a long-range magnetic order, examination of the entropy release at the measured temperature can also help to estimate the possibility of the system to establish a long-range magnetic order at low temperatures. Beside the above mentioned macroscopic probes, more local probes such as muon spin relaxation and nuclear magnetic resonance are usually in play to detect the possible spin freezing or order. Neutron diffraction is also used to detect magnetic ordering.

While the absence of the long-range magnetic ordering is the first step to check, it is still not very satisfactory to establish a system as a QSL, while in principle, disorder effects can also give rise to such ground states without a long-range magnetic ordering. Another aspect that defines the QSL state is the fractional spin excitations, which might also be the key point to identifying this state more confidently. For instance, spinons predicted within the RVB model are worth seeking. They are fermionic quasiparticles carrying fractional spins with their own dispersion expected to give low-lying excitations and can be eventually used to identify the QSL state.

This review will mainly focus on the inorganic systems of the frustrated lattices, such as pyrochlore, triangular, honeycomb, and kagome compounds. Several review articles are already written on this topic [6–11], while this particular one aims to bring together the published works and to present the ongoing discussion of the QSL state emerging under high pressure. However, owing to the fact that the QSLs are Mott insulators and in light of theoretical proposals [12] that they are the parent states of the high-temperature superconductivity, organic conductors also are promising candidates to search for. Moreover, external pressure has already been successfully used to tune these systems to a superconducting state, albeit the high-temperature superconductivity could not be achieved. Although, it is still experimentally challenging to prove whether the Mott insulator ground state is QSL, the lack of magnetic ground states have been reported by nuclear magnetic resonance (NMR) measurements for several organic charge transfer salts. The QSL state and its evolution with external and/or chemical pressure have been discussed extensively in several review articles [13,14].

In this review, we would like to discuss the search of QSL state from another perspective, from a rather indirect route. Within the search of QSL candidates, many others also come into light that eventually are proven to be not a QSL. On the other hand, they already are very close to the conditions in which are searched for the realization of the QSL state. Perhaps a fine tuning in certain parameters, such as magnetic exchange interactions, lattice parameters, etc., can push these closer to the QSL state. Here, we look into the external pressure as this tuning mechanism. Pressure is generally accepted to be a clean tuning parameter of structural properties and electronic correlations. While it can be compared to the chemical doping effects in some cases, it allows one to eliminate the additional disorder introduced via chemical doping. In this review, we want to discuss the recent progress of the high-pressure studies, especially on the frustrated magnets, systems that are often studied in the search of the QSL candidates. We want to focus on the following questions: How does the pressure affect the frustrated systems? Which phases can we tune? Can we tune magnetic interactions directly? How do chemical and external pressure differ? Finally, can we use external pressure as a pathway to realize QSLs?

2. Pyrochlore Lattice

The pyrochlore lattice is a prime example for frustrated magnetism in three dimensions. While classical spin ice states are realized in $Ho_2Ti_2O_7$ and $Dy_2Ti_2O_7$ [15,16], $Yb_2Ti_2O_7$ is a candidate system for a quantum spin ice ground state. Here, the magnetic monopoles, obeying an ice rule, become long-range entangled. Necessary conditions for such a quantum mechanical state are small spin quantum numbers and quantum fluctuations within the degenerated ground state manifold. $Yb_2Ti_2O_7$ hosts a minimal $S = 1/2$ spin of the crystal field Kramers doublet (Yb_3^{+11}). Strong quantum fluctuations are mediated by anisotropic exchange interactions and an XY g-tensor [17]. Considering the magnetic ground state, a sample-dependency possibly induced by disorder (e.g., excess magnetic ions in the stuffed pyrochlore lattice) might explain different reported results. In particular, ordered ferromagnetic ground states were reported for single-crystal samples, while polycrystalline powders showed no indication of a spin freezing. Kermarrec et al. [17] combined muon spin relaxation (μSR) and neutron diffraction measurements under pressure to explore the low-temperature ground state of $Yb_{2+x}Ti_{2-x}O_7$. Figure 1a shows a pressure-dependent phase diagram. Upon cooling, the paramagnetic state vanishes and most of the Yb magnetic moments were found to be in a fast fluctuating regime even down to low temperatures, reminiscent of a QSL state. Under hydrostatic pressure, pristine samples undergo a transition from this nonmagnetic ground state to a splayed ice-like ferromagnet (the magnetic moments are sketched in Figure 1a). Figure 1b shows the pressure dependency of a developing magnetic fraction upon cooling, leading to the magnetically ordered phase. By applying pressure a freezing of magnetic moments, increasing the magnetic fractions is observed. Additionally, the freezing temperature (defined as a fraction of 50% frozen out magnetic moments) increases with pressure. In contrast, in the stuffed compound ($x = 0.046$), no transition is observed up to the maximal pressure of 2.41 GPa. This study shows how fragile the balance of anisotropic exchange in the quantum spin ice Hamiltonian on the pyrochlore lattice can be against external or chemical pressure. Remarkably, the lattice structure is not expected to change noticeably in the low-pressure range. A structural phase transition is observed only above 29 GPa [18]. This illustrates that pressure is an effective tool to tune magnetic exchange interactions directly.

Chemical pressure on $Yb_2X_2O_7$ ($X = $ Sn, Ti, Ge) was probed on polycrystalline samples [19]. While the $X = $ Ti and Sn samples order into a ferromagnet at 0.13 K and 0.25 K, respectively, $Yb_2Ge_2O_7$ exhibits an antiferromagnetic ground state below 0.62 K. Different to the physical pressure, the lattice parameter increases for the Sn compound and decreases for the smaller Ge^{4+} ion. In general, a decreasing Curie–Weiss temperature was found with increasing lattice parameter.

Note that there are other high-pressure studies on the pyrochlore lattice ($A_2B_2O_7$, A = Eu, Dy; B = Ti, Zr) [20].

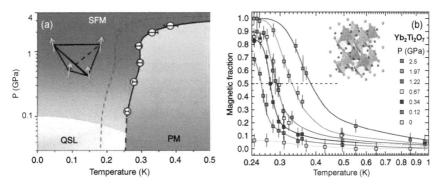

Figure 1. Results of μSR experiments on $Yb_{2+x}Ti_{2-x}O_7$: (**a**) The pressure-dependent phase diagram hosts three phases in the experimental accessed range. Under ambient pressure, the compound shows a ground state with large quantum fluctuations reminiscent of a quantum spin liquid (QSL) which freezes under pressure in favour of a canted ferromagnet. The dashed purple line marks the hypothetical transition for the stuffed compound $x = 0.046$. (**b**) As the ordered phase is approached, the magnetic fraction gets enhanced due to pressure. The freezing temperature (dashed line) is furthermore increased at the high-pressure side. Graphs are reproduced from Reference [17].

2.1. $Tb_2Ti_2O_7$

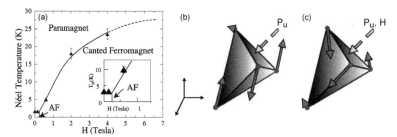

Figure 2. (**a**) The magnetic phase diagram of single crystalline $Tb_2Ti_2O_7$ under hydrostatic pressure $P_i = 2.4$ GPa and uniaxial stress $P_u = 0.3$ GPa: Under pressure, an ordered AFM ground state forms below T_N. With a magnetic field of 0.6 T, the AFM order gets lifted in favour of a canted ferromagnet. (**b**) Spin structure of Tb tetrahedron under pressure: The AFM structure (red arrows) is formed under a pressure of $P_i = 2.4$ GPa and uniaxial stress $P_u = 0.3$ GPa along the [011] axis (green arrow). (**c**) Canted ferromagnet for an additional field $H = 4$ T parallel to P_u: Blue spheres indicate Tb sites; principal axes of the cubic cell consisting of 4 Tb tetrahedra are indicated by black arrows. Graphs are reproduced from Reference [21].

The low-temperature ground state of the spin liquid compound $Tb_2Ti_2O_7$ has been extensively studied but remains intriguing. With an antiferromagnetic Curie–Weiss temperature of $\theta_{CW} = -13$ K (respectively -19 K, including crystal field contribution) and Ising-like 5 μ_b Tb^{3+} spins of the crystal-field doublet, the compound, however, shows no onset of static order down to at least 43 mK [22–26]. Interestingly, antiferromagnetic short-range correlations are set below 50 K [22]. Various suggestions to explain the missing Néel state have been given (see Reference [27] and references herein): a quantum spin ice state [28], structural distortion [29], or magnetoelastic excitations [30]. Mirebeau et al. extensively investigated the enigmatic nonmagnetic ground state by disturbing it by means of pressure and magnetic field [21,31–33]. Here, pressure is used to destabilize the balance of superexchange, crystal-field interactions, and dipolar coupling between neighbouring Tb^{3+} cations. With a decreasing lattice constant of 1% at 8.6 GPa and preserved $Fd\bar{3}m$ symmetry, the effect of pressure on the crystal structure was found to be rather small [31,34]. Neutron diffraction on polycrystalline samples under pressure reveals a complex antiferromagnetic structure below $T_N = 2.1$ K, coexisting with the spin liquid ground state. Bragg peaks of this antiferromagnetic structure are observed for a surprisingly low pressure of 1.5 GPa [31]. Interestingly, the Néel temperature seems to be insensitive to hydrostatic pressure but depends on applied strain [21,32]. Moreover, experiments on single crystals [21] show that a combination of isotropic and uniaxial pressures is crucial for disturbing the spin liquid ground state since hydrostatic pressure alone does not introduce a magnetic order in the single crystals.

Combining a magnetic field parallel to uniaxial stress of $P_u = 0.3$ GPa along the [011] direction and an isotropic pressure component of $P_i = 2.4$ GPa leads to suppression of the antiferromagnetic structure in favor of a canted ferromagnet (see the phase diagram in Figure 2a). Simultaneously, T_N gets increased under magnetic field. Mirebeau et al. argue that, while the dipolar interaction is only weakly affected by pressure, superexchange is strongly influenced. The effect of pressure on the spin structure of the Tb tetrahedron is shown in Figure 2b,c, with and without magnetic field parallel to P_u, respectively. Here, the isotropic compound increases exchange energy about $\frac{dJ}{JdP} = 0.07$ GPa^{-1}. A uniaxial strain along the [011] axis lifts the geometrical frustration by compressing 1/3 of Tb–Tb bonds of about 0.3% (for $P = 0.3$ GPa) and by decreasing the remaining 2/3 by about 0.1%. Due to this lifting of frustration, magnetic order is introduced. With increase of magnetic field, the spins are further reoriented in the canted ferromagnetic structure (Figure 2c).

Compared to physical pressure, the effect of chemical pressure on the ground state of $Tb_2Ti_2O_7$ is different [27,33,35]. A negative chemical pressure for the case of replacing titanium by the larger tin leads to lattice expansion. Despite antiferromagnetic interactions at higher temperatures leading to a Curie–Weiss temperature between −11 to −12 K, the compound shows a ferromagnetic contribution in 37% of the Tb^{3+} spins below a transition temperature of 0.87 K. Together with an antiferromagnetic "two in, two out" ice rule, these ferromagnetic domains form the magnetic ground state. Possibly, the compound orders because of the weakened antiferromagnetic exchange compared to the pure $Tb_2Ti_2O_7$ sample and because of a distortion of the local crystal field [33]. A positive chemical pressure is reached by substituting titanium with germanium, leading to a smaller lattice constant [27]. The contraction was found to be larger (2%) compared to a physical pressure of 8.6 GPa (1% cf. Reference [21]). The reduced Tb–Tb distance results in a stronger antiferromagnetic exchange as indicated by a higher Curie–Weiss temperature of −19.2 K compared to Ti and Sn compounds. Different from the physical counterpart, positive chemical pressure here induces short-ranged ferromagnetic correlations, coexisting with the liquid-like correlations, as observed by neutron scattering [27]. Similar to $Tb_2Ti_2O_7$, no long-range order is observed down to 20 mK.

2.2. Dichalcogenides

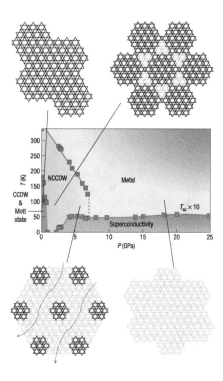

Figure 3. Phase diagram of 1T-TaS_2 under pressure: Pressure decreases the swelling of the planes related to the David-star pattern (sketched above and below the diagram). In the light grey areas, deformations are reduced or completely suppressed. Mott phase and CCDW state are suppressed over 0.8 GPa, as the NCCDW state stabilizes between 1–7 GPa. Here, hexagonal domains are formed (see sketch). As the pressure is further increased, the superconducting state develops (with a T_c of about 5 K). Superconductivity remains stable up to 25 GPa, with a metallic state above T_c. Graphs are reproduced from Reference [36].

Transition metal dichalcogenides (TMDs) attract great attention [37]. Due to the complex interplay of charge, spin, and orbital degrees of freedom, they display a rich phase diagram ranging from charge density waves (CDW), superconductivity, Mott physics, and possibly QSLs [38].

The two-dimensional 1T-TaS$_2$ exhibits multi charge density wave ground states [39], which can be continuously manipulated via external stimuli, such as temperature, chemical [40,41] or optical doping [39], disorder [42], and hydrostatic pressure [36,43]. Under ambient pressure, the compound shows metallic behavior in the high-temperature range. Below 550 K, an incommensurate charge density wave (ICCDW) superlattice develops. As resistivity increases, the ICCDW undergoes a transition to a nearly commensurate order (NCCDW) around 350 K. The low temperature range is governed by Mott physics in the commensurate charge density wave (CCDW) [36]. Thus, a simple model is the Mott–Hubbard Hamiltonian:

$$H = - \sum_{<i,j>,\sigma} t_{i,j} \left(c_{i\sigma}^{\dagger} c_{j\sigma} + h.c. \right) + U \sum_i n_i (n_i - 1). \tag{1}$$

The first term accounts for a hopping between two adjacent lattice sites $< i,j >$ and can be interpreted as kinetic energy. t is the transfer integral, which gives the hopping probability between two sites and is therefore proportional to the atomic overlap and effectively the bandwidth W. The creation and annihilation operators of an electron with spin σ at site i are written as $c_{i\sigma}$ and $c_{i\sigma}^{\dagger}$, respectively. The second term introduces an on-site Coulomb repulsion U if site i is fully occupied.

Here, pressure acts as a natural tuning parameter of the CDW states by affecting the transfer integral, t, and on-site Coulomb repulsion, U, without introducing chemical disorder. A phase diagram reproduced from Reference [36] is shown in Figure 3. With an increase of t and a decrease of U, Mott physics melt away at a pressure of about 1 GPa, giving rise to a transition to the NCCDW state, which persists up to 7 GPa, and finally entering to a metallic state. The high-pressure ground state (over 3 GPa) is superconducting (T_c of about 5 K) at least up to 25 GPa. While the Mott state CCDW clearly competes with the superconducting region [36,43], the coexistence of NCCDW and superconductivity is intriguing. At present, a macroscopic picture of superconductivity in the NCCDW phase is not fully clear. One proposed suggestion is that the superconducting phase forms within the metallic interdomain spaces of the CDW, which become connected as the CDW domains smear out under pressure [36]. On the other hand, an XRD study under pressure [43] suggests that the distance between the CDW domains decreases while domain boundaries remain sharp, meaning no interconnected metallic regions are formed. According to this picture, the whole NCCDW structure must form a single coherent superconducting phase.

By substituting sulfur by the isovalent selenium, a Mott-insulator-to-metal transition is observed [40,41]. Recent studies [41] show a melting of the Mott state CCDW due to the formation of a pseudogap, suggesting the importance of disorder and further inducing superconductivity [40]. The superconducting ground states is then suppressed in favor of a CCDW in the 1T-TaSe$_2$ compound.

In summary, 1T-TaS$_{2-x}$Se$_x$ is an interesting system to explore quantum phase transitions and various exotic states of matter. Especially, the possibility of a QSL ground state remains intriguing [38]. Optical infrared or Raman measurements at low temperatures and under pressure could give further information about the CDW phases.

3. Triangular Lattice

The highly frustrated triangular lattice hosts a rich phase diagram of magnetic phases and QSL ground states. Focusing on the case of a Heisenberg magnetic exchange, we want to show the tunability of magnetic phases on this lattice. The general model for an anisotropic Heisenberg antiferromanget on the triangular lattice can be written as

$$\mathcal{H} = J \sum_{<i,j>} \mathbf{S_i} \cdot \mathbf{S_j} + J' \sum_{<i,j'>} \mathbf{S_i} \cdot \mathbf{S_{j'}}, \tag{2}$$

where J and J' are the magnetic exchange interactions along the horizontal and diagonal bonds, respectively, and S_i, S_j, and $S_{j'}$ give the spin-1/2 operators at sites i, j, and j', respectively [44].

3.1. Cs_2CuCl_4

The two-dimensional Heisenberg magnet, Cs_2CuCl_4 [45], posses $S = 1/2$ spins with slightly different exchange values J and J' ($J'/J = 0.3$) along the b-direction (horizontal bonds) and for the interchain coupling (diagonal bonds), respectively [46]. Furthermore, Dzyaloshinskii–Moriya (DM), an interplane exchange interaction, was found to be important [47,48]. Below $T_N = 0.62$ K, an incommensurate spiral ground state is formed (DM spiral), with the spiral along the b-axis [49]. A magnetic field was shown to lift this confinement of the spins in the bc-plane in favour of a commensurate coplanar AFM ground state within the ab-plane [48]. Recently, the tunability of the spin Hamiltonian via external pressure and magnetic field was demonstrated, combining high-pressure electron spin resonance (ESR), radio frequency susceptibility measurements [44]. Here, Heisenberg exchange is continuously enhanced under pressure, leading to an increase of J'/J by 12% at 1.8 GPa, as determined by the ESR (cf. Figure 4a,b) Most importantly, the interchain coupling J' gets enhanced with increasing pressure. Due to the tuned exchange interactions, new phases are emerging under magnetic field (see Figure 4c,d). The magnetic field favors more classical phases; therefore, suppressing the DM spiral at about 2.2–2.6 T, a coplanar order is stabilized in the ab-plane. Due to the enhanced J', a non-coplanar frustrated phase becomes stable at around 6.9 T. The emerging magnetic anomalies at 9.2 and 9.8 T under a pressure of 1.8 GPa are interpreted as double-cone and single-cone order, respectively (see Reference [44] and references herein). Furthermore, the fully polarized ferromagnetic high-field phase is shifted from around 9 T at ambient pressure to 11.5 T at 1.8 GPa.

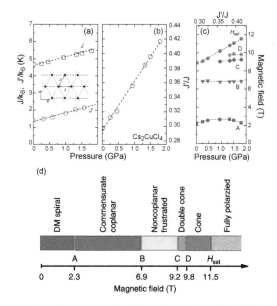

Figure 4. (**a,b**) Pressure-dependent magnetic exchange interactions J and J' along the horizontal and diagonal bonds, respectively (see inset of the crystallographic structure): As the pressure is decreased, the Heisenberg exchange is continuously altered. (**c**) Simultaneously increasing a magnetic field ($H \parallel b$) at $T = 350$ mK unveils five different magnetic phases beside the DM spiral. (**d**) Proposed phase diagram under pressure (1.8 GPa) and magnetic field. Graphs are reproduced from Reference [44].

3.2. YbMgGaO$_4$

YbMgGaO$_4$ [50] consists of $S = 1/2$ Yb^{3+} spins on triangular lattices, separated by nonmagnetic Mg^{2+} and Ga^{3+} ions (Figure 5a–c). The edge-sharing YbO$_6$ octahedra in $R\bar{3}m$ are characterized by equal Yb–O distances and two equal angles, the Yb–O–Yb bridging angle α and the O–Yb–O angle β. Structural randomness is induced by a random distribution of the Mg^{2+} and Ga^{3+} ions. The material became of recent interest as a QSL candidate, when an absence of magnetic order down to at least 50 mK was shown [51], although possibly a weak spin freezing takes place at around 100 mK. More interesting, a magnetic continuum at low temperatures is possibly related to gapless spinons or a nearest-neighbour RVB state (see the references in Reference [52]). Focusing on combined μSR, XRD, and DFT studies [52], we are going to review the pressure dependency of the intriguing ground state. XRD shows no changes of the crystal symmetry up to pressures as high as 10 GPa. However, the Yb–O distances are shrinked by about 0.6% at 2.6 GPa. Importantly, the angles α and β are weakly decreasing with a change of about 0.07° at 2.6 GPa and 0.2° at 10 GPa. Figure 5d shows the temperature dependency of the zero-field μSR spectra at ambient conditions and under pressures as high as 2.6 GPa. Of interest here is the increase of the zero-field muon relaxation rate below 4 K. This was interpreted as the onset of spin–spin correlations, which are fully developed around 0.8 K, leading to a dynamic spin state in YbMgGaO$_4$. Comparing the ambient and high-pressure data, the external pressure seems to have no effect on the development of such a ground state. While comparably strong pressure-induced structural changes have been proven in other compounds to have an effect on magnetic couplings, here, the structural randomness, which is not affected by pressure, seems to be crucial in stabilizing the QSL ground state.

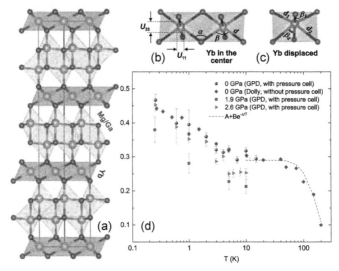

Figure 5. (**a**) Crystallographic structure of YbMgGaO$_4$: Edge-sharing YbO$_6$ octahedra are separated by slabs of nonmagnetic Mg^{2+} and Ga^{3+} ions. (**b**) YbO$_6$ octahedra with characteristic parameters. (**c**) Trigonal distortion of the YbO$_6$ octahedra induced by mixing of Mg^{2+} and Ga^{3+} sides: Importantly, no changes of the structural symmetry are observed under pressure. (**d**) The temperature dependence of the zero-field muon relaxation rate at different pressures shows the onset of spin–spin correlations below 4 K, fully developed at 0.8 K. Reproduced from Reference [52].

4. Honeycomb Lattice

Recently frustrated magnets on the honeycomb lattice have attracted attention [8,53–55]. Among them, the Kitaev systems are of particular interest because (a) they are highly relevant for theoretical considerations, as Kitaev's model is directly solvable, and (b) rich physics in terms of exotic quasiparticles, e.g., Majorana fermions [56,57]. Besides the iridate systems [55], like Li_2IrO_3 and Na_2IrO_3, α-$RuCl_3$ [58,59] has been extensively studied.

4.1. α-$RuCl_3$

α-$RuCl_3$ implements the $4d^5$ transition metal ruthenium (Ru^{3+}) with $\lambda_{Ru} \approx 0.15$ eV [60,61] on a nearly ideal honeycomb lattice. It renders an edge-sharing geometry with an octahedral cage of chlorine ions. Different polymorphs with an ABC-like stacking of the ab honeycomb layers along the c axis (α-$RuCl_3$), and one-dimensional face-sharing $RuCl_6$ chains (β-$RuCl_3$) are known. Regarding the structural symmetry, different space groups ($P3_112$, $R\bar{3}$, and $C2/m$) were reported (see the discussion in Reference [55]). This seems to be explained by a majority of samples showing stacking faults with a mixing of the ABC (rhombohedral $R\bar{3}$) and AB (monoclinic $C2/m$) stacking patterns. Due to an only weak van der Waals attraction between the layers, the system is quasi-two-dimensional in the ab-plane and susceptible to a mixing of the two stacking patterns [62,63].

Looking closer at the magnetic properties, a transition within the ab-plane from a paramagnetic phase, obeying a Curie–Weiss law with $\theta_{CW} \approx +40$ K and an effective magnetic moment of about 2.3 μ_B [64], to an antiferromagnetic zigzag order is observed [63]. As well, susceptibility measurements expose an anisotropy ($\chi_{ab} > \chi_c$) [65]. While the cleanest crystals with a minimum of stacking faults show a Néel temperature of about $T_N \approx 7$ to 8 K, T_N is found to increase up to 10 to 14 K with higher amount of ABC and AB mixing, indicating different order temperatures of these patterns [62,63].

Under ambient pressure, α-$RuCl_3$ is considered as a spin-orbit assisted $j_{eff} = 1/2$ Mott insulator with a Mott gap of about 1.1 eV [66–70]. Specific heat measurements [71] show a pressure sensitivity of the Néel transition and indicate a suppression of the magnetic order at around 0.7 GPa (in accordance with the theoretical, predicted coupling constants [60]). Resistivity experiments up to 140 GPa clearly prove the persistence of the insulating state [71]. Further NMR and magnetization studies demonstrated the vanishing magnetic order under pressure together with a strongly reduced susceptibility and the absence of the low-energy fluctuations [72]. The pressure-induced state was further investigated by optical infrared spectroscopy and *ab initio* DFT calculations [64] and magnetization measurements combined with high-resolution x-ray diffraction experiments [73].

Here, we want to focus on the spectroscopic study of Reference [64]. First, the phononic part of the optical spectrum with the pronounced 320 cm^{-1} mode was investigated to elaborate on a possible structural transition under pressure. The $T = 10$ K optical conductivity is shown in Figure 6a. In addition to the main mode (peak 1, which is assigned to a trio with mostly in-plane contributions at $\omega_0 \approx 321$, 322, and 326 cm^{-1}), a weaker out-of-plane mode is located at around 290 cm^{-1}. Generally, a hardening together with a broadening of the peaks under pressure is observed. Over 0.7 GPa, the main-mode peak 1 splits while another distinct resonance (peak 2) emerges. As the pressure is increased, both peaks further experience a hardening due to lattice contraction. Since phonon modes are generally strongly dependent on the lattice symmetry, the observed splitting is a direct evidence for a symmetry breaking over a pressure of 0.7 GPa.

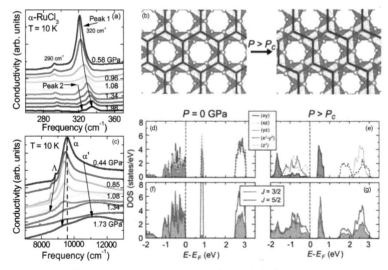

Figure 6. Pressure-dependent optical spectroscopy of α-RuCl₃ [64]: (**a**) Splitting of the phononic contributions (peak 1) is a direct evidence for a symmetry breaking over a pressure of 0.7 GPa. (**b**) The ambient pressure honeycomb (left-hand side) transforms to a dimerized structure under pressure (right-hand side). (**c**) The electronic ground state is studied by the pressure dependency of the Mott gap α. As the pressure increases, the gap gets suppressed, indicating a collapse of Kitaev magnetism and a breakdown of the j_{eff} picture. (**d–g**) Results of the GGA+SOC+U calculations, showing the breakdown of the j_{eff} picture (see text for further explanations).

DFT calculations suggest that parallel dimerization of neighbouring Ru sites set in at high pressure. The homogeneous ambient-pressure honeycomb structure transforms to a triclinic $P\bar{1}$ structure (Figure 6b). As a result, a pressure-driven structural transition at $P > 1$ GPa to a triclinic, dimerized structure is established by the spectroscopic experiment, well in accordance with the calculations.

To further understand the dimerized phase, the electronic part of the optical spectrum is investigated in the region of the optical gap α under pressure. The general suppression of the α peak strongly indicates a collapse of Kitaev interactions above 0.7 GPa. In this frame, a breakdown of the j_{eff} picture in accordance with the suppressed magnetic susceptibility above 0.7 GPa [72] is most probable. Note that, according to transport measurements, the suppression of the α peak is not interpreted as a closure of the optical gap. Instead, α-RuCl₃ stays well in the Mott insulation region at least up to 140 GPa [71].

The orbital-dependent density of states (DOS) was calculated including spin-orbit coupling and electronic correlations (GGA+SOC+U) with $U = 1.5$ eV giving insight to the observed optical excitations. The results are shown in Figure 6d–g. For the undimerized structure ($P = 0$ GPa, left-hand side panels), the relativistic j_{eff} picture is validated. In Figure 6d. the lower lying t_{2g} DOS with the expected orbital contributions is splitted from the higher e_g manifold and further gapped by SOC and U. Thus, the narrow peak at around 1 eV represents the single t_{2g} hole at each site residing in the upper Hubbard band. Projecting on the atomic J orbitals (Figure 6f) shows a mixture of the lower Hubbard band $J = 5/2$ ($j_{eff} = 1/2$) with the $J = 3/2$ ($j_{eff} = 3/2$) states, whereas the upper Hubbard band with mainly $J = 5/2$ contribution is clearly distinguishable. Here, virtual hopping through the ligands via hopping channel t_2 induces a strong ferromagnetic Kitaev exchange. In the high-pressure phase ($P > P_c$, right-hand side panels), the electronic density of states changes (cf. Figure 6e). The d_{xy} orbitals split into a bonding contribution which is lower in energy and an antibonding part at higher energy. This destroys the j_{eff}, states as can be seen in the atomic projection Figure 6g. At the former upper Hubbard band, now a pronounced mixture of $J = 3/2$ and $J = 5/2$ states emerges. With this breakdown of the j_{eff} picture, we can explain the vanishing magnetic ground state under pressures of around 1 GPa [71,72] as a result of the formation of pseudocovalent bonds in the dimerized structure. In the high-pressure phase, the direct Ru–Ru hopping path t_3 is enhanced along these bonds. This leads to a large antiferromagnetic Heisenberg exchange located on the dimers, further destabilizing and suppressing the magnetic low-temperature zigzag order. Consistently, the computed magnetic moments are completely suppressed in the high-pressure structure. Instead of increasing the t_2 channel towards a dominant Kitaev regime, hydrostatic pressure promotes the direct t_3 hopping. Results, therefore, conclude that the high-pressure nonmagnetic state of α-RuCl$_3$ is a valence-bond crystal and excludes a transition to a Kitaev QSL in the dimerized structure.

4.2. Iridates

The 5d honeycomb iridates A$_2$IrO$_3$ (A=Li, Na) [74,75] have gained attraction as Kitaev candidates [54,55,76,77] and were considered in terms of the Heisenberg–Kitaev model and with additional off-diagonal contribution. Due to strong spin-orbit coupling, they are located in the relativistic Mott insulating limit, hosting j_{eff} spins of the magnetic Ir ions [78,79]. Compared to α-RuCl$_3$ (4d vs. 5d electronic configuration), a stronger SOC in addition to a weaker Coulomb repulsion is generally expected. While in principle a band-insulating picture featuring quasimolecular orbitals is imaginable to explain the insulating nature of these systems [80], the Mott insulating picture is well backed by experiments and the majority of theoretical approaches. Additional trigonal crystal field splitting of the $j_{eff} = 3/2$ quartet can be sufficient high compared to SOC. This might induce a splitting of the SO-exciton and a mixing of the $j_{eff} = 3/2$ quartet with the $j_{eff} = 1/2$ doublet [78,81,82].

The three polymorphs of the Li-iridates are different in structure. While the α type is comparable to Na$_2$IrO$_3$ with a layered honeycomb lattice, the β variant shows a more three-dimensional hyper honeycomb. The γ type is characterized by a stripy-honeycomb structure. All types show an Ir–O–Ir edge-sharing geometry, allowing a nearly ideal 90° bond. The centered Ir^{4+} ion is coordinated by a cage of six O^{2-} ions. The Ir ions generate a 5d^5 electronic structure with a single t_{2g} hole [83].

4.2.1. α-Li$_2$IrO$_3$

The lattice structure of α-Li$_2$IrO$_3$ (Figure 7a) resembles a honeycomb of edge-sharing IrO$_6$ octahedra (Ir–O–Ir bond angles of around 95° with 5.7% difference in bond length) with a centered Li ion as the buffer element and shows a $C2/m$ monoclinic symmetry [84]. Magnetic susceptibility characterization indicates a Curie–Weiss temperature of −33 K, with an effective moment of $\mu_{eff} = 1.83(5)\ \mu_B$. A Néel transition at $T_N \approx 15$ K is further observed in accordance with specific heat measurements [76]. The resulting antiferromagnetic ground state is of an incommensurate counter-rotating type in the Ir plane [84].

X-ray diffraction studies together with DFT calculations [83] showed a structural phase transition under pressure at $P_c = 3.8$ GPa, from the monoclinic $C2/m$ to a dimerized triclinic $P\bar{1}$ structure. With an increasing Z1 bond, the dimerization is taking place either in the X1 or in the Y1 Ir–Ir bond of the honeycomb, resulting in two possible order patterns (cf. Figure 7b,c) in the high-pressure phase. Furthermore, it was elaborated that different properties balance, whether an energy gain by forming a magnetic order or a dimerization is higher. These are the size of the buffer ion and the electronic configuration of the metal species. The latter influences the strength of spin-orbit coupling, electronic correlations, and Hund's rule coupling. Larger interactions (SOC, U, and J_H) protect Kitaev physics by ensuring Ising-like spins and by inhibiting dimerization [83]. A larger buffer ion, in principle, inhibits dimerization by an additional hardening of the lattice. Thus, it was argued that α-Li$_2$IrO$_3$ is the intermediate case between Na$_2$IrO$_3$ and Li$_2$RuO$_3$. Indeed, Li$_2$RuO$_3$ already dimerizes at ambient conditions [85], whereas Na$_2$IrO$_3$ (discussed below) with a larger center ion size (Na vs. Li) is expected to dimerize at higher pressure.

Furthermore, studies using multiple X-ray techniques—X-ray powder diffraction (XRD) investigating the crystal structure, resonant inelastic X-ray scattering (RIXS) unveiling the electronic structure, and X-ray absorption spectroscopy (XAS) probing SOC under hydrostatic pressure together with DFT calculations—give additional insights [86]. At low pressures of about 0.1 GPa, the X-ray powder diffraction shows a gradual elongation of the honeycomb, where two long bonds (3.08 Å) and four short bonds (2.92 Å), still within the C2/m symmetry, are formed. The XAS data suggest a strongly decreasing SOC up to 1.1 GPa, saturating at around 2.8 GPa. Interestingly, the RIXS spectra in Figure 7d indicate a pressure dependence of the crystal field excitations. At low pressure, the SO-exciton (peak A) with corresponding energy of $\frac{3\lambda}{2} \approx 0.72$ eV is clearly identified. While this excitation should, in principle, be splitted due to reasonable trigonal crystal field splitting $\Delta_{Tr} \approx 0.11$ eV (see Figure 7e), a substructure is not resolved within the resolution of the setup. Under pressure, the corresponding intensity gets suppressed and peak A slightly shifts to lower energies. A new peak B develops at around 1.4 eV and gets intensified under pressure. This was interpreted as an increase of trigonal crystal field splitting over spin-orbit coupling. Consequently, Clancy et al. argued that the relativistic j_{eff} picture breaks down, even at low pressures of around 0.1 GPa, in favour of a localized pseudospin approach or an itinerant quasimolecular orbital (QMO) model (Figure 7f). At around 3 GPa, powder diffraction identified a first-order structural transition in accordance with Reference [83]. Here, a pronounced transfer of spectral weight from peak A to peak B is observed. Fitting of Peak B unveiled a two-peak structure related to the possible transitions to a QMO picture.

Note that there is a pressure-dependent optical study on this compound [87].

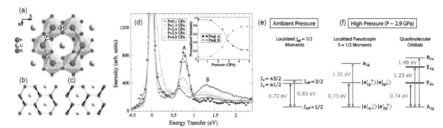

Figure 7. (**a**) Below P_c, the honeycomb of α-Li$_2$IrO$_3$ realizes symmetry-equivalent X1, Y1, and distinct Z1 bonds. (**b**,**c**) The high-pressure dimerized phase consists of two degenerate order patterns. Reproduced from Reference [83]. (**d**) Pressure-dependent resonant inelastic X-ray scattering (RIXS) on α-Li$_2$IrO$_3$ [86]. At ambient pressure, the SO-exciton (peak A) can be clearly identified according to the level structure of the j_{eff} model (**e**). At higher pressure (around 1.4 GPa), a second contribution (peak B) emerges, related to an enhancement of Δ_{Tr}. Below 2 GPa, the two peaks A and B can be fitted to the transitions of a pseudospin model or an itinerant quasimolecular orbital (QMO) state equivalently well. In the high-pressure phase, peak B shows a two-peak structure with a contribution at around 1.6 eV, interpreted as a transition from the j_{eff} picture to a QMO state (**f**).

4.2.2. β-Li$_2$IrO$_3$

The β-polymorph forms a hyper-honeycomb structure of edge-sharing IrO$_6$ octahedra with nearly identical Ir–O–Ir bonds (0.2% difference) and angles of around 94.5°. It extends Kitaev physics in three dimensions [88]. This relativistic Mott insulator has effective moments of $\mu_{eff} = 1.6(1)\ \mu_B$, and the magnetic susceptibility shows a positive Curie–Weiss temperature of 40 K, which may stem from ferromagnetic Kitaev couplings [89]. A transition to a noncollinear or incommensurate ground state at 38 K is seen under rather strong fluctuations. Furthermore, an unusual cusp in specific heat measurements [88] indicates a second-order transition. The low ferromagnetic Curie–Weiss temperature was interpreted as an effective cancellation of two competitive and nearly degenerate ferromagnetic (possibly Kitaev exchange) and antiferromagnetic ground states. These observations were interpreted as proximity to a Kitaev QSL [88]. Furthermore, Raman spectroscopy indicates signatures of fractionalized excitations [90], similar to α-RuCl$_3$.

A relative weak magnetic field of 3 T polarizes the compound with 0.35 μ_B/Ir [88] and induces strongly correlated ferromagnetic zigzag chains. Magnetic resonant X-ray scattering [91] shows a thermal driven crossover from a paramagnetic behavior in this quantum correlated (quantum paramagnetic) state. The field-induced moments were traced by X-ray magnetic circular dichroism (XMCD) [88] to be suppressed under pressure at 1 GPa and vanished over 2 GPa, while the compound remained insulating. Finally, this was interpreted as a rearrangement of the j_{eff} moments. By applying pressure without external magnetic field, the order temperature shifts first from around 38 K to 15 K [88]. Under further increase, β-Li$_2$IrO$_3$ undergoes an electronic/magnetic phase transition at 1.5 GPa, as observed by X-ray absorption near edge structure (XANES) measurements [92], without breaking the lattice symmetry. Probing the 5d holes, SOC was found to be reduced but remains important. *Ab initio* calculations indicate a dominant Dzyaloshinskii–Moriya regime under pressure, pushed away from the pure j_{eff} limit. However, the compound still remains in a relativistic Mott picture with an enlarged mixture between $j_{eff} = 3/2$ and $j_{eff} = 1/2$ states [88,93]. The new ground state remains intriguing [94]. Around 4 GPa further, a phase transition to a monoclinic C2/m symmetry was observed [92]. The compound dimerizes under a compression of X and Y Ir–Ir bonds, compared to the Z bonds.

4.2.3. γ−Li$_2$IrO$_3$

The zigzag chains in γ−Li$_2$IrO$_3$ host $j_{eff} = 1/2$ spins with $\mu_{eff} = 1.6\ \mu_B$ in a noncoplanar, counter-rotating pattern [95,96]. A transition temperature of about 38 K and a strongly anisotropic susceptibility, which rather does not allow a determined Curie–Weiss temperature, were observed. The underlying lattice is of Cccm symmetry. Equally to β-Li$_2$IrO$_3$, Raman spectroscopy shows signatures of fractionalized excitations [90]. Resonant X-ray scattering (RXS) measurements under pressure find an abrupt suppression of the spiral magnetic order at 1.4 GPa without indications of a changed lattice symmetry and point out a continuous reduction of the unit cell volume [97]. This non-magnetic pressure state remains of further interest.

4.2.4. Na$_2$IrO$_3$

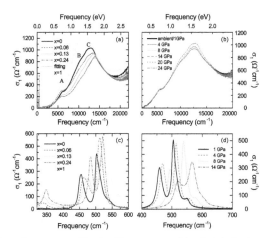

Figure 8. Infrared spectroscopy combined with pressure and isoelectric doping studies on Na$_2$IrO$_3$ of Reference [98]: The effect of Li doping ((Na$_{1-x}$Li$_x$)$_2$IrO$_3$ for $x \leq 0.24$ and $x = 1$ (represented by α-Li$_2$IrO$_3$)) on the electronic spectra (**a**) and phononic part (**c**) is shown. (**a**) While the SO-exciton (peak A) is only marginally affected by Li doping, the direct hopping between $j_{eff} = 1/2$ orbitals (peak B) gets suppressed. A blueshift of the intersite $j_{eff} = 3/2 \rightarrow j_{eff} = 1/2$ transition (peak C) for $x \leq 0.24$ Li doping indicates a increasing U_{eff}/t. (**c**) The phononic part shows a hardening upon increasing the Li doping. Further, the effect of hydrostatic pressure on the electronic spectra (**b**) and on the phononic part (**d**) is shown. (**b**) Under pressure U_{eff}/t is slightly lowered, indicated by a redshift of all features while the lattice is contracted (**d**).

The layered honeycomb of Na_2IrO_3 is similar to α-Li_2IrO_3, with edge-sharing IrO_6 octahedra and Na as buffer ion ($C2/m$ space group). Ir–O–Ir bond angles reach from 98° to 99.4° [99]. High-temperature moments of $\mu_{eff} = 1.79\ \mu_B$ are determined together with a Curie–Weiss temperature of -125 K [76]. Below 15–18 K, an antiferromagnetic zigzag order is observed with 0.22(1) μ_B/Ir [99,100]. Spin-wave excitations in this magnetic order were studied by inelastic neutron scattering experiments and compared with theoretical considerations [99], showing the importance of higher-order coupling contributions on the honeycomb lattice [101].

Optical studies established a mostly temperature-independent onset of a Mott gap at around 340 meV [102]. Electronic features below 3 eV are assigned to be transitions, belonging to the Ir 5d t_{2g} multiplets, and above mostly to charge transfer transition from O 2p to Ir 5d t_{2g}. A clear absorption edge is visible, resulting in an effective on-site Coulomb repulsion of about 1.5 eV. This seems to match the LDA+SOC+U calculated DOS with $U = 3$ eV and $J_H = 0.6$ eV [102]. The j_{eff} picture is clearly valid as pointed out by resonant inelastic X-ray scattering. These studies show a splitted but pronounced SO-exciton, concluding that trigonal distortions are weaker compared to a strong SOC (110 meV and 0.4–0.5 eV, respectively) [78].

Hydrostatic pressure and isoelectric doping studies on Na_2IrO_3 using infrared spectroscopy and synchrotron X-ray diffraction were performed by Hermann et al. [98]. The optical conductivity under ambient pressure for an as-grown sample is displayed in Figure 8a (black line). The d–d contributions peak A (0.7 eV), peak B (1.2 eV), and peak C (1.6 eV) (Figure 8a) are assigned as follows: intrasite $j_{eff} = 3/2 \rightarrow j_{eff} = 1/2$, intersite $j_{eff} = 1/2 \rightarrow j_{eff} = 1/2$, and intersite $j_{eff} = 3/2 \rightarrow j_{eff} = 1/2$ transitions, respectively, and thus probing the Mott insulating picture directly (peak A) and Kitaev correlations indirectly (peak B and C). The first intersite excitation peak B reveals a Mott gap with $U_{eff} = 1.2$ eV, in accordance with Reference [102]. On the low-energy side, phononic excitations contribute (Figure 8c). While the $C2/m$ symmetry hosts 18 infrared-active modes, five were resolved by the experiment in the undoped sample at ambient pressure.

The effect of Li doping, discriminating Na, with $(Na_{1-x}Li_x)_2IrO_3$ for $x \leq 0.24$ and $x = 1$ (represented by α-Li_2IrO_3) is analyzed in detail. Focusing on the electronic part (Figure 8a), the intrasite contribution (peak A) remains mostly stable upon doping, with only a slight redshift, while the intersite excitation (peak C) shifts to higher energies. This indicates an increasing ratio of U_{eff}/t, while SOC and a distortion of the crystal field remain only marginally affected. Further, it corresponds to a shift towards the Mott insulating side. According to theoretical predictions [103], the decreasing spectral weight of the $j_{eff} = 1/2 \rightarrow j_{eff} = 1/2$ intersite transition B with increasing amount of doping was related to enhanced Kitaev couplings due to a suppression of the direct Ir–Ir hopping channel. This emphasizes a proximity of the $x = 0.24$ compound to the Kitaev limit. Further X-ray measurements show that the chemical pressure upon Li doping only affects a contraction of the ab-plane. The c direction remains nearly constant because only in-plane Na sites are affected for a sufficient low doping concentration. This naturally tunes the Ir–O–Ir bond angle, crucially influencing Kitaev magnetism. However, α-Li_2IrO_3 is located not so deep in the Mott insulating state, as indicated by a redshifted absorption edge (Figure 8a), therefore resulting in a lower ratio of U_{eff}/t, backed by previous studies [60]. The similar position and shape of transition A prove a comparable SOC and distortion of the crystal field in Na_2IrO_3 and α-Li_2IrO_3 [55].

On the phononic part (Figure 8c), increasing the Li doping is expressed as a hardening of the in-plane modes due to compression of the ab-plane. In addition to this chemical pressure effect, the phonon modes are intrinsically affected by the contribution of Li. The observed modes, previously without any Na contribution, were simulated to have an increasing contribution of Li discriminating Ir. The lowest mode at around 350 cm^{-1} was found to be purely Li based, in accordance to the optical spectra.

Hermann et al. further compared the effect of chemical pressure to the physical hydrostatic one. The optical conductivity for pressurized samples is shown in Figure 8b,d. On the electronic part (Figure 8b), hydrostatic pressure over 8 GPa leads to a decreasing intensity of the absorption edge, while the intrasite transition (peak A) remains nearly unaffected. A slight redshift of all features upon increasing pressure is observed. U_{eff}/t is therefore only slightly adjusted between 8 GPa and 24 GPa. The nearly unchanged intrasite contribution indicates a smooth monotonic contraction without disturbing the crystal field symmetry. In addition, the phononic contributions (Figure 8d) were found to experience a monotonic hardening while the damping increases with pressure. Overall, no indications for a breaking of lattice symmetry are found. Further X-ray measurements show that, additionally to the *ab*-plane, the *c* direction is contracted upon pressurizing. This naturally explains the different affected intensities of peak B and peak C. While transition C $j_{eff} = 3/2 \rightarrow j_{eff} = 1/2$ is influenced by changes of the Ir–O–Ir bond angle, the direct hopping of peak B is nearly unchanged through the smooth contraction of the *ab*-plane. Thus, it was argued that external hydrostatic pressure, in contrast to the chemical counterpart, drives the compound away from the Kitaev limit.

5. Kagome Lattice

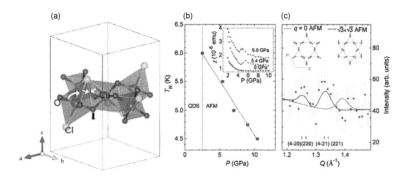

Figure 9. (**a**) Pressure effect on the crystal structure of Herbertsmithite. While the $R\overline{3}m$ symmetry is preserved under pressure, trigonal distortion is added by tilting the CuO_4-plane. (**b**) Phase diagram under pressure: At 2.1 GPa, a transition from a QSL to an AFM ordered phase takes place, which can be seen in the susceptibility data (inset). (**c**) Neutron diffraction pattern under pressure. The local spin structure under pressure resembles a $\sqrt{3} \times \sqrt{3}$ type. After Reference [104].

Herbertsmithite $ZnCu_3(OH)_6Cl_2$ is one of the most highlighted QSL model systems, crystallizing in the highly frustrated Kagome lattice (see Reference [105] and references herein). Cu Kagome planes ($S = 1/2$) are separated by Zn^{2+} ions realizing a highly frustrated system. Note that there is a prone to disorder induced by mixing of Cu and Zn ions [106]. Besides strong antiferromagnetic interaction ($J = 190$ K) due to the 120° Cu bonds, no signature of magnetic order was found down to the lowest temperatures of around 50 mK [107,108]. Furthermore, Dzyaloshinskii–Moriya interaction was found to be important [109]. Spin freezing was observed under magnetic field [110] and pressure [104], on which we want to discuss now in detail. Under pressure, the $R\overline{3}m$ symmetry is preserved up to at least 5 GPa. However, Cu–O–Cu bond angles are non-monotonically affected. First, for $P < 0.25$ GPa, a linear increase is observed, followed by a decrease up to 5.1 GPa. The Cu–O bond distance was found to decrease linearly at the low pressure side of $P < 0.25$ GPa and to be pressure independent above. Figure 9a shows the pressure effect on the crystal structure of Herbertsmithite. Most importantly, the CuO_4-plane tilts respectively to the Cl–Cu–Cl axis, inducing a trigonal distortion. Interestingly, the ratio of DM and Heisenberg interaction in Herbertsmithite is lowered under pressure. A quantum phase transition from the QSL ground state to an ordered AFM phase is observed at 2.5 GPa with $T_N = 6$ K as a peak in susceptibility measurements (Figure 9b). In the AFM ordered state, T_N is further decreased under pressure and explained by a decrease of Heisenberg interaction (15% from 2.5 GPa to 5.1 GPa). The AFM structure is of type $\sqrt{3} \times \sqrt{3}$ (see Figure 9c).

6. Spin 1/2 Dimer Systems

We also want to highlight pressure-dependent studies on the spin dimer system TlCuCl₃ [111,112]. This magnetic insulator host dimerized $S = 1/2$ moments of Cu^{2+} ions confined by strong AFM interaction. The formation of spin dimers leads to a quantum disordered phase at ambient pressure. Between the singlet ground state ($S = 0$) and the first excited triplet state ($S = 1$), there is a small gap of about 0.7 meV for spin excitations. Pressure [113–117], magnetic field [118], and impurity doping were shown to generate AFM order [119]. We are going to review the pressure-induced phase transition in detail. In a simple picture, the interdimer coupling can be increased by external pressure, closing the spin gap [120]. Rüegg et al. found a quantum phase transition (QCP) at $P = 0.107$ GPa and a power law increase of T_N in the AFM phase [116]. Spin dimer formation can be destroyed by the suppression of quantum fluctuations or the reduction of thermal fluctuations, both leading to magnetic order. Figure 10 summarizes both phase transitions as probed by inelastic neutron scattering [117]. First, we focus on the QPT (Figure 10a). As the pressure is increased, the spin gap is suppressed and finally closed with a temperature-dependent pressure p_c. In the ordered Néel state (right-hand side), two types of excitations are observed, namely the conventional Goldstone mode or spin wave (Figure 10a grey symbols) and, remarkably, the longitudinal Higgs mode (Figure 10a red symbols) [121]. Decreasing the pressure softens the Higgs mode, and finally as dimer-based quantum fluctuations destroy magnetic order, the system becomes gapped under p_c. While, transverse excitation remains gapped (0.38 meV at p_c) over the full pressure range, they can be well distinguished from the gapless longitudinal modes at p_c. Now, we want to focus on the classical phase transition (Figure 10b). At high temperatures, thermal fluctuations gap all observed modes. Having the temperature as a tuning parameter and a fixed pressure of 0.175 GPa, the ordered state emerges below T_N as the longitudinal mode becomes gapless at T_N. Here, we see that quantum and thermal melting of the ordered phase are affecting the neutron spectra in a qualitatively very similar way. Finally, a full phase diagram covering quantum critical and classical critical region is shown in Figure 10c.

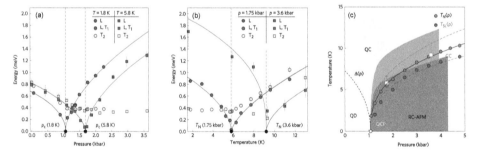

Figure 10. Quantum and classical phase transition from a dimer $S = 1/2$ state to magnetic order in TlCuCl₃ observed by inelastic neutron scattering: (**a,b**) Evolution of mode energies for pressure and temperature as tuning parameters. At the quantum critical point, QPT takes place: (**a**) Transverse magnetic modes (T) or Goldstone mode of the ordered phase remains gapped while the longitudinal Higgs mode (L) is gapless at the QCP with a temperature-dependent p_c. Taking temperature as the tuning parameter (**b**) and a fixed pressure, the ordered state emerges at a pressure-dependent T_N. A qualitatively similar evolution of the mode gaps for classical and quantum phase transition is observed. The results are summarized in the phase diagram (**c**) showing the quantum disordered state (QD) and induced magnetic phase (RC-AFM). Quantum critical and classical critical regions are indicated as QC, and CC, respectively. Grey spheres show the power-law behavior of $T_N(p)$, while blues symbols ($T_{SL}(p)$) denote the limit of classical scaling. Reproduced from Reference [117].

7. Summary

In summary, external pressure can be a very powerful tool to tune the electronic, magnetic, and structural parameters opening a new route in the investigations of QSLs and candidates. For instance, small perturbations to the crystal structure can drive the geometrical frustration factor towards a favourable state. More complicated but possible is a direct tuning of the exchange coupling (Cs_2CuCl_4). Moreover, unwanted magnetic interactions and fluctuations can be suppressed, leaving room for realization of a pure QSL state. However, it is often difficult to predict the influence of external pressure on magnetic properties of a candidate system. In fact, we have seen few systems where the pressure induces spin freezing rather than a liquid state (for instance, in $Tb_2Ti_2O_7$ or $Yb_2Ti_2O_7$) or an unfavourable modification of the crystal structure (α-$RuCl_3$). Promising are candidates where a magnetic order vanishes before the structural transition as in the case of the iridate systems. Therefore, there is no easy answer for whether external pressure is always a pathway to introduce spin liquid physics in the candidate systems. Albeit that the end results are unpredictable, exotic states of matter in the vicinity of the QSL state can be investigated.

Author Contributions: T.B. wrote the manuscript. T.B. and E.U. discussed the content. All authors have read and agreed to the published version of the manuscript.

Funding: This research was funded by Deutsche Forschungsgemeinschaft through grant No. DR228/52-1. and the "Margarete von Wrangell Habilitation Programm" by the Ministry of Sciences, Research, and Arts in Baden Württemberg.

Acknowledgments: We thank Weiwu Li, Seulki Roh, Andrej Pustogow, Artem Pronin, and Martin Dressel for fruitful discussions. T.B. acknowledges funding from Deutsche Forschungsgemeinschaft through grant No. DR228/52-1. E.U. acknowledges the support by "Margarete von Wrangell Habilitation Programm" by the Ministry of Sciences, Research, and Arts in Baden Württemberg.

Conflicts of Interest: The authors declare no conflict of interest.

References

1. Pesin, D.; Balents, L. Mott physics and band topology in materials with strong spin–orbit interaction. *Nat. Phys.* **2010**, *6*, 376–381. [CrossRef]
2. Wen, X.G. Colloquium: Zoo of quantum-topological phases of matter. *Rev. Mod. Phys.* **2017**, *89*, 041004. [CrossRef]
3. Anderson, P.W. Resonating valence bonds: A new kind of insulator? *Mater. Res. Bull.* **1973**, *8*, 153–160. [CrossRef]
4. Kitaev, A. Anyons in an exactly solved model and beyond. *Ann. Phys.* **2006**, *321*, 2–111. [CrossRef]
5. Nasu, J.; Udagawa, M.; Motome, Y. Vaporization of Kitaev Spin Liquids. *Phys. Rev. Lett.* **2014**, *113*, 197205. [CrossRef]
6. Savary, L.; Balents, L. Quantum spin liquids: A review. *Rep. Prog. Phys.* **2016**, *80*, 016502. [CrossRef]
7. Zhou, Y.; Kanoda, K.; Ng, T.K. Quantum spin liquid states. *Rev. Mod. Phys.* **2017**, *89*, 025003. [CrossRef]
8. Hermanns, M.; Kimchi, I.; Knolle, J. Physics of the Kitaev Model: Fractionalization, Dynamic Correlations, and Material Connections. *Annu. Rev. Condens. Matter Phys.* **2018**, *9*, 17–33. [CrossRef]
9. Takagi, H.; Takayama, T.; Jackeli, G.; Khaliullin, G.; Nagler, S.E. Concept and realization of Kitaev quantum spin liquids. *Nat. Rev. Phys.* **2019**, *1*, 264–280. [CrossRef]
10. Knolle, J.; Moessner, R. A Field Guide to Spin Liquids. *Annu. Rev. Condens. Matter Phys.* **2019**, *10*, 451–472. [CrossRef]
11. Wen, J.; Yu, S.L.; Li, S.; Yu, W.; Li, J.X. Experimental identification of quantum spin liquids. *Npj Quantum Mater.* **2019**, *4*, 12. [CrossRef]
12. Anderson, P.W. The Resonating Valence Bond State in La_2CuO_4 and Superconductivity. *Science* **1987**, *235*, 1196. [CrossRef] [PubMed]
13. Powell, B.J.; McKenzie, R.H. Quantum frustration in organic Mott insulators: From spin liquids to unconventional superconductors. *Rep. Prog. Phys.* **2011**, *74*, 056501. [CrossRef]
14. Dressel, M.; Pustogow, A. Electrodynamics of quantum spin liquids. *J. Phys. Condens. Matter* **2018**, *30*, 203001. [CrossRef] [PubMed]

15. Morris, D.J.P.; Tennant, D.A.; Grigera, S.A.; Klemke, B.; Castelnovo, C.; Moessner, R.; Czternasty, C.; Meoptissner, M.; Rule, K.C.; Hoffmann, J.U.; et al. Dirac Strings and Magnetic Monopoles in the Spin Ice $Dy_2Ti_2O_7$. *Science* **2009**, *326*, 411–414. [CrossRef] [PubMed]

16. Bramwell, S.T.; Gingras, M.J.P. Spin Ice State in Frustrated Magnetic Pyrochlore Materials. *Science* **2001**, *294*, 1495–1501. [CrossRef] [PubMed]

17. Kermarrec, E.; Gaudet, J.; Fritsch, K.; Khasanov, R.; Guguchia, Z.; Ritter, C.; Ross, K.A.; Dabkowska, H.A.; Gaulin, B.D. Ground state selection under pressure in the quantum pyrochlore magnet $Yb_2Ti_2O_7$. *Nat. Commun.* **2017**, *8*, 14810. [CrossRef]

18. Mishra, A.K.; Poswal, H.K.; Sharma, S.M.; Saha, S.; Muthu, D.V.S.; Singh, S.; Suryanarayanan, R.; Revcolevschi, A.; Sood, A.K. The study of pressure induced structural phase transition in spin-frustrated $Yb_2Ti_2O_7$ pyrochlore. *J. Appl. Phys.* **2012**, *111*, 033509. [CrossRef]

19. Dun, Z.L.; Lee, M.; Choi, E.S.; Hallas, A.M.; Wiebe, C.R.; Gardner, J.S.; Arrighi, E.; Freitas, R.S.; Arevalo-Lopez, A.M.; Attfield, J.P.; et al. Chemical pressure effects on magnetism in the quantum spin liquid candidates $Yb_2X_2O_7$ (X=Sn, Ti, Ge). *Phys. Rev. B* **2014**, *89*, 064401. [CrossRef]

20. Rittman, D.R.; Turner, K.M.; Park, S.; Fuentes, A.F.; Yan, J.; Ewing, R.C.; Mao, W.L. High-pressure behavior of $A_2B_2O_7$ pyrochlore (A=Eu, Dy; B=Ti, Zr). *J. Appl. Phys.* **2017**, *121*, 045902. [CrossRef]

21. Mirebeau, I.; Goncharenko, I.N.; Dhalenne, G.; Revcolevschi, A. Pressure and Field Induced Magnetic Order in the Spin Liquid $Tb_2Ti_2O_7$ as Studied by Single Crystal Neutron Diffraction. *Phys. Rev. Lett.* **2004**, *93*, 187204. [CrossRef] [PubMed]

22. Gardner, J.S.; Dunsiger, S.R.; Gaulin, B.D.; Gingras, M.J.P.; Greedan, J.E.; Kiefl, R.F.; Lumsden, M.D.; MacFarlane, W.A.; Raju, N.P.; Sonier, J.E.; et al. Cooperative Paramagnetism in the Geometrically Frustrated Pyrochlore Antiferromagnet $Tb_2Ti_2O_7$. *Phys. Rev. Lett.* **1999**, *82*, 1012–1015. [CrossRef]

23. Gingras, M.; den Hertog, B.; Faucher, M.; Gardner, J.; Dunsiger, S.; Chang, L.; Gaulin, B.; Raju, N.; Greedan, J. Thermodynamic and single-ion properties of Tb^{3+} within the collective paramagnetic-spin liquid state of the frustrated pyrochlore antiferromagnet $Tb_2Ti_2O_7$. *Phys. Rev. B* **2000**, *62*, 6496–6511. [CrossRef]

24. Gardner, J.S.; Keren, A.; Ehlers, G.; Stock, C.; Segal, E.; Roper, J.M.; Fåk, B.; Stone, M.B.; Hammar, P.R.; Reich, D.H.; et al. Dynamic frustrated magnetism in $Tb_2Ti_2O_7$ at 50 mK. *Phys. Rev. B* **2003**, *68*, 180401. [CrossRef]

25. Lhotel, E.; Paulsen, C.; de Réotier, P.D.; Yaouanc, A.; Marin, C.; Vanishri, S. Low-temperature magnetization in geometrically frustrated $Tb_2Ti_2O_7$. *Phys. Rev. B* **2012**, *86*, 020410. [CrossRef]

26. Legl, S.; Krey, C.; Dunsiger, S.R.; Dabkowska, H.A.; Rodriguez, J.A.; Luke, G.M.; Pfleiderer, C. Vibrating-Coil Magnetometry of the Spin Liquid Properties of $Tb_2Ti_2O_7$. *Phys. Rev. Lett.* **2012**, *109*, 047201. [CrossRef]

27. Hallas, A.M.; Cheng, J.G.; Arevalo-Lopez, A.M.; Silverstein, H.J.; Su, Y.; Sarte, P.M.; Zhou, H.D.; Choi, E.S.; Attfield, J.P.; Luke, G.M.; et al. Incipient Ferromagnetism in $Tb_2Ge_2O_7$: Application of Chemical Pressure to the Enigmatic Spin-Liquid Compound $Tb_2Ti_2O_7$. *Phys. Rev. Lett.* **2014**, *113*, 267205. [CrossRef]

28. Molavian, H.R.; Gingras, M.J.P.; Canals, B. Dynamically Induced Frustration as a Route to a Quantum Spin Ice State in $Tb_2Ti_2O_7$ via Virtual Crystal Field Excitations and Quantum Many-Body Effects. *Phys. Rev. Lett.* **2007**, *98*, 157204. [CrossRef]

29. Bonville, P.; Mirebeau, I.; Gukasov, A.; Petit, S.; Robert, J. Tetragonal distortion yielding a two-singlet spin liquid in pyrochlore $Tb_2Ti_2O_7$. *Phys. Rev. B* **2011**, *84*, 184409. [CrossRef]

30. Fennell, T.; Kenzelmann, M.; Roessli, B.; Mutka, H.; Ollivier, J.; Ruminy, M.; Stuhr, U.; Zaharko, O.; Bovo, L.; Cervellino, A.; et al. Magnetoelastic Excitations in the Pyrochlore Spin Liquid $Tb_2Ti_2O_7$. *Phys. Rev. Lett.* **2014**, *112*, 017203. [CrossRef]

31. Mirebeau, I.; Goncharenko, I.N.; Cadavez-Peres, P.; Bramwell, S.T.; Gingras, M.J.P.; Gardner, J.S. Pressure-induced crystallization of a spin liquid. *Nature* **2002**, *420*, 54–57. [CrossRef] [PubMed]

32. Mirebeau, I.; Goncharenko, I.N. $Tb_2Ti_2O_7$: A 'spin liquid' single crystal studied under high pressure and high magnetic field. *J. Phys. Condens. Matter* **2005**, *17*, S771–S782. [CrossRef]

33. Mirebeau, I.; Apetrei, A.; Rodríguez-Carvajal, J.; Bonville, P.; Forget, A.; Colson, D.; Glazkov, V.; Sanchez, J.P.; Isnard, O.; Suard, E. Ordered Spin Ice State and Magnetic Fluctuations in $Tb_2Sn_2O_7$. *Phys. Rev. Lett.* **2005**, *94*, 246402. [CrossRef]

34. Kumar, R.S.; Cornelius, A.L.; Somayazulu, M.; Errandonea, D.; Nicol, M.F.; Gardner, J. High pressure structure of $Tb_2Ti_2O_7$ pyrochlore at cryogenic temperatures. *Phys. Status Solidi B* **2007**, *244*, 266–269. [CrossRef]

35. Matsuhira, K.; Hinatsu, Y.; Tenya, K.; Amitsuka, H.; Sakakibara, T. Low-Temperature Magnetic Properties of Pyrochlore Stannates. *J. Phys. Soc. Jpn.* **2002**, *71*, 1576–1582. [CrossRef]

36. Sipos, B.; Kusmartseva, A.F.; Akrap, A.; Berger, H.; Forró, L.; Tutiš, E. From Mott state to superconductivity in 1T-TaS$_2$. *Nat. Mater.* **2008**, *7*, 960–965. [CrossRef]

37. Novoselov, K.S.; Mishchenko, A.; Carvalho, A.; Castro Neto, A.H. 2D materials and van der Waals heterostructures. *Science* **2016**, *353*, aac9439. [CrossRef]

38. Law, K.T.; Lee, P.A. 1T-TaS$_2$ as a quantum spin liquid. *Proc. Natl. Acad. Sci. USA* **2017**, *114*, 6996–7000. [CrossRef]

39. Han, T.R.T.; Zhou, F.; Malliakas, C.D.; Duxbury, P.M.; Mahanti, S.D.; Kanatzidis, M.G.; Ruan, C.Y. Exploration of metastability and hidden phases in correlated electron crystals visualized by femtosecond optical doping and electron crystallography. *Sci. Adv.* **2015**, *1*, e1400173. [CrossRef]

40. Liu, Y.; Ang, R.; Lu, W.J.; Song, W.H.; Li, L.J.; Sun, Y.P. Superconductivity induced by Se-doping in layered charge-density-wave system 1T-TaS$_{2-x}$Se$_x$. *Appl. Phys. Lett.* **2013**, *102*, 192602. [CrossRef]

41. Gao, J.; Park, J.W.; Kim, K.; Song, S.K.; Chen, F.; Luo, X.; Sun, Y.; Yeom, H.W. Pseudogap and weak multifractality in disordered Mott charge-density-wave insulator. *arXiv* **2019**, arXiv:1904.04508.

42. Lahoud, E.; Meetei, O.N.; Chaska, K.B.; Kanigel, A.; Trivedi, N. Emergence of a Novel Pseudogap Metallic State in a Disordered 2D Mott Insulator. *Phys. Rev. Lett.* **2014**, *112*, 206402. [CrossRef]

43. Ritschel, T.; Trinckauf, J.; Garbarino, G.; Hanfland, M.; v. Zimmermann, M.; Berger, H.; Büchner, B.; Geck, J. Pressure dependence of the charge density wave in 1T-TaS$_2$ and its relation to superconductivity. *Phys. Rev. B* **2013**, *87*, 125135. [CrossRef]

44. Zvyagin, S.A.; Graf, D.; Sakurai, T.; Kimura, S.; Nojiri, H.; Wosnitza, J.; Ohta, H.; Ono, T.; Tanaka, H. Pressure-tuning the quantum spin Hamiltonian of the triangular lattice antiferromagnet Cs$_2$CuCl$_4$. *Nat. Commun.* **2019**, *10*, 1064. [CrossRef] [PubMed]

45. Balents, L. Spin liquids in frustrated magnets. *Nature* **2010**, *464*, 199–208. [CrossRef] [PubMed]

46. Zvyagin, S.A.; Kamenskyi, D.; Ozerov, M.; Wosnitza, J.; Ikeda, M.; Fujita, T.; Hagiwara, M.; Smirnov, A.I.; Soldatov, T.A.; Shapiro, A.Y.; et al. Direct Determination of Exchange Parameters in Cs$_2$CuBr$_4$ and Cs$_2$CuCl$_4$: High-Field Electron-Spin-Resonance Studies. *Phys. Rev. Lett.* **2014**, *112*, 077206. [CrossRef] [PubMed]

47. Coldea, R.; Tennant, D.A.; Habicht, K.; Smeibidl, P.; Wolters, C.; Tylczynski, Z. Direct Measurement of the Spin Hamiltonian and Observation of Condensation of Magnons in the 2D Frustrated Quantum Magnet Cs$_2$CuCl$_4$. *Phys. Rev. Lett.* **2002**, *88*, 137203. [CrossRef] [PubMed]

48. Starykh, O.A.; Katsura, H.; Balents, L. Extreme sensitivity of a frustrated quantum magnet: Cs$_2$CuCl$_4$. *Phys. Rev. B* **2010**, *82*, 014421. [CrossRef]

49. Coldea, R.; Tennant, D.; Cowley, R.; McMorrow, D.; Dorner, B.; Tylczynski, Z. Neutron scattering study of the magnetic structure of Cs$_2$CuCl$_4$. *J. Phys. Condens. Matter* **1996**, *8*, 7473–7491. [CrossRef]

50. Li, Y.; Chen, G.; Tong, W.; Pi, L.; Liu, J.; Yang, Z.; Wang, X.; Zhang, Q. Rare-Earth Triangular Lattice Spin Liquid: A Single-Crystal Study of YbMgGaO$_4$. *Phys. Rev. Lett.* **2015**, *115*, 167203. [CrossRef]

51. Li, Y.; Adroja, D.; Biswas, P.K.; Baker, P.J.; Zhang, Q.; Liu, J.; Tsirlin, A.A.; Gegenwart, P.; Zhang, Q. Muon Spin Relaxation Evidence for the U(1) Quantum Spin-Liquid Ground State in the Triangular Antiferromagnet YbMgGaO$_4$. *Phys. Rev. Lett.* **2016**, *117*, 097201. [CrossRef]

52. Majumder, M.; Simutis, G.; Collings, I.E.; Orain, J.C.; Dey, T.; Li, Y.; Gegenwart, P.; Tsirlin, A.A. Persistent spin dynamics in the pressurized spin-liquid candidate YbMgGaO$_4$. *arXiv* **2019**, arXiv:1902.07749.

53. Jackeli, G.; Khaliullin, G. Mott Insulators in the Strong Spin-Orbit Coupling Limit: From Heisenberg to a Quantum Compass and Kitaev Models. *Phys. Rev. Lett.* **2009**, *102*, 017205. [CrossRef]

54. Trebst, S. Kitaev Materials. *arXiv* **2017**, arXiv:1701.07056.

55. Winter, S.M.; Tsirlin, A.A.; Daghofer, M.; van den Brink, J.; Singh, Y.; Gegenwart, P.; Valentí, R. Models and materials for generalized Kitaev magnetism. *J. Phys. Condens. Matter* **2017**, *29*, 493002. [CrossRef]

56. Banerjee, A.; Lampen-Kelley, P.; Knolle, J.; Balz, C.; Aczel, A.A.; Winn, B.; Liu, Y.; Pajerowski, D.; Yan, J.; Bridges, C.A.; et al. Excitations in the field-induced quantum spin liquid state of α-RuCl$_3$. *Npj Quantum Mater.* **2018**, *3*, 8. [CrossRef]

57. Do, S.H.; Park, S.Y.; Yoshitake, J.; Nasu, J.; Motome, Y.; Kwon, Y.S.; Adroja, D.T.; Voneshen, D.J.; Kim, K.; Jang, T.H.; et al. Majorana fermions in the Kitaev quantum spin system α-RuCl$_3$. *Nat. Phys.* **2017**, *13*, 1079–1084. [CrossRef]

58. Plumb, K.W.; Clancy, J.P.; Sandilands, L.J.; Shankar, V.V.; Hu, Y.F.; Burch, K.S.; Kee, H.Y.; Kim, Y.J. α-RuCl$_3$: A spin-orbit assisted Mott insulator on a honeycomb lattice. *Phys. Rev. B* **2014**, *90*, 041112. [CrossRef]

59. Johnson, R.D.; Williams, S.C.; Haghighirad, A.A.; Singleton, J.; Zapf, V.; Manuel, P.; Mazin, I.I.; Li, Y.; Jeschke, H.O.; Valentí, R.; et al. Monoclinic crystal structure of α-RuCl$_3$ and the zigzag antiferromagnetic ground state. *Phys. Rev. B* **2015**, *92*, 235119. [CrossRef]

60. Winter, S.M.; Li, Y.; Jeschke, H.O.; Valentí, R. Challenges in design of Kitaev materials: Magnetic interactions from competing energy scales. *Phys. Rev. B* **2016**, *93*, 214431. [CrossRef]

61. Kim, H.S.; Catuneanu, A.; Kee, H.Y. Kitaev magnetism in honeycomb RuCl3 with intermediate spin-orbit coupling. *Phys. Rev. B* **2015**, *91*, 241110. [CrossRef]

62. Banerjee, A.; Bridges, C.A.; Yan, J.Q.; Aczel, A.A.; Li, L.; Stone, M.B.; Granroth, G.E.; Lumsden, M.D.; Yiu, Y.; Knolle, J.; et al. Proximate Kitaev quantum spin liquid behaviour in a honeycomb magnet. *Nat. Mater.* **2016**, *15*, 733–740. [CrossRef]

63. Cao, H.B.; Banerjee, A.; Yan, J.Q.; Bridges, C.A.; Lumsden, M.D.; Mandrus, D.G.; Tennant, D.A.; Chakoumakos, B.C.; Nagler, S.E. Low-temperature crystal and magnetic structure of α-RuCl$_3$. *Phys. Rev. B* **2016**, *93*, 134423. [CrossRef]

64. Biesner, T.; Biswas, S.; Li, W.; Saito, Y.; Pustogow, A.; Altmeyer, M.; Wolter, A.U.B.; Büchner, B.; Roslova, M.; Doert, T.; et al. Detuning the honeycomb of α-RuCl$_3$: Pressure-dependent optical studies reveal broken symmetry. *Phys. Rev. B* **2018**, *97*, 220401. [CrossRef]

65. Sears, J.A.; Songvilay, M.; Plumb, K.W.; Clancy, J.P.; Qiu, Y.; Zhao, Y.; Parshall, D.; Kim, Y.J. Magnetic order in α-RuCl$_3$: A honeycomb-lattice quantum magnet with strong spin-orbit coupling. *Phys. Rev. B* **2015**, *91*, 144420. [CrossRef]

66. Fletcher, J.M.; Gardner, W.E.; Fox, A.C.; Topping, G. X-ray, Infrared, and Magnetic Studies of α- and β-Ruthenium Trichloride. *J. Chem. Soc. A* **1967**, *0*, 1038–1045. [CrossRef]

67. Binotto, L.; Pollini, I.; Spinolo, G. Optical and transport properties of the magnetic semiconductor α-RuCl$_3$. *Phys. Stat. Sol. (b)* **1971**, *44*, 245–252. [CrossRef]

68. Guizzetti, G.; Reguzzoni, E.; Pollini, I. Spin-polarized electron tunneling. *Phys. Lett. A* **1979**, *70*, 34–36. [CrossRef]

69. Sandilands, L.J.; Tian, Y.; Reijnders, A.A.; Kim, H.S.; Plumb, K.W.; Kim, Y.J.; Kee, H.Y.; Burch, K.S. Spin-orbit excitations and electronic structure of the putative Kitaev magnet α-RuCl$_3$. *Phys. Rev. B* **2016**, *93*, 075144. [CrossRef]

70. Sandilands, L.J.; Sohn, C.H.; Park, H.J.; Kim, S.Y.; Kim, K.W.; Sears, J.A.; Kim, Y.J.; Noh, T.W. Optical probe of Heisenberg-Kitaev magnetism in α-RuCl$_3$. *Phys. Rev. B* **2016**, *94*, 195156. [CrossRef]

71. Wang, Z.; Guo, J.; Tafti, F.F.; Hegg, A.; Sen, S.; Sidorov, V.A.; Wang, L.; Cai, S.; Yi, W.; Zhou, Y.; et al. Pressure-induced melting of magnetic order and emergence of a new quantum state in α-RuCl$_3$. *Phys. Rev. B* **2018**, *97*, 245149. [CrossRef]

72. Cui, Y.; Zheng, J.; Ran, K.; Wen, J.; Liu, Z.X.; Liu, B.; Guo, W.; Yu, W. High-pressure magnetization and NMR studies of α-RuCl$_3$. *Phys. Rev. B* **2017**, *96*, 205147. [CrossRef]

73. Bastien, G.; Garbarino, G.; Yadav, R.; Martinez-Casado, F.J.; Beltrán Rodríguez, R.; Stahl, Q.; Kusch, M.; Limandri, S.P.; Ray, R.; Lampen-Kelley, P.; et al. Pressure-induced dimerization and valence bond crystal formation in the Kitaev-Heisenberg magnet α-RuCl$_3$. *Phys. Rev. B* **2018**, *97*, 241108. [CrossRef]

74. Kimchi, I.; Analytis, J.G.; Vishwanath, A. Three-dimensional quantum spin liquids in models of harmonic-honeycomb iridates and phase diagram in an infinite-D approximation. *Phys. Rev. B* **2014**, *90*, 205126. [CrossRef]

75. Kimchi, I.; Coldea, R.; Vishwanath, A. Unified theory of spiral magnetism in the harmonic-honeycomb iridates α, β, and γ Li$_2$IrO$_3$. *Phys. Rev. B* **2015**, *91*, 245134. [CrossRef]

76. Singh, Y.; Manni, S.; Reuther, J.; Berlijn, T.; Thomale, R.; Ku, W.; Trebst, S.; Gegenwart, P. Relevance of the Heisenberg-Kitaev Model for the Honeycomb Lattice Iridates A$_2$IrO$_3$. *Phys. Rev. Lett.* **2012**, *108*, 127203. [CrossRef]

77. Chaloupka, J.; Jackeli, G.; Khaliullin, G. Kitaev-Heisenberg Model on a Honeycomb Lattice: Possible Exotic Phases in Iridium Oxides A$_2$IrO$_3$. *Phys. Rev. Lett.* **2010**, *105*, 027204. [CrossRef]

78. Gretarsson, H.; Clancy, J.P.; Liu, X.; Hill, J.P.; Bozin, E.; Singh, Y.; Manni, S.; Gegenwart, P.; Kim, J.; Said, A.H.; et al. Crystal-Field Splitting and Correlation Effect on the Electronic Structure of A$_2$IrO$_3$. *Phys. Rev. Lett.* **2013**, *110*, 076402. [CrossRef]

79. Li, Y.; Foyevtsova, K.; Jeschke, H.O.; Valentí, R. Analysis of the optical conductivity for A_2IrO_3(A=Na, Li) from first principles. *Phys. Rev. B* **2015**, *91*, 161101. [CrossRef]

80. Kim, B.H.; Shirakawa, T.; Yunoki, S. From a Quasimolecular Band Insulator to a Relativistic Mott Insulator in t_{2g}^5 Systems with a Honeycomb Lattice Structure. *Phys. Rev. Lett.* **2016**, *117*, 187201. [CrossRef]

81. Kim, B.J.; Ohsumi, H.; Komesu, T.; Sakai, S.; Morita, T.; Takagi, H.; Arima, T. Phase-Sensitive Observation of a Spin-Orbital Mott State in Sr_2IrO_4. *Science* **2009**, *323*, 1329–1332. [CrossRef]

82. Sohn, C.H.; Kim, H.S.; Qi, T.F.; Jeong, D.W.; Park, H.J.; Yoo, H.K.; Kim, H.H.; Kim, J.Y.; Kang, T.D.; Cho, D.Y.; et al. Mixing between $J_{eff} = \frac{1}{2}$ and $\frac{3}{2}$ orbitals in Na_2IrO_3: A spectroscopic and density functional calculation study. *Phys. Rev. B* **2013**, *88*, 085125. [CrossRef]

83. Hermann, V.; Altmeyer, M.; Ebad-Allah, J.; Freund, F.; Jesche, A.; Tsirlin, A.A.; Hanfland, M.; Gegenwart, P.; Mazin, I.I.; Khomskii, D.I.; et al. Competition between spin-orbit coupling, magnetism, and dimerization in the honeycomb iridates: α-Li_2IrO_3 under pressure. *Phys. Rev. B* **2018**, *97*, 020104. [CrossRef]

84. Williams, S.C.; Johnson, R.D.; Freund, F.; Choi, S.; Jesche, A.; Kimchi, I.; Manni, S.; Bombardi, A.; Manuel, P.; Gegenwart, P.; et al. Incommensurate counterrotating magnetic order stabilized by Kitaev interactions in the layered honeycomb α-Li_2IrO_3. *Phys. Rev. B* **2016**, *93*, 195158. [CrossRef]

85. Kimber, S.A.J.; Mazin, I.I.; Shen, J.; Jeschke, H.O.; Streltsov, S.V.; Argyriou, D.N.; Valentí, R.; Khomskii, D.I. Valence bond liquid phase in the honeycomb lattice material Li_2RuO_3. *Phys. Rev. B* **2014**, *89*, 081408. [CrossRef]

86. Clancy, J.P.; Gretarsson, H.; Sears, J.A.; Singh, Y.; Desgreniers, S.; Mehlawat, K.; Layek, S.; Rozenberg, G.K.; Ding, Y.; Upton, M.H.; et al. Pressure-driven collapse of the relativistic electronic ground state in a honeycomb iridate. *Npj Quantum Mater.* **2018**, *3*, 35. [CrossRef]

87. Hermann, V.; Ebad-Allah, J.; Freund, F.; Jesche, A.; Tsirlin, A.A.; Gegenwart, P.; Kuntscher, C.A. Optical signature of the pressure-induced dimerization in the honeycomb iridate α-Li_2IrO_3. *Phys. Rev. B* **2019**, *99*, 235116. [CrossRef]

88. Takayama, T.; Kato, A.; Dinnebier, R.; Nuss, J.; Kono, H.; Veiga, L.S.I.; Fabbris, G.; Haskel, D.; Takagi, H. Hyperhoneycomb Iridate β-Li_2IrO_3 as a Platform for Kitaev Magnetism. *Phys. Rev. Lett.* **2015**, *114*, 077202. [CrossRef]

89. Katukuri, V.M.; Yadav, R.; Hozoi, L.; Nishimoto, S.; van den Brink, J. The vicinity of hyper-honeycomb β-Li_2IrO_3 to a three-dimensional Kitaev spin liquid state. *Sci. Rep.* **2016**, *6*, 29585. [CrossRef]

90. Glamazda, A.; Lemmens, P.; Do, S.H.; Choi, Y.S.; Choi, K.Y. Raman spectroscopic signature of fractionalized excitations in the harmonic-honeycomb iridates β- and γ-Li_2IrO_3. *Nat. Commun.* **2016**, *7*, 12286. [CrossRef]

91. Ruiz, A.; Frano, A.; Breznay, N.P.; Kimchi, I.; Helm, T.; Oswald, I.; Chan, J.Y.; Birgeneau, R.J.; Islam, Z.; Analytis, J.G. Correlated states in β-Li_2IrO_3 driven by applied magnetic fields. *Nat. Commun.* **2017**, *8*, 961. [CrossRef]

92. Veiga, L.S.I.; Etter, M.; Glazyrin, K.; Sun, F.; Escanhoela, C.A.; Fabbris, G.; Mardegan, J.R.L.; Malavi, P.S.; Deng, Y.; Stavropoulos, P.P.; et al. Pressure tuning of bond-directional exchange interactions and magnetic frustration in the hyperhoneycomb iridate β-Li_2IrO_3. *Phys. Rev. B* **2017**, *96*, 140402. [CrossRef]

93. Kim, H.S.; Kim, Y.B.; Kee, H.Y. Revealing frustrated local moment model for pressurized hyperhoneycomb iridate: Paving the way toward a quantum spin liquid. *Phys. Rev. B* **2016**, *94*, 245127. [CrossRef]

94. Majumder, M.; Manna, R.S.; Simutis, G.; Orain, J.C.; Dey, T.; Freund, F.; Jesche, A.; Khasanov, R.; Biswas, P.K.; Bykova, E.; et al. Breakdown of Magnetic Order in the Pressurized Kitaev Iridate β-Li_2IrO_3. *Phys. Rev. Lett.* **2018**, *120*, 237202. [CrossRef]

95. Modic, K.A.; Smidt, T.E.; Kimchi, I.; Breznay, N.P.; Biffin, A.; Choi, S.; Johnson, R.D.; Coldea, R.; Watkins-Curry, P.; McCandless, G.T.; et al. Realization of a three-dimensional spin–anisotropic harmonic honeycomb iridate. *Nat. Commun.* **2014**, *5*, 4203. [CrossRef]

96. Biffin, A.; Johnson, R.D.; Kimchi, I.; Morris, R.; Bombardi, A.; Analytis, J.G.; Vishwanath, A.; Coldea, R. Noncoplanar and Counterrotating Incommensurate Magnetic Order Stabilized by Kitaev Interactions in γ-Li_2IrO_3. *Phys. Rev. Lett.* **2014**, *113*, 197201. [CrossRef]

97. Breznay, N.P.; Ruiz, A.; Frano, A.; Bi, W.; Birgeneau, R.J.; Haskel, D.; Analytis, J.G. Resonant x-ray scattering reveals possible disappearance of magnetic order under hydrostatic pressure in the Kitaev candidate γ-Li_2IrO_3. *Phys. Rev. B* **2017**, *96*, 020402. [CrossRef]

98. Hermann, V.; Ebad-Allah, J.; Freund, F.; Pietsch, I.M.; Jesche, A.; Tsirlin, A.A.; Deisenhofer, J.; Hanfland, M.; Gegenwart, P.; Kuntscher, C.A. High-pressure versus isoelectronic doping effect on the honeycomb iridate Na_2IrO_3. *Phys. Rev. B* **2017**, *96*, 195137. [CrossRef]

99. Choi, S.K.; Coldea, R.; Kolmogorov, A.N.; Lancaster, T.; Mazin, I.I.; Blundell, S.J.; Radaelli, P.G.; Singh, Y.; Gegenwart, P.; Choi, K.R.; et al. Spin Waves and Revised Crystal Structure of Honeycomb Iridate Na_2IrO_3. *Phys. Rev. Lett.* **2012**, *108*, 127204. [CrossRef]

100. Ye, F.; Chi, S.; Cao, H.; Chakoumakos, B.C.; Fernandez-Baca, J.A.; Custelcean, R.; Qi, T.F.; Korneta, O.B.; Cao, G. Direct evidence of a zigzag spin-chain structure in the honeycomb lattice: A neutron and X-ray diffraction investigation of single-crystal Na_2IrO_3. *Phys. Rev. B* **2012**, *85*, 180403. [CrossRef]

101. Foyevtsova, K.; Jeschke, H.O.; Mazin, I.I.; Khomskii, D.I.; Valentí, R. *Ab initio* analysis of the tight-binding parameters and magnetic interactions in Na_2IrO_3. *Phys. Rev. B* **2013**, *88*, 035107. [CrossRef]

102. Comin, R.; Levy, G.; Ludbrook, B.; Zhu, Z.H.; Veenstra, C.N.; Rosen, J.A.; Singh, Y.; Gegenwart, P.; Stricker, D.; Hancock, J.N.; et al. Na_2IrO_3 as a Novel Relativistic Mott Insulator with a 340-meV Gap. *Phys. Rev. Lett.* **2012**, *109*, 266406. [CrossRef]

103. Li, Y.; Winter, S.M.; Jeschke, H.O.; Valentí, R. Electronic excitations in γ-Li_2IrO_3. *Phys. Rev. B* **2017**, *95*, 045129. [CrossRef]

104. Kozlenko, D.P.; Kusmartseva, A.F.; Lukin, E.V.; Keen, D.A.; Marshall, W.G.; de Vries, M.A.; Kamenev, K.V. From Quantum Disorder to Magnetic Order in an $s = 1/2$ Kagome Lattice: A Structural and Magnetic Study of Herbertsmithite at High Pressure. *Phys. Rev. Lett.* **2012**, *108*, 187207. [CrossRef]

105. Norman, M.R. Colloquium: Herbertsmithite and the search for the quantum spin liquid. *Rev. Mod. Phys.* **2016**, *88*, 041002. [CrossRef]

106. Puphal, P.; Bolte, M.; Sheptyakov, D.; Pustogow, A.; Kliemt, K.; Dressel, M.; Baenitz, M.; Krellner, C. Strong magnetic frustration in $Y_3Cu_9(OH)_{19}Cl_8$: a distorted kagome antiferromagnet. *J. Mater. Chem. C* **2017**, *5*, 2629–2635. [CrossRef]

107. Mendels, P.; Bert, F.; de Vries, M.A.; Olariu, A.; Harrison, A.; Duc, F.; Trombe, J.C.; Lord, J.S.; Amato, A.; Baines, C. Quantum Magnetism in the Paratacamite Family: Towards an Ideal Kagome Lattice. *Phys. Rev. Lett.* **2007**, *98*, 077204. [CrossRef]

108. Helton, J.S.; Matan, K.; Shores, M.P.; Nytko, E.A.; Bartlett, B.M.; Yoshida, Y.; Takano, Y.; Suslov, A.; Qiu, Y.; Chung, J.H.; et al. Spin Dynamics of the Spin-1/2 Kagome Lattice Antiferromagnet $ZnCu_3(OH)_6Cl_2$. *Phys, Rev, Lett,* **2007**, *98*, 107204. [CrossRef]

109. Zorko, A.; Nellutla, S.; van Tol, J.; Brunel, L.C.; Bert, F.; Duc, F.; Trombe, J.C.; de Vries, M.A.; Harrison, A.; Mendels, P. Dzyaloshinsky-Moriya Anisotropy in the Spin-1/2 Kagome Compound $ZnCu_3(OH)_6Cl_2$. *Phys. Rev. Lett.* **2008**, *101*, 026405. [CrossRef]

110. Jeong, M.; Bert, F.; Mendels, P.; Duc, F.; Trombe, J.C.; de Vries, M.A.; Harrison, A. Field-Induced Freezing of a Quantum Spin Liquid on the Kagome Lattice. *Phys. Rev. Lett.* **2011**, *107*, 237201. [CrossRef]

111. Takatsu, K.i.; Shiramura, W.; Tanaka, H. Ground States of Double Spin Chain Systems $TlCuCl_3$, NH_4CuCl_3 and $KCuBr_3$. *J. Phys. Soc. Jpn.* **1997**, *66*, 1611–1614. [CrossRef]

112. Oosawa, A.; Ishii, M.; Tanaka, H. Field-induced three-dimensional magnetic orderingin the spin-gap system $TlCuCl_3$. *J. Phys. Condens. Matter* **1999**, *11*, 265–271. [CrossRef]

113. Oosawa, A.; Fujisawa, M.; Osakabe, T.; Kakurai, K.; Tanaka, H. Neutron Diffraction Study of the Pressure-Induced Magnetic Ordering in the Spin Gap System $TlCuCl_3$. *J. Phys. Soc. Jpn.* **2003**, *72*, 1026–1029. [CrossRef]

114. Tanaka, H.; Goto, K.; Fujisawa, M.; Ono, T.; Uwatoko, Y. Magnetic ordering under high pressure in the quantum spin system $TlCuCl_3$. *Physica B* **2003**, *329-333*, 697–698. [CrossRef]

115. Goto, K.; Fujisawa, M.; Ono, T.; Tanaka, H.; Uwatoko, Y. Pressure-induced Magnetic Quantum Phase Transition from Gapped Ground State in $TlCuCl_3$. *J. Phys. Soc. Jpn.* **2004**, *73*, 3254–3257. [CrossRef]

116. Rüegg, C.; Furrer, A.; Sheptyakov, D.; Strässle, T.; Krämer, K.W.; Güdel, H.U.; Mélési, L. Pressure-Induced Quantum Phase Transition in the Spin-Liquid $TlCuCl_3$. *Phys. Rev. Lett.* **2004**, *93*, 257201. [CrossRef]

117. Merchant, P.; Normand, B.; Krämer, K.W.; Boehm, M.; McMorrow, D.F.; Rüegg, C. Quantum and classical criticality in a dimerized quantum antiferromagnet. *Nat. Phys.* **2014**, *10*, 373. [CrossRef]

118. Rüegg, C.; Cavadini, N.; Furrer, A.; Güdel, H.U.; Krämer, K.; Mutka, H.; Wildes, A.; Habicht, K.; Vorderwisch, P. Bose–Einstein condensation of the triplet states in the magnetic insulator $TlCuCl_3$. *Nature* **2003**, *423*, 62–65. [CrossRef]

119. Oosawa, A.; Ono, T.; Tanaka, H. Impurity-induced antiferromagnetic ordering in the spin gap system TlCuCl$_3$. *Phys. Rev. B* **2002**, *66*, 020405. [CrossRef]

120. Matsumoto, M.; Normand, B.; Rice, T.M.; Sigrist, M. Field- and pressure-induced magnetic quantum phase transitions in TlCuCl$_3$. *Phys. Rev. B* **2004**, *69*, 054423. [CrossRef]

121. Rüegg, C.; Normand, B.; Matsumoto, M.; Furrer, A.; McMorrow, D.F.; Krämer, K.W.; Güdel, H.U.; Gvasaliya, S.N.; Mutka, H.; Boehm, M. Quantum Magnets under Pressure: Controlling Elementary Excitations in TlCuCl$_3$. *Phys. Rev. Lett.* **2008**, *100*, 205701. [CrossRef] [PubMed]

Review

Mechanisms of Pressure-Induced Phase Transitions by Real-Time Laue Diffraction

Dmitry Popov [1],*, Nenad Velisavljevic [1,2] and Maddury Somayazulu [1]

[1]   High Pressure Collaborative Access Team, X-ray Science Division, Argonne National Laboratory, Lemont, IL 60439, USA; hpcat-director@anl.gov (N.V.); zulu@anl.gov (M.S.)

[2]   Physics Division-Physical & Life Sciences Directorate, Lawrence Livermore National Laboratory, Livermore, CA 94550, USA

*   Correspondence: dpopov@anl.gov

Received: 30 November 2019; Accepted: 11 December 2019; Published: 14 December 2019

Abstract: Synchrotron X-ray radiation Laue diffraction is a widely used diagnostic technique for characterizing the microstructure of materials. An exciting feature of this technique is that comparable numbers of reflections can be measured several orders of magnitude faster than using monochromatic methods. This makes polychromatic beam diffraction a powerful tool for time-resolved microstructural studies, critical for understanding pressure-induced phase transition mechanisms, by in situ and in operando measurements. The current status of this technique, including experimental routines and data analysis, is presented along with some case studies. The new experimental setup at the High-Pressure Collaborative Access Team (HPCAT) facility at the Advanced Photon Source, specifically dedicated for in situ and in operando microstructural studies by Laue diffraction under high pressure, is presented.

Keywords: Laue diffraction; high pressure; mechanisms of phase transitions

1. Introduction

White beam (polychromatic) Laue diffraction is a powerful experimental tool for studying mechanisms of pressure-induced phase transitions. Use of the full white beam spectrum allows for fast data collection, which then provides both spatially and time-resolved microstructural information to be gained simultaneously from a sample in Diamond Anvil Cell (DAC), with spatial resolution down to microns and time resolution down to seconds [1,2]. The alternative, using monochromatic beam diffraction, would not provide comparable time resolution, even with high X-ray energies, due to the need to rotate the sample while collecting X-ray images [3,4]. This makes total data collection time across the sample in a DAC substantially longer.

To some extent, the mechanisms of transitions can be determined by studying samples recovered at ambient pressure [5–7]. However, such samples may be substantially altered during the recovery process or undergo reversible phase transformations as pressure is released. In contrast, polychromatic beam diffraction makes it possible to determine the morphology, deformation and orientation relations of co-existing parental and product phases in real time in situ. Time-resolved measurements are vital for revealing details of phase transitions, which may be controlled by kinetics, and therefore require the fastest techniques of data collection.

Despite the fairly wide use of Laue diffraction for characterization of materials, including in situ studies of materials under external stress [8–10], the application of this technique for high-pressure DAC studies requires additional experimental development and consideration. The purpose of this review article is to summarize the currently available polychromatic beam diffraction experimental techniques at high pressure and data analysis capabilities at HPCAT, as well as to present some recent case studies.

2. Experimental Facilities and Procedures

There are multiple synchrotron beamlines dedicated to Laue micro-diffraction [10–13]. However, currently these facilities are not specifically optimized for measurements on materials at high pressures. The first experimental setup specifically dedicated to high-pressure Laue micro-diffraction was developed at the 16BMB beamline of the Advanced Photon Source (APS) [2] (Figure 1). This setup is mounted on a granite table in order to maintain mechanical stability and positioning of the x-ray beam spot on the sample. A fast PerkinElmer area detector (PerkinElmer Optoelectronics) is positioned with a detector arm and, therefore, can be oriented in either transmitted or 90° geometry. The X-ray polychromatic beam from the bending magnet is focused down to $2\mu m^2$ with KB-mirrors. The highest limit of X-ray energy at the sample position is adjustable by changing the tilt of the mirrors. For measurements in transmitted geometry, the highest possible energy is typically set at ~90keV, while in 90° geometry, the highest energy limit is around 35keV. The sample for high-pressure studies in a DAC is mounted on top of a fast x/y table, which in turn is mounted on top of an elevation stage. These stages are used to collect a series of two-dimensional (2D) translational scans across the sample. With a vertical rotational stage, sample orientation can be optimized to obtain the highest number of reflections. A Si 111 channel-cut monochromator provides monochromatic beam switchable with the polychromatic beam in a matter of minutes. This monochromatic beam is mainly used to identify powdered product phases after destructive phase transitions, to calibrate sample to detector distance and detector tilt in transmitted geometry using the CeO_2 standard. Detector calibration in 90° geometry is done using the Laue diffraction pattern from a Si single crystal and the known energy of one reflection on this pattern measured with the monochromator [14].

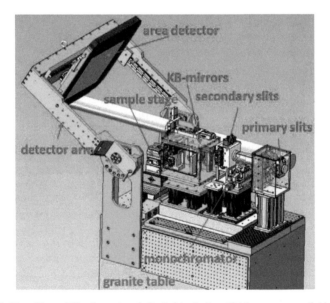

Figure 1. Outline of Laue diffraction setup dedicated to studies at high pressure available at 16BMB beamline of Advanced Photon Source [2].

Measurements in a DAC introduce some specific requirements to the sample environment (Figure 2). Strong Laue reflections from diamonds can damage the area detector. To protect the detector during data collection a detector mask is used: small pieces of lead are placed on a kapton foil in order to block the reflections from the diamonds. Positions of strong diamond reflections are pre-determined by collecting an X-ray image with strong absorbers inserted in the incident beam to reduce intensities of these reflections to the level at which they will not damage the detector. A metal grid is placed on the kapton foil during the collecting of this image; this introduces a clear 'imprint' on the background of the pattern. Using the grid as a reference the pieces of lead are placed to the positions on the kapton foil coinciding with the strong diamond reflections. Before the detector mask is positioned, the sample is aligned in the X-ray beam and co-linear with the axis of the rotational stage by doing absorption scans across the sample with photodiode. A movable lead shield is used to protect the detector during this process and is removed for data collection.

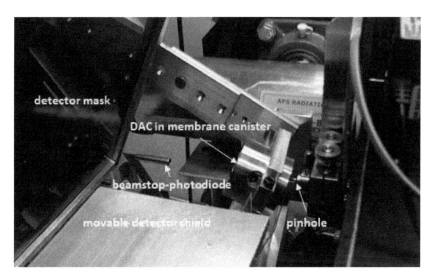

Figure 2. Sample environment of the Laue diffraction setup available at 16BMB beamline of Advanced Photon Source [2].

A series of 2D scans is collected across the sample undergoing a phase transition, while sample pressure is varied remotely using a gas membrane system. Microstructural changes during the transition are observed in real time using ImageJ software [15], and pressure can be fine-tuned based on the observed changes, such that the most important details of the transition are not lost. As the sample position may shift during pressure change, it has to be periodically re-centered on the rotation axis during data collection, in order to maintain a constant spatial sample-to-detector relationship. This procedure requires absorption scans across the sample at different angular positions of the sample [16,17]. Therefore, a detector mask is not sufficient to eliminate unwanted diffraction spots, and the movable detector shield is used instead. During the experiment, pressure is continuously monitored, and is also measured before and after the transition using an off-line Ruby fluorescence system [18] that is available inside the experimental hutch.

For the measurements in 90° geometry, a panoramic DAC with X-ray transparent gasket is used. The incident beam passes through the diamond, while the diffracted beams pass through the gasket material. In transmitted geometry, both incident and diffracted beams pass through diamonds. For the measurements in this configuration typically DACs with total opening of ~60° are used. The DACs are tilted vertically by ~25° with a sample holder, and the area detector is tilted vertically by 30°. This configuration optimizes the amount of reciprocal space that can be accessed and allows us to collect sufficient number of reflections for indexing them.

Phase transitions may proceed very rapidly, and in order to observe details of phase transitions, pressure has to be changed in small steps. In the case of strongly displacive transitions, the initial parent single crystal sample breaks up into smaller crystals or powder-like crystalline aggregates. Due to this process, the sample becomes heavily deformed, which introduces additional difficulties in data analysis. In practice, one must therefore follow the process until this deformation sets in and makes analysis difficult. Using Re gaskets in the transmitted geometry currently pressure rate can be as small as ~0.2 GPa/hour and can therefore allow us to appropriately capture the transition.

Measurements in diamond anvil cells require sufficiently small samples, typically smaller than 100 μm. Therefore, if the sample is bigger, a small part has to be separated from it. On the other hand, Laue diffraction is sensitive only to single crystals with sizes comparable to or larger than X-ray beams, so the original samples must be either single crystals or poly crystals with crystal size typically at least in the micron range. If the sample is too big to be put into a DAC but it is a single crystal with strong cleavage, it can be mechanically split into smaller crystals that still have good quality for the measurements. However, if the mechanical disintegration of the sample destroys the single crystals, an alternative approach may be considered using a laser micro-machining system [19]. The sufficiently small sample is carefully put in the center of gasket hole of a DAC using Axis Pro Micro Support micromanipulator (Supplementary Materials, Figure S1). Positioning of the sample right in the center of gasket hole is crucial to minimize diffraction signal from gasket material and to avoid sample-gasket interaction during pressure changes.

3. Data Analysis

Analysis of the high-pressure Laue diffraction data includes two major steps: indexation of diffraction patterns [14,20], and mapping of reflections. By systematic indexing of diffraction spots, single crystals can be identified, and therefore reflections from parent and daughter phases can be clearly distinguished. Furthermore, by going through systematic indexation, orientations of crystals can be determined, which in turn can be used to find orientation relations between coexisting parent and product phases. Relative orientations of single crystals can be determined from one Laue diffraction pattern with an angular precision down to ~0.02°, which makes Laue diffraction a powerful tool for characterizing twining [21]. By application of the polychromatic beam diffraction, one can overcome a major challenge to recognize domain structures formed after phase transitions: the newly formed domains typically have pseudo symmetrical translational lattices with very low mismatch with respect to the lattice of parental phase [22]. As a Laue diffraction pattern is typically collected in a matter of a few seconds, such domain structures can be recognized in real time with a resolution reaching down to seconds. At the same time, determining twin relationships with the same level of precision using a monochromatic beam requires sample rotations in very small steps and correspondingly large data collection times.

Software for indexing and mapping of Laue reflections was developed in-house by the lead author of this paper D. Popov, in Python [23]. This program can be made available upon request, as it is developed for broader distribution. The indexation routine is tailored toward simultaneous indexation of reflections from multiple crystals or crystallites that can coexist on the same diffraction pattern. Two strong diffraction spots that could belong to the same crystal are selected first. Typically, two reflections from the same 'zone line' are selected, which is a good indication that reflections originate from the same crystal. Based on the positions of these diffraction spots and their indices, the orientation matrix of the crystal can be calculated [20]. However, in many cases, the indices are not known, and therefore the software is used to explore and converge on the possible indices for calculating the orientation matrix and indexing all other reflections using this matrix. Orientation matching to substantially bigger number of reflections than multiple orientations is first exhausted. The range of possible indices is determined based on 2θ angles of the selected two reflections and the highest limit of X-ray energy. After the diffraction spots from a crystal are determined, they are removed from further data analysis, and reflections from other crystals are indexed in the same repetitive way. The peak search function available in Fit2d [24] software is used for determining the positions of reflections.

The newly developed program can even detect a rather unlikely situation in which the same set of reflections can be indexed by multiple crystal orientations. In such a case, the data has to be recollected in order to obtain a broader set of reflections from the sample. A larger set of reflections can typically be obtained by using DACs with larger openings, varying sample orientation, and reducing the sample-to-detector distance. For example, one should avoid situations where only one 'zone line' from a crystal is present in the diffraction pattern, since, in those cases, there are two possible orientation variants that satisfy the diffraction condition. These two orientations are related by a mirror plane perpendicular to the zone axis.

Crystal orientation is determined from reflections picked up on X-ray images using a peak search algorithm which may not be able to recognize all diffraction spots. More reliably, all the reflections from a single crystal can be visually recognized. For this purpose, positions of all possible reflections from a single crystal are predicted and shown on X-ray image, by the new program, based on orientation matrix, highest X-ray energy limit, and some lower limit of d-values (Figure 3a). Software Dioptas [25], widely implemented in the high-pressure area, is used to visualize X-ray images with marks of predicted reflection positions and to look through series of such diffraction patterns collected during a 2D scan.

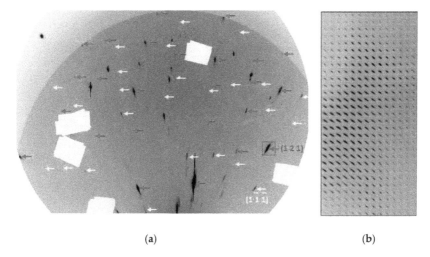

(a)   (b)

Figure 3. Results of identification and mapping of two ω-Zr crystals using Laue diffraction data collected at 34IDE beamline of APS with X-ray beam focused down to $500 \times 500 \text{nm}^2$ at pressure of 5.16GPa. (**a**) Predicted positions of reflections from the two crystals of ω-Zr (shown by different colors, slightly shifted to the right not to overlap with the observed reflections) on a Laue diffraction pattern assuming X-ray energy limit of 30keV and the smallest d-values of 0.55Å (positions shown in red) and 0.65Å (positions shown in yellow). (**b**) Composite frame of area denoted by red rectangle in (**a**) presenting a map of (121) reflection. Composite frames of areas around some other reflections from the same crystal are presented in the Supplementary Materials (Figure S2).

For mapping the single crystal, the new program combines all images collected during the 2D scan of the crystal into a composite frame in the same order as those images were collected during the scan (Figure 3b). Such composite frames reproduce the shapes of single crystals and, at the same time, variations in the shapes and positions of diffraction spots across the sample are also clearly visible, indicating deformation of the crystals. The composite images are currently visualized with Fit2d.

In general, the microstructural changes during phase transitions are continuous, and as such, in order to analyze the 2D scans, one has to distinguish between changes that could be attributed to spatial inhomogeneity of the sample and those caused by temporal changes during the transition. Typically, there is variation in the diffraction spot positions across both the parental and product single crystals caused by their deformation. At the same time, rotation of the crystals due to deformation of the entire sample during a translational scan causes such a variation as well. Pressure increase may slightly mis-orient the DAC during a translational scan, also resulting in variation of diffraction spot positions across the single crystals. The shapes of diffraction spots may also vary across parent crystals coexisting with product phase due to the inhomogeneity of their deformation. However, if the crystals are deformed during a 2D translational scan, they exhibit the same kind of variation. Maps of reflections across the sample are defined by crystal morphology, but the maps can also be affected by decreasing parent fraction or growth of product phase during the scan. The variations due to sample inhomogeneity are distinguished by comparing concurrent translational scans in order to find reproducible features. For instance, the composite frame presented in Figure 3b was extracted from a scan repeated twice. The composite frame from the second scan is identical, indicating that there were no observable changes in the sample within the time interval of 12 h required to collect both of the scans. Therefore, the composite frame clearly reproduces the shape of the crystal, and no changes to its morphology are indicated. More composite frames of reflections from the same crystal are presented in Supplementary Materials (Figure S2).

4. Examples

The mechanism of the α → β transition in Si was studied using Laue diffraction [1]. In Figure 4, maps of coexisting parent and product phases at the onset of transition are presented (Figure 4a), along with map of the rest of parent α-Si single crystal (Figure 4b).

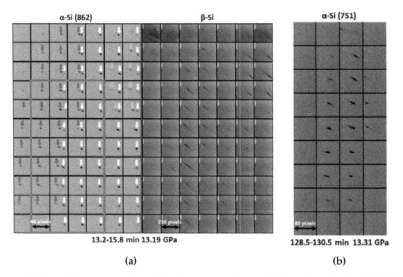

Figure 4. Maps of the parent and product phases across the α → β phase transition in Si [1]. Time intervals of the 2D scans since the beginning of the data collection routine are shown. The pressure values have been interpolated based on the pressures measured before and after the transition. Step size was 5 μm. (**a**) Composite frames of reflections from three slightly mis-oriented crystals of α-Si (shown in different colors) and β-Si. The blue rectangle denotes the overlap area with the map presented in (**b**). (**b**) Map of the rest of parent α-Si crystal when the transition was nearly over.

Microstructural changes that evolve over longer time periods in comparison to typical acquisition times of one translational scan can be better analyzed and viewed by combining composite frames from consecutive scans. As examples, movies (Movie S1a,b) compiled from composite frames shown in Figure 4 are presented in the Supplementary Materials. Movie S1a captures the splitting of the original single crystal of α-Si at the onset of transition into three crystals that are slightly mis-oriented with respect to one another, while the newly formed β-Si is located at the interface of these crystals. In Movie S1b, the disintegration of the rest of parental single crystal is shown.

The results of study of the mechanism of the α → β phase transition in Si [1] presented in Movie S1a indicate that the interface between the parental and product phases is highly incoherent. While the α-phase produced very sharp reflections, the β-phase produced highly broadened 'streaky' reflections, indicating that crystals of this phase were heavily deformed. Most likely, this is due to the large volume collapse accompanying this transition, which then resulted in large lattice mismatch between the parent and daughter phases. The areas of β-Si were strongly elongated domains formed parallel to a <110> direction of the parent α-Si phase. This may also indicate that the nucleation of the product phase could be controlled by defects introduced along the cleavage planes during the sample preparation. It is interesting to note that the single crystals of α-Si were slightly bent as the β-phase started growing. This is indicated by the systematic shift of positions of reflections from α-Si across the sample reproducible on the composite frames from adjacent translational scans. Toward the end of the transition, the rest of the parental single crystal also exhibited the same kind of deformation (Movie S1b). The remaining α-Si crystal was strongly elongated parallel to the same <110> direction as the β-Si areas on the onset of transition.

Another example is our recent study of zirconium (Zr) metal. Zr has previously been observed to exhibit unusual grain enlargement across the α → ω phase transition [26]. The parent phase is a nano-crystalline aggregate, while the product phase is much grainier. The relative change in grain size structure was inferred from the 'spotty' diffraction lines recorded with a monochromatic beam on the daughter ω-phase of Zr in comparison to the continuous diffraction lines from the parent α-phase of Zr. The grain enlargement in Zr is observed under high pressure and room temperature conditions, as opposed to typical grain enlargement processes that are required to occur at elevated temperatures to overcome the activation barriers for diffusion. Currently, both the driving force and the mechanism of this phenomenon remain unknown. Measurements with a polychromatic beam that is 500×500 nm^2 size demonstrated that the newly formed ω-Zr crystals may have an irregular morphology (Figure 3b).

Using Laue diffraction, the grain enlargement process in Zr was studied in real time [27]. Initially, we did not observe any Laue reflections from the fine-grained parent α-phase. As Zr started to undergo a transition to ω-phase, we started to observe the appearance of strong diffraction spots. The intensities of these reflections increased gradually along with increasing pressure. Mapping of the ω-Zr reflections clearly demonstrated that the ω-Zr crystals nucleated and grew in the nano-crystalline matrix of α-Zr. In the Supplementary Materials, a movie combined from maps of an ω-Zr reflection is presented as an example (Movie S2). The sizes of the newly formed ω-Zr crystals gradually increased along with the increase in pressure, while the intensities of the ω-Zr reflections gradually decreased towards the edges of the crystals away from the nucleation points. These observations indicate a grain enlargement mechanism that involves formation and movement of highly angular grain boundaries, distinguishing the enlargement from a recovery process in which stored energy of dislocations is released without movement of the grain boundaries. In the case of recovery, intensities of ω-Zr reflections on the composite frames would increase simultaneously over a single grain area. This is followed by gradual sharpening of the ω-Zr reflections. The observed ω-Zr reflections stayed at the same level of broadening during the grain enlargement process indicating that there was no substantial texture development involved in this process. The majority of ω-Zr reflections exhibited only positive shifts of their intensities, indicating that ω-Zr crystals mainly grew at a cost to α-phase, and not at cost to each other, as would be the case if the enlargement process had a similar nature to the widely known grain coarsening process.

5. Future Developments

In the future, faster area detectors and translational stages will be implemented in order to improve the time resolution of this technique, while improvement of its spatial resolution will require smaller incident beams. Improvement of both spatial and time resolution will provide better details of the phase transition mechanisms and interface developments thereof. For example, implementation of

X-ray polychromatic beams focused down below 100 nm [12] may provide details of the mechanism of the pressure induced grain enlargement observed across the $\alpha \to \omega$ phase transition in Zr.

A clear advantage of DAC compared to other stress generation devices is that the sample can be studied under hydrostatic compression or alternately under shear. Hydrostatic pressure using either He or Ne as a transmitting medium can be contrasted with studies with other media (such as silicone oil for example), thereby modifying the shear forces.

Supplementary Materials: The following are available online at http://www.mdpi.com/2073-4352/9/12/672/s1, Figure S1: Axis Pro Micro Support micromanipulator available in sample preparation laboratory of HPCAT, Figure S2: maps of reflections (indices are shown) from a crystal of ω-Zr, Movies S1a and S1b: series of composite frames of reflections from coexisting α- and β-Si [1] at the onset of $\alpha \to \beta$ transition (S1a) and when the transition was nearly over (S1b), arrows of different colors in S1a denote reflections from different crystals of α-Si, rectangle in S1a denotes overlap area with S1b, Movie S2: series of composite frames of a reflection from an ω-Zr crystal growing during the $\alpha \to \omega$ transition [27], time intervals since starting of data collection are shown in the movies.

Author Contributions: Conceptualization, D.P. and N.V.; software, D.P.; writing—original draft preparation, D.P. and N.V.; writing—review and editing, M.S.

Funding: This research was funded by DOE-NNSA's Office of Experimental Sciences. The Advanced Photon Source was funded by the U.S. Department of Energy (DOE) Office of Science.

Acknowledgments: This work was performed at HPCAT (Sector 16), Advanced Photon Source (APS), Argonne National Laboratory. HPCAT operations are supported by DOE-NNSA's Office of Experimental Sciences. The Advanced Photon Source is a U.S. Department of Energy (DOE) Office of Science User Facility operated for the DOE Office of Science by Argonne National Laboratory under Contract No. DE-AC02-06CH11357. Part of this work was performed under the auspices of the U.S. Department of Energy by Lawrence Livermore National Laboratory under Contract DE-AC52-07NA27344.

Conflicts of Interest: The authors declare no conflict of interest.

References

1. Popov, D.; Park, C.; Kenney-Benson, C.; Shen, G. High pressure Laue diffraction and its application to study microstructural changes during the $\alpha \to \beta$ phase transition in Si. *Rev. Sci. Instrum.* **2015**, *86*, 072204. [CrossRef] [PubMed]
2. Popov, D.; Sinogeikin, S.; Park, C.; Rod, E.; Smith, J.; Ferry, R.; Kenney-Benson, C.; Velisavljevic, N.; Shen, G. New Laue micro-diffraction setup for real-time in situ microstructural characterization of materials under external stress. In *Advanced Real Time Imaging II. The Minerals, Metals & Materials Series*; Nakano, J., Chris Pistorius, P., Tamerler, C., Yasuda, H., Zhang, Z., Dogan, N., Wang, W., Saito, N., Webler, B., Eds.; Springer: Cham, Germsny, 2019; pp. 43–48. [CrossRef]
3. Shen, G.; Mao, H.K. High-pressure studies with x-rays using diamond anvil cells. *Rep. Prog. Phys.* **2017**, *80*, 016101. [CrossRef] [PubMed]
4. McMahon, M.I. High-pressure crystallography. *Top. Curr. Chem.* **2012**, *315*, 69–110. [CrossRef] [PubMed]
5. Lorenz, H.; Lorenz, B.; Kuhne, U. The kinetics of cubic boron nitride formation in the system BN-Mg$_3$N$_2$. *J. Mater. Sci.* **1988**, *23*, 3254–3257. [CrossRef]
6. Dupas-Bruzek, C.; Sharp, T.G.; Rubie, D.C.; Durham, W.B. Mechanisms of transformation and deformation in Mg$_{1.8}$Fe$_{0.2}$SiO$_4$ olivine and wadsleyite under non-hydrostatic stress. *Phys. Earth Planet. Inter.* **1998**, *108*, 33–48. [CrossRef]
7. Kerschhofer, L.; Rubie, D.C.; Sharp, T.G.; McConnell, J.D.C.; Dupas-Bruzek, C. Kinetics of intracrystalline olivine–ringwoodite transformation. *Phys. Earth Planet. Inter.* **2000**, *121*, 59–76. [CrossRef]
8. Ice, G.E.; Pang, J.W.L. Tutorial on x-ray microLaue diffraction. *Mater. Charact.* **2009**, *60*, 1191–1201. [CrossRef]
9. Cornelius, T.W.; Thomas, O. Progress of in situ synchrotron X-ray diffraction studies on the mechanical behavior of materials at small scales. *Prog. Mater. Sci.* **2018**, *94*, 384–434. [CrossRef]
10. Robach, O.; Kirchlechner, C.; Micha, J.S.; Ulrich, O.; Biquard, X.; Geaymond, O.; Castelnau, O.; Bornert, M.; Petit, J.; Berveiller, S.; et al. Laue microdiffraction at the ESRF. In *Strain and Dislocation Gradients from Diffraction*; Barabash, R., Ice, G., Eds.; Imperial College Press: London, UK, 2014; pp. 156–204.
11. Ulrich, O.; Biquard, X.; Bleuet, P.; Geaymond, O.; Gergaud, P.; Micha, J.S.; Robach, O.; Rieutord, F. A new white beam x-ray microdiffraction setup on the BM32 beamline at the European Synchrotron Radiation Facility. *Rev. Sci. Instrum.* **2011**, *82*, 033908. [CrossRef] [PubMed]

12. Liu, W.; Ice, G.E. X-ray Laue diffraction microscopy in 3D at the Advanced Photon Source. In *Strain and Dislocation Gradients from Diffraction*; Barabash, R., Ice, G., Eds.; Imperial College Press: London, UK, 2014; pp. 53–81.

13. Tamura, N.; Kunz, M.; Chen, K.; Celestre, R.S.; MacDowell, A.A.; Warwick, T. A superbend X-ray microdiffraction beamline at the advanced light source. *Mater. Sci. Eng. A* **2009**, *524*, 28–32. [CrossRef]

14. Tischler, J.Z. Reconstructing 2D and 3D X-ray orientation maps from white-beam Laue. In *Strain and Dislocation Gradients from Diffraction*; Barabash, R., Ice, G., Eds.; Imperial College Press: London, UK, 2014; pp. 358–375.

15. ImageJ. Available online: https://imagej.nih.gov/ij/ (accessed on 12 December 2019).

16. Smith, J.S.; Desgreniers, S. Selected techniques in diamond anvil cell crystallography: Centring samples using X-ray transmission and rocking powder samples to improve X-ray diffraction image quality. *J. Synchrotron Radiat.* **2009**, *16*, 83–96. [CrossRef] [PubMed]

17. Smith, J.S.; Rod, E.A.; Shen, G. Fly scan apparatus for high pressure research using diamond anvil cells. *Rev. Sci. Instrum.* **2019**, *90*, 015116. [CrossRef]

18. Mao, H.K.; Xu, J.; Bell, P.M. Calibration of the Ruby pressure gauge to 800-kbar under quasi-hydrostatic conditions. *J. Geophys. Res. Solid Earth* **1986**, *91*, 4673–4676. [CrossRef]

19. Hrubiak, R.; Sinogeikin, S.; Rod, E.; Shen, G. The laser micro-machining system for diamond anvil cell experiments and general precision machining applications at the High Pressure Collaborative Access Team. *Rev. Sci. Instrum.* **2015**, *86*, 072202. [CrossRef] [PubMed]

20. Chung, J.S.; Ice, G.E. Automated indexing for texture and strain measurement with broad-bandpass x-ray microbeams. *J. Appl. Phys.* **1999**, *86*, 5249–5255. [CrossRef]

21. Barabash, R.; Barabash, O.; Popov, D.; Shen, G.; Park, C.; Yang, W. Multiscale twin hierarchy in NiMnGa shape memory alloys with Fe and Cu. *Acta Mater.* **2015**, *87*, 344–349. [CrossRef]

22. Hahn, T.; Klapper, H. Twinning of crystals. In *International Tables for Crystallography*, 2nd ed.; Chapter, 3.3; Authier, A., Ed.; John Wiley and Sons Limited: Hoboken, NJ, USA, 2013; Volume D, pp. 413–487. [CrossRef]

23. Python. Available online: https://www.python.org/ (accessed on 12 December 2019).

24. Hammersley, A.P.; Svensson, S.O.; Hanfland, M.; Fitch, A.N.; Hausermann, D. Two-dimensional detector software: From real detector to idealized image or two-theta scan. *J. High Press. Res.* **1996**, *14*, 235–248. [CrossRef]

25. Prescher, C.; Prakapenka, V. DIOPTAS: A program for reduction of two-dimensional X-ray diffraction data and data exploration. *J. High Press. Res.* **2015**, *35*, 223–230. [CrossRef]

26. Velisavljevic, N.; Chesnut, G.N.; Stevens, L.L.; Dattelbaum, D.M. Effects of interstitial impurities on the high pressure martensitic α to ω structural transformation and grain growth in zirconium. *J. Phys. Condens. Matter* **2011**, *23*, 125402. [CrossRef] [PubMed]

27. Popov, D.; Velisavljevic, N.; Liu, W.; Hrubiak, R.; Park, C.; Shen, G. Real time study of grain enlargement in zirconium under room-temperature compression across the α to ω phase transition. *Sci. Rep.* **2019**, *9*, 15712. [CrossRef] [PubMed]

Review

Pressure-Induced Phase Transitions in Sesquioxides

Francisco Javier Manjón [1,*], Juan Angel Sans [1], Jordi Ibáñez [2] and
André Luis de Jesús Pereira [3]

1   Instituto de Diseño para la Fabricación y Producción Automatizada, MALTA-Consolider Team, Universitat
    Politècnica de València, 46022 Valencia, Spain; juasant2@upv.es
2   Institute of Earth Sciences Jaume Almera, MALTA-Consolider Team, Consell Superior d'Investigacions
    Científiques (CSIC), 08028 Barcelona, Catalonia, Spain; jibanez@ictja.csic.es
3   Grupo de Pesquisa de Materiais Fotonicos e Energia Renovável—MaFER, Universidade Federal da Grande
    Dourados, Dourados 79825-970, MS, Brazil; andreljpereira@gmail.com
*   Correspondence: fjmanjon@fis.upv.es; Tel.: +34-699-133-078

Received: 12 November 2019; Accepted: 26 November 2019; Published: 28 November 2019

Abstract: Pressure is an important thermodynamic parameter, allowing the increase of matter density by reducing interatomic distances that result in a change of interatomic interactions. In this context, the long range in which pressure can be changed (over six orders of magnitude with respect to room pressure) may induce structural changes at a much larger extent than those found by changing temperature or chemical composition. In this article, we review the pressure-induced phase transitions of most sesquioxides, i.e., A_2O_3 compounds. Sesquioxides constitute a big subfamily of ABO_3 compounds, due to their large diversity of chemical compositions. They are very important for Earth and Materials Sciences, thanks to their presence in our planet's crust and mantle, and their wide variety of technological applications. Recent discoveries, hot spots, controversial questions, and future directions of research are highlighted.

Keywords: sesquioxides; high pressure; phase transitions

---

1. Introduction

The family of sesquioxides (SOs), i.e., compounds with A_2O_3 stoichiometry, is very important from an applied point of view, since SOs play a vital role in the processing of ceramics as additives, grain growth inhibitors and phase stabilizers. They also have potential applications in nuclear engineering and as hosts for optical materials for rare-earth (RE) phosphors. Besides, SOs with corundum-like structure, like corundum (α-Al_2O_3), hematite (α-Fe_2O_3) and eskolaite (α-Cr_2O_3), are relevant to Earth and planetology Sciences, because they are minerals present in the Earth's crust or in meteorites. Therefore, knowledge and understanding of phase transitions (PTs) occurring at high pressure (HP) in SOs is very important for Physics, Chemistry and Earth and Materials Sciences.

SOs are mainly formed by materials featuring an A cation of valence 3+, so that the two A cations compensate the total negative valence (6−) of three O atoms acting with valence 2−. Among the cations featuring 3+ valence, we find all group-13 and group-15 elements, some transition metal (TM) elements and all RE elements. Additionally, we can also find SOs where cations show an average 3+ valence. In this context, SOs with cations showing both 2+ and 4+ valences (both valences present in the A_2O_3 compound), are named mixed-valence (MV) SOs. The different compositions of these compounds provide a very rich environment ranging from molecular or zero-dimensional (0D), one-dimensional (1D) and two-dimensional (2D) solids to the conventional (bulk) three-dimensional (3D) solids; all of them showing radically different properties and behaviors under compression.

After more than 50 years of exploration of pressure-induced PTs in SOs, we present in this work a review of the PTs of SOs at HP, including those occurring at high temperatures (HTs), at low temperatures like room temperature (RT), and even at room pressure (RP). We have structured the present paper in five main sections that are devoted to RE-SOs, group-13-SOs, group-15-SOs, TM-SOs, and MV-SOs.

Finally, before attempting to present our results, we must comment that several reviews of the PTs in SOs at HP have been already published. We can cite the review on oxides of Liu and Basset in 1986 [1], of Adachi and Imanaka in 1998 [2], of Smyth et al. in 2000 [3], of Zinkevich in 2007 [4] and of Manjón and Errandonea in 2009 [5]. The present review goes beyond previous studies of SOs by expanding the number of families studied, by discussing the most recent results published in the last decade, and by suggesting pressure-based studies still to be performed.

2. High-Pressure Phase Transitions in Sesquioxides

2.1. Rare-Earth Sesquioxides

RE-SOs are highly interesting and versatile for different types of applications, since the atomic radius can be finely tuned along the lanthanide family, thus enabling a wide range of technological applications, including light emitters (lasers and improved phosphors), catalysts, and high-dielectric constant (high-k) gates.

RE-SOs, and in particular lanthanide SOs (Ln_2O_3; Ln = La to Lu, including Y and Sc), usually crystallize at room conditions in either the A-, B-, or C-type structures, depending on the RE atomic size. Large cations from La to Nd usually crystallize in the trigonal A-type structure (s.g. $P\bar{3}m1$), medium-size cations from Sm to Gd tend to show a monoclinic B-type structure (s.g. $C2/m$), and small cations from Tb to Lu, including Sc and Y, tend to adopt a cubic C-type structure (s.g. $Ia\bar{3}$). At HTs, two additional phases, hexagonal H (s.g. $P6_3/mmc$) and cubic X (s.g. $Im\bar{3}m$) structures, have also been found in Ln_2O_3 [6]. Zinkevich has reviewed temperature-induced PTs in RE-SOs and found that a C→B→A sequence of PTs usually occurs upon an increase in temperature [4]. Since the density and cation coordination of the A-, B- and C-type structures increase in the sequence C-B-A, C→B, B→A, and even direct C→A transitions are expected on increasing pressure (upstroke), while inverse PTs are expected on decreasing pressure (downstroke) [7,8]. Curiously, it has been found that the cationic distances in the three structures of the RE-SOs at room conditions are similar to those in the bcc, fcc and hcp structures of the RE metals, thus suggesting that a relationship must exist between them [9,10].

The effect of HP on Ln_2O_3 has been extensively studied by many research groups, mainly by X-ray diffraction (XRD), Raman scattering (RS) and photoluminescence measurements. The HP behavior of C-type compounds has been studied in Lu_2O_3 [11–14], Yb_2O_3 [15–18], Tm_2O_3 [19,20], Er_2O_3 [11,21,22], Ho_2O_3 [23–26], Dy_2O_3 [27,28], Gd_2O_3 [11,29–36], Eu_2O_3 [11,37–43], Sm_2O_3 [11,32,44,45], Sc_2O_3 [46–48], Y_2O_3 [11,32,49–60], and recently on Tb_2O_3 [61]. Examples of pressure-driven C→B, B→A, direct C→A, and even C→A+B transitions have been shown to occur in a number of RE-SOs. As example, Lu_2O_3 was found to show either a C→B PT around 12 GPa without B→A PT up to 47 GPa [12,13], or a C→A+B PT above 25 GPa, with a recovery of a single B phase on downstroke from 47 GPa [14]. Similarly, Yb_2O_3 was found to exhibit a C→B PT above 13 GPa, and no further PT was found up to 20 GPa [15], but an A phase was found above 19.6 GPa that transforms into the B phase below 12 GPa [16]. The A phase in coexistence with the B phase was confirmed by RS measurements up to 45 GPa [17]. The C→B→A sequence of PTs was also found in Y_2O_3, the most studied RE-SO at HP, around 12 and 19 GPa [49], and has been confirmed in many studies [60]. In this context, it must be stressed that the pressure-induced PT sequence in Y_2O_3 has been shown to be affected by the nature of the sample, since different sequences of PTs have been observed in nanocrystals [33]. It must be stressed that direct C→A PT, with A→B PT on downstroke from 44 GPa, has been observed in Sm_2O_3 [44]. This sequence of PTs has also been observed in Gd_2O_3 [31]. Other PTs in RE-SOs have been found at HP and HT. In particular, an orthorhombic Gd_2S_3-type structure (s.g. *Pnma*) has been found in C-type Y_2O_3 above 8 GPa [57].

The HP behavior of B-type compounds has been also studied in Y_2O_3 [62], Eu_2O_3 [63], and Sm_2O_3 [64,65]. A reversible pressure-induced B→A PT has been observed in all B-type compounds. Additionally, the Gd_2S_3-type structure has been found in B-type Sc_2O_3 above 19 GPa upon laser heating [47]. Table 1 summarizes the experimental PT pressures observed in C- and B-type RE-SOs.

Table 1. Experimental data of the phase transition pressures (in GPa) for Ln_2O_3 and related sesquioxides having a C- or B-type phase at room conditions. The measurement technique (Tech.) and the pressure-transmitting medium (PTM) are also provided. Comp. stands for compound, MEW stands for the 16:3:1 mixture of methanol:ethanol:water and Polyethy. stands for polyethylene glycol.

Comp.	$P_{C \to A}$	$P_{C \to B}$	$P_{B \to A}$	Tech.	PTM	Reference
Sc_2O_3		36		XRD	MEW	[46]
		25–28		XRD	Neon	[48]
Lu_2O_3	14			XRD	N_2	[11]
		12.7–18.2		XRD	Silicon Oil	[12]
		17.0–21.4		XRD	ME	[13]
Yb_2O_3		13		XRD	Silicon Oil	[15]
	30–47	14	11.9	XRD	MEW	[14,16]
	17.0			XRD	N_2	[11]
	20.6			RS	ME	[18]
Tm_2O_3		7		XRD	MEW	[19]
		12		XRD	MEW	[20]
Er_2O_3		9.9–16.3		XRD	Silicon Oil	[66]
	14			XRD	N_2	[11]
	17.8–20.0	13.6	17.8–23.5	XRD	He	[22]
		6.6–12.7	22.5–42.0	XRD	Silicon Oil	
Y_2O_3	13.0			XRD	N_2	[11]
	15.0–25.6	15.0–25.6	14.3–17.5	XRD	MEW	[57]
			23.5–44.0	XRD	Neon	[62]
		13	24.5	XRD	No PTM	[54]
Ho_2O_3		8.9–16.3	14.8–26.4	XRD	Silicon Oil	[24]
		9.5		XRD	MEW	[19]
		9.5–16.0		XRD	MEW	[23]
		8.8		XRD	MEW	[20]
Dy_2O_3		7.7–18.8	10.9–26.6	XRD	Silicon Oil	[28]
Tb_2O_3		7	12	XRD	MEW	[61]
Gd_2O_3	7.0–15.0		6.8	XRD	MEW	[31]
	8.9–14.8	2.5	8.9	XRD	Ar	[33]
	7.0			XRD	N_2, Ar, He	[11]
	8.6–12.5		5.1	XRD	Silicon Oil	[34]
	12.0			XRD	MEW	[19]
Eu_2O_3			4.7	XRD	Silicon Oil	[63]
	5.0–13.1			XRD	Silicon Oil	[40]
	6.0			XRD	N_2	[11]
	5.7–12.9			XRD	MEW	[41,42]
Sm_2O_3			3.2–3.9	XRD	Polyethy.	[64]
	7.5–12.5		4.7	XRD	Silicon Oil	[44]
	4.0			XRD	N_2, Ar	[11]
	4.2		2.5	XRD	Silicon Oil	[45]

The HP behavior of A-type compounds has been studied in Nd_2O_3 [67,68], Ce_2O_3 [69], and La_2O_3 [11]. A-type Nd_2O_3 has been observed to suffer a reversible PT above 27 GPa towards a monoclinic s.g. *P2/m* phase [67]. However, other work has suggested a 1st-order isostructural PT above 10 GPa [68]. This behavior is in contrast with that of Ce_2O_3, which seems to be stable in the A-type phase up to 70 GPa. Curiously, A-type Ce_2O_3 shows anomalies in the compressibility of the lattice parameters around 20 GPa, which could be related to an isostructural PT similar to that of Nd_2O_3. Finally, A-type La_2O_3 seems not to be stable above 4 GPa, following a PT to a hexagonal superlattice and eventually to a distorted monoclinic structure with a group–subgroup relationship with the hexagonal one [11].

It must be mentioned that RE-SOs and their pressure-induced PTs have also been investigated theoretically, and the simulations of the bandgap, volume, bulk modulus and its pressure derivative for the C, B and A phases, as well as their PT pressures, have been reported and found in rather good agreement with experiment [70–83]. Theoretical studies have found that the C→A and C→B PTs result in a decrease in the unit cell volume per formula unit at the transition pressure of the order of 10% ± 2%, in good agreement with experiments [81]. On the other hand, the B→A PT results in a decrease in the unit cell volume per formula unit at the transition pressure of the order of less than 2%, in good agreement with experiments [76]. Figures 1–3 show a comparison of experimental data (from Table 1) and theoretical data for the C→A, C→B and B→A PT pressures in bulk RE-SOs as a function of the atomic number of the lanthanide, Z_{Ln}, and of the ionic radius of the RE cation. In general, a good agreement between theoretical and experimental data can be found for all PTs. As regards the C→A PT, the pressure of the transition increases along the lanthanide series with the increase in atomic number, i.e., with the decrease in the ionic radius. It must be noted that no work has reported the C→A PT for Sc_2O_3, and that the reported C→A PT in materials with a small ionic radius, like Lu_2O_3 and Yb_2O_3, are doubtful. Regarding the C→B PT, the pressure of the transition also increases along the lanthanide series with the increase in atomic number or the decrease in the ionic radius. Unfortunately, no theoretical data exist for the C→B PT along the whole RE-SOs series; consequently, theoretical data for the C→A PTs are plotted for comparison with experimental data of the C→B PTs. It must be noted that the C→B PT is not observed in Sm_2O_3 and Eu_2O_3, since these two SOs undergo a direct C→A PT. Finally, as regards the B→A PT, the pressure of the transition once again increases along the lanthanide series with the increase in atomic number or the decrease in the ionic radius. As observed, no experimental data for the B→A PT has been reported in Sc_2O_3, Lu_2O_3 and Tm_2O_3. This will require new experiments up to 40 GPa in Lu_2O_3 and Tm_2O_3 and above 80 GPa in Sc_2O_3.

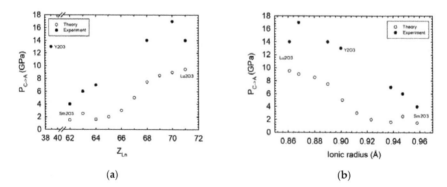

(a)   (b)

Figure 1. Experimental (solid symbols) and theoretical (empty symbols) data for the C→A phase transition pressures in Ln_2O_3 and related sesquioxides as a function of (**a**) the atomic number, Z_{Ln}, and (**b**) the ionic radius. Experimental data are from Table 1, while theoretical data are from reference [81]. Ionic radii of Shannon for cation 6-fold coordination were considered.

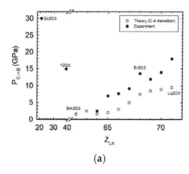 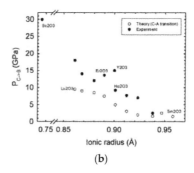

(a) (b)

Figure 2. Experimental (solid symbols) and theoretical (empty symbols) data for the C→B phase transition pressures in Ln_2O_3 and related sesquioxides as a function of (**a**) the atomic number, Z_{Ln}, and (**b**) the ionic radius. Experimental data are from Table 1, while theoretical data correspond to those of the C→A phase transition in reference [81], since no theoretical data for the C→B phase transition are known to our knowledge. Ionic radii of Shannon for cation 6-fold coordination were considered.

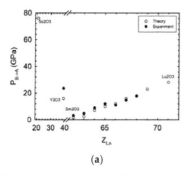 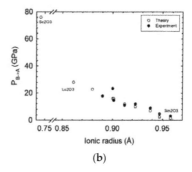

(a) (b)

Figure 3. Experimental (solid symbols) and theoretical (empty symbols) data for the B→A phase transition pressures for Ln_2O_3 and related sesquioxides as a function of (**a**) the atomic number, Z_{Ln}, and (**b**) the ionic radius. Experimental data are from Table 1, while theoretical data are from reference [76]. Ionic radii of Shannon for cation 6-fold coordination were considered.

In summary, there are many HP studies in cubic C-type RE-SOs that allow us to conclude that this structure undergoes a pressure-induced irreversible 1st-order PT either to a hexagonal A-type structure or to an intermediate monoclinic B-type structure. It is also observed that the A-type structure of RE-SOs usually reverts to the B-type structure on decreasing pressure, but the B-type structure remains metastable at room conditions and does not revert to the C-type structure unless it is conveniently heated. The irreversibility of the C→A and C→B PTs means that there is a considerable kinetic energy barrier between the C-type and the A- and B-type structures, due to the reconstructive nature of these PTs. On the other hand, the reversibility of the B→A PT is consequence of the weak 1st-order nature of this PT, as expected, because A- and B-type structures obey a group-subgroup relationship. In this context, HP studies of B-type RE-SOs show a pressure-induced reversible weak 1st-order PT to the A-type structure. It must be noted that some PTs for RE-SOs with a small RE ionic radius have yet to be reported. Besides, a pressure-induced PT to the Gd_2S_3 structure has been found from some C- and B-type RE-SOs at HT. Finally, we must comment that HP studies on trigonal A-type RE-SOs are quite scarce and do not allow us to conclude which is the typical HP phase of A-type RE-SOs. Consequently, more work has to be done in A-type RE-SOs, either studying C- or B-type RE-SOs at higher pressures or studying in better detail RE-SOs crystallizing in the A phase at room conditions, in order to understand the behavior of the trigonal structure and its pressure-induced PTs.

2.2. Group-13 Sesquioxides

Group-13-SOs are highly interesting compounds, due to their chemical and physical properties, which enable their application as catalyst, lasers and light emitting materials [84]. HP studies in group-13-SOs have been conducted on the five compounds. Al_2O_3, also known as alumina, is by far the most studied compound [85–99], followed by In_2O_3 [100–109], Ga_2O_3 [107,110–114], and B_2O_3 [115–121]. Finally, Tl_2O_3 has been the least studied compound [122].

The interest in alumina comes from its high stability, making it interesting for HP science applications, especially as windows in shockwave experiments [123]. However, its main application occurs when alumina is doped with a small amount of Cr^{3+}. $Al_2O_3:Cr^{3+}$, also known as ruby, presents fluorescence lines that are pressure-dependent and is used as a pressure standard in static HP experiments using anvil cells [124]. Corundum (α-Al_2O_3) is a mineral of alumina that is found in abundance in the earth mantle incorporated into enstatite ($MgSiO_3$) [99]. Given that both minerals might adopt similar structures at ultra-HPs, their study is interesting from a geophysical point of view. In fact, the most stable phase of alumina at room conditions is the corundum structure (s.g. $R\bar{3}c$), where coordination of all Al atoms is six-fold and of O atoms is four-fold [94]. The unit cell has two formula units, with O atoms forming a slightly distorted, hexagonal, close-packed structure [88,125]. The same structure is also found in other SO minerals, such as hematite (Fe_2O_3), eskolaite (Cr_2O_3), karelianite (V_2O_3), and also Ti_2O_3 and Ga_2O_3 [125]. In this structure, cations occupy positions $(0, 0, z)$ and oxygens occupy positions $(x, 0, \frac{1}{4})$ and, if both z and x coordinates were 1/3, the corundum-type structure would have a perfect hexagonal close-packed packing [125].

β-Al_2O_3 is another very interesting polymorph of alumina from a technological application point of view. This material is a non-stoichiometric compound ($Na_2O_x \cdot Al_2O_3$, $5 \le x \le 11$) and not a simple variation of alumina [126,127]. Due to its layered crystalline structure, β-Al_2O_3 has a high ionic conductivity and has been studied for solid state battery applications [126,127]. We have not found any HP study of this material.

Theoretical studies of corundum predicted a pressure-induced PT to the Rh_2O_3-II structure (s.g. *Pbcn*) between 78 and 91 GPa [87,92,96,98] and a second PT to an orthorhombic perovskite (Pv) structure (s.g. *Pbnm*) at 223 ± 15 GPa [87,92,96,98,128]. Other theoretical studies predicted that the post-perovskite (PPv) $CaIrO_3$-type structure (s.g. *Cmcm*) would be more stable than the Pv structure at pressures higher than 150 GPa [85,99,129]. However, RT measurements performed up to 175 GPa under hydrostatic and non-hydrostatic conditions showed no PT [88,89]. The PT to the Rh_2O_3-II structure was only observed at HP (>95 GPa) and HT (>1200 K) [86,97] once the energy supplied by temperature allows the overcoming of the kinetic barrier of this 1st-order PT [86,89,96,97,128]. The Rh_2O_3-II structure is directly related to corundum structure; nevertheless, it presents a great distortion in the AlO_6 octahedron [97]. Lin et al. reported that the Rh_2O_3-II-type Al_2O_3 (Cr^{3+} doped) remains stable up to 130 GPa and 2000 K [97]. However, studies using shock waves have shown that a decrease by one order of magnitude in resistivity occurs at 130 GPa [123] that it is coincident with an increase in the density of Al_2O_3 [130], thus indicating a possible PT. This hypothesis was supported by the discovery of a PPv $CaIrO_3$-type polymorph of Fe_2O_3 [131] and $MgSiO_3$ [132]. In 2005, Oganov and Ono were finally able to synthesize the PPv $CaIrO_3$-type phase of Al_2O_3 at ~ 170 GPa and 2500 K [94,129]. They showed that the Rh_2O_3-II → $CaIrO_3$-type PT is a 1st-order PT with ~ 5% of volume decrease [94,129]. The possibility of obtaining this structure at HP and HT was important to understand the physical properties of the elements present in the D″ layer of the earth mantle [129]. This layer is basically composed of a solid solution of $MgSiO_3$, Fe_2O_3 and Al_2O_3. Since, under the HP and HT conditions of the Earth mantle, all these materials are isostructural, their solubility may be facilitated [94].

When synthesized as nanoparticles smaller than 20 nm, alumina tends to crystallize into a cubic structure known as a phase γ (s.g. $Fd\bar{3}m$) [133]. HP experiments at room temperature observed an amorphization of γ-Al_2O_3 around 5 GPa [133]. On the other hand, it was also observed that the γ-Al_2O_3 → $R\bar{3}c$ PT, which occurs at RP near 1473 K, decreases to ~ 1023 K when γ-Al_2O_3 is pressurized at

1 GPa [134]. In addition, HP-HT experiments have revealed other phases of alumina, like polymorphs of RE-SOs B- and C-type, as well as polymorphs λ and μ [135,136]; however, no HP studies on these phases have been reported to our knowledge.

The search for new high hardness materials, such as Al_2O_3, has attracted the attention of many research groups for SOs, including indium oxide (In_2O_3) and gallium oxide (Ga_2O_3) [106]. In_2O_3 is a semiconductor with high potential for many transparent oxide applications, including touch and photovoltaic devices, thermoelectric and gas sensors [137]. At room conditions, In_2O_3 crystallizes in the cubic bixbyite structure (s.g. $Ia\bar{3}$), also known as the C-type RE-SO structure (c-In_2O_3) [106]. In this structure there are two types of In atoms (they are surrounded by oxygen in the octahedral and trigonal prismatic coordination) located at positions 8b and 24d and one type of O located at position 48e [138]. In the 1960s, a PT from c-In_2O_3 to a metastable corundum-like rhombohedral phase (rh-In_2O_3; s.g. $R\bar{3}c$) was reported at HP and HT (6.5 GPa and 1000 °C) [106,139,140]. This PT was also observed through shockwave compression experiments between 15 and 25 GPa [141]. Some HP-XRD studies of both bulk and nanoparticle samples claimed that the c-In_2O_3 → rh-In_2O_3 PT occurs at RT when the sample is exposed to pressures between 12 and 25 GPa [103,104]. This result is controversial since other studies using HP-XRD and HP-RS measurements did not clearly observe this PT up to 30 GPa [100,101]. On the other hand, García-Domene et al. reported that bulk c-In_2O_3 actually undergo a PT to an Rh_2O_3-II-type structure (o1-In_2O_3; s.g. *Pbcn*) orthorhombic phase at pressures above 31 GPa at RT [101]. This structure was also obtained at 7 GPa after laser heating at 1500 °C [106]. This result is consistent with the prediction of theoretical calculations that the Rh_2O_3-II-type phase is the most stable In_2O_3 structure between 8 GPa and 36 GPa [101]. On top of that, looking for a possible $CaIrO_3$-type PPv phase in In_2O_3, Yusa et al. observed another PT from Rh_2O_3-II-type to α-Ga_2S_3-type structure (s.g. *Pnma*) above 40 GPa and 2000 K [107]. A reduction of 7%–8% in volume was observed, unlike the reduction of 2% expected for a Rh_2O_3-II-type → $CaIrO_3$-type transition [107].

Releasing the pressure after obtaining the Rh_2O_3-II-type structure in In_2O_3, García-Domene et al. demonstrated that, at 12 GPa, o1-In_2O_3 undergoes a PT to a previously unknown phase [101]. The combination of experimental and theoretical results helped to conclude that, upon decreasing the pressure to 12.1 GPa, the In_2O_3 recrystallizes in a distorted bixbyite-like structure (o3-In_2O_3; s.g. *Pbca*), isostructural to Rh_2O_3-III [101,108]. Decreasing the pressure below 1 GPa, o3-In_2O_3 undergoes another PT to the metastable rh-In_2O_3 structure. The metastable rh-In_2O_3 phase at room conditions remains stable up to 15 GPa and above this pressure shows a reversible PT to o3-In_2O_3, which is stable up to 25 GPa [101]. The observation of the *Pbca* phase took place only when decreasing pressure using original bulk c-In_2O_3 samples was related to kinetic barriers in the c-In_2O_3 → o3-In_2O_3 PT, that cannot be overcome at RT during upstroke [101]. Similar studies in c-In_2O_3 nanoparticles performed up to 44 GPa did not clearly show these PTs [100,104,105]. Qi et al. reported that, when compressing 6 nm c-In_2O_3 nanoparticles, an irreversible PT to the rh-In_2O_3 phase occurs between 20–25 GPa [104]. Nowadays, it is now possible to synthesize In_2O_3 nanoparticles at room conditions with the rh-In_2O_3 metastable phase at relatively low temperatures (250 to 500 °C) [142–144]. Sans et al. studied rh-In_2O_3 nanoparticles (10–30 nm) up to 30 GPa using XRD and RS measurements [109]. XRD measurements did not show any PT up to 30 GPa, but RS measurements showed peaks of the o3-In_2O_3 phase above 20 GPa [109]. One reason for this discrepancy was attributed to the sensitivity of RS measurements to local structural changes, which is responsible for the observation of PTs at lower pressures in RS compared to XRD measurements [109]. Another reason might be related to the use of less hydrostatic pressure-transmitting medium in RS (MEW) than in XRD (Ar) measurements [109].

Regarding Ga_2O_3, its most stable structure at room conditions is the monoclinic phase (s.g. *C2/m*), known as β-Ga_2O_3 [145,146]. However, depending on the pressure, temperature and atmosphere conditions, it is possible to find Ga_2O_3 in the α, β, γ, σ and ε phases [111,147]. Compared to Al_2O_3 and In_2O_3, HP studies on Ga_2O_3 are scarcer. In 1965, Remeika and Marezio reported the synthesis of α-Ga_2O_3 (s.g. $R\bar{3}c$), which is isostructural to α-Al_2O_3, by pressurizing β-Ga_2O_3 to 4.4 GPa, after heating to 1000 °C and quenching to room conditions [147]. Lipinska-Kalita et al. reported that, upon pressing nanoparticles (~15 nm) of β-Ga$_2$O$_3$ homogeneously dispersed in a host silicon oxide-based glass matrix, it is possible to observe a PT to α-Ga$_2$O$_3$ at 6 GPa [111]. In that study, the β-Ga$_2$O$_3$ → α-Ga$_2$O$_3$ PT is not completed up to 15 GPa and it is not clear whether the PT is induced by glass matrix densification or is an intrinsic behavior of the β-Ga$_2$O$_3$ nanoparticles [111]. This work led Machon et al. to study the behavior of β-Ga$_2$O$_3$ microparticles by XRD and RS measurements with different pressure-transmitting media [113]. They observed a PT to the α phase above 20 GPa, unlike what was observed in reference [111]. A highly disordered α-Ga$_2$O$_3$ was recovered at RP after a fast decompression from 25 GPa and a mixture of α and β phases were recovered when decompression was performed slowly [113]. The β-Ga$_2$O$_3$ → α-Ga$_2$O$_3$ PT can be considered to be of 1st-order and involves a change in O^{2-} ions packing from cubic to hexagonal, accompanied by a change in Ga^{3+} coordination from four to six [111,113,148].

The β-Ga$_2$O$_3$ → α-Ga$_2$O$_3$ PT was not clearly observed in an experiment performed up to 65 GPa without pressure-transmitting medium [106]. Instead, XRD measurements at 65 GPa showed only a diffraction pattern similar to a material with low crystallinity or amorphous. After laser heating (~2300 K), a recrystallization to an orthorhombic phase Rh$_2$O$_3$-II-type (s.g. *Pbcn*) was observed [106]. This phase remained stable up to 108 GPa and 2500 K. During decompression, the Rh$_2$O$_3$-II-type → α-Ga$_2$O$_3$ PT was observed between 32 and 21 GPa, being the α phase metastable at RP [106]. These results differ from other published by Lipinska-Kalite et al. up to 70 GPa with (N$_2$) and without pressure-transmitting medium [110]. In that study, a β-Ga$_2$O$_3$ → α-Ga$_2$O$_3$ PT was clearly observed, which starts at 6.5–7.0 GPa (3 GPa) and completes at 40 GPa (30 GPa) when measured with (without) pressure-transmitting medium [110]. Through XRD measurements, Lipinska-Kalita et al. observed no evidence of pressure-induced amorphization or deterioration of the diffractogram up to 70 GPa [110]. The β-Ga$_2$O$_3$ → α-Ga$_2$O$_3$ PT was also observed at 16 GPa in shockwave experiments [149]. Wang et al. conducted HP studies on β-Ga$_2$O$_3$ nanoparticles (14 nm) at RT and found the PT to α-Ga$_2$O$_3$ above 13.6 GPa [114]. Above this pressure, the α phase remains stable up to 64.9 GPa and stays metastable upon returning to RP [114].

Boron sesquioxide (B$_2$O$_3$) is one of the most important oxides in applications involving vitreous systems. In glass form, B$_2$O$_3$ has a low melting point (750 K) and a structure similar to vitreous SiO$_2$ and H$_2$O [117]. When pressing the glassy material, it is possible to observe a structural rearrangement around 3 GPa that was considered a low- to high-density vitreous PT [118]. Furthermore, starting from the glassy material, it is possible to obtain B$_2$O$_3$ crystals at HP and HT. At room conditions, the most stable structure of B$_2$O$_3$ is a trigonal structure (s.g. *P3$_1$21*) known as B$_2$O$_3$-I, where B is in the center of triangles of O atoms [115–117,119,121]. Despite its stability at room conditions, working with this material is not simple due to its strong hygroscopic behavior, capable of decomposing the sample in a few minutes [121]. Above 3.5 GPa and 800 K, B$_2$O$_3$-I undergoes a PT to an orthorhombic structure, known as B$_2$O$_3$-II (s.g. *Ccm2$_1$*), composed of distorted BO4 tetrahedra [116,120]. The B$_2$O$_3$-I → B$_2$O$_3$-II transition leads to an increase of B and O coordination from 3 to 4, and from 2 to an intermediate between 2 and 3, respectively, and a volume change of ~27% [117,120,150].

The last group-13-SO is thalium oxide (Tl$_2$O$_3$). This material is very interesting from a technological point of view as it has potential for interesting applications such as solar cell electrode, optical communication and HT superconductors [151–154]. Its most stable phase at room conditions is c-Tl$_2$O$_3$, a bixbyite body-centered cubic phase isostructural to c-In$_2$O$_3$. A corundum-like structure (s.g. $R\bar{3}c$) was also observed at HP (6.5 GPa) and HT (500-600 °C) [140]. So far, only one paper reports the properties of the c-Tl$_2$O$_3$ at HP, probably due to the toxicity of this material [122]. Theoretical calculations predict a c-Tl$_2$O$_3$ → Rh$_2$O$_3$-II-type (s.g. *Pbcn*) PT at 5.8 GPa and a second Rh$_2$O$_3$-II-type → α-Gd$_2$S$_3$-type (s.g. *Pnma*) PT at 24.2 GPa [122]. However, ADXRD measurements at RT, performed using MEW as pressure-transmitting medium, showed only an irreversible pressure-induced amorphization at 25.2 GPa, which was a consequence of kinetic barriers that cannot be overcome at RT [122]. In addition, theoretical calculations predicted a mechanical instability of c-Tl$_2$O$_3$ above 23.5 GPa [122]. Future work involving HP and HT is needed to better understand the behavior of Tl$_2$O$_3$ at HP.

To finish this section, we summarize in Table 2 the phase transition pressures observed in group-13-SOs.

Table 2. Experimental data of the phase transition pressures (in GPa) for group-13 sesquioxides. The letters A, B, C, D, E, F, G, H, and I refer to the structures $R\bar{3}c$, $Rh_2O_3 - II$-type, $CaIrO_3$-type, $Ia\bar{3}$, $\alpha - Ga_2S_3$-type, $Pbca$, $C2/m$, $P3_121$, and $Ccm2_1$, respectively. Between parentheses are the transition temperatures (when necessary). SW stands for shockwave experiments and NP for nanoparticles.

	$P_{A\to B}$	$P_{A\to C}$			Reference
	95 (1200 K)				[86]
	96 (>2000 K)				[97]
Al_2O_3		150 (>1500 K)			[94]
		170 (2500 K)			[127]

	$P_{D\to A}$	$P_{D\to B}$	$P_{B\to E}$	$P_{B\to F}$	Reference
	6.5 (1273 K)				[140]
	15–25 (SW)				[141]
In_2O_3		31			[101]
		7 (>1700 K)			[106]
			40		[107]
				12	[101]

	$P_{G\to A}$	$P_{G\to B}$			Reference
	4.4 (1273 K)				[147]
	20–29				[113]
Gd_2O_3	16 (SW)				[149]
	16.4–39.2 (NP)				[114]
		65 (2300 K)			[106]

	$P_{H\to I}$				Reference
B_2O_3	3.5 (800 K)				[116,120]

	$P_{D\to A}$				Reference
Tl_2O_3	6.5 (>720 K)				[140]

2.3. Transition-Metal Sesquioxides

TM-SOs belong to a highly heterogeneous family of compounds, as their fundamental (electronic, magnetic) properties strongly depend on the number, and therefore on the spin state, of d electrons in the A^{3+} cation. TM-SOs include, among others, the ubiquitous ferric oxide (Fe_2O_3), the commonly-used green pigment Cr_2O_3, or the relatively rare Ti_2O_3 oxide, found in the famous Allende meteorite as the mineral tistarite [155]. Thus, each of these compounds has its own particularities and, regarding the HP structural properties, its specific behavior upon compression.

Among all the TMs, excluding the f-block lanthanide and actinide series and including group-3 and group-12 elements, in this section we will only consider A_2O_3 compounds with at least one d electron. Thus, Sc_2O_3, Y_2O_3 and La_2O_3 are not included here, since they have been discussed together with the RE-SOs in Section 2.1. Among the rest of possible TM-SOs, only the following compounds have been found to be stable at room conditions: Ti_2O_3, V_2O_3, Cr_2O_3, Mn_2O_3, Fe_2O_3, Co_2O_3, Ni_2O_3, Rh_2O_3 and Au_2O_3. Other compounds, like Hf_2O_3 and Zr_2O_3, might be metastable at HP [156].

With the exception of Mn_2O_3, the most stable form of all TM-SOs at room conditions corresponds to the rhombohedral structure of the corundum. As can be seen in Figure 4a, the lattice parameters of the corundum-like structure adopted by most of TM-SOs are barely reduced by increasing the number of d electrons in the A^{3+} cation. This result simply reflects the slight reduction in the atomic radius of the cations with increasing atomic number. However, as can be seen in Figure 4b, the c/a ratio of these compounds displays a well-known anomaly around Ti_2O_3 and V_2O_3 that deserves further attention. In particular, V_2O_3 has highly remarkable properties and behaves, upon compression, in a highly different fashion than other corundum-type TM-SOs [157]. In turn, Mn_2O_3 crystallizes in the so-called bixbyite structure (s.g. 206, $Ia\bar{3}$), which takes the name of the mineral form of Mn_2O_3. The absence of the corundum structure in Mn_2O_3 at room conditions has been attributed to the Jahn-Teller distortion associated with the d^4 Mn^{3+} ion [158].

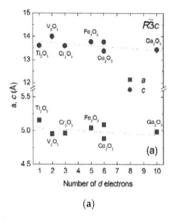

(a)

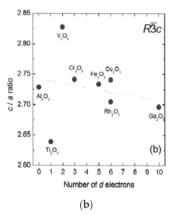

(b)

Figure 4. (**a**) Lattice parameters, a and c, of different corundum-type 3d-TM-SOs. The plot shows the similarity between the lattice parameters of these compounds. Corundum-type Ga_2O_3 (with 10 d electrons) is also included. (**b**) c/a ratio for different corundum-type TM-SOs. Al_2O_3 (with no d electrons) and Ga_2O_3 (with 10 d electrons) are also included. Note the well-known anomaly around Ti_2O_3 and V_2O_3.

Most TM-SOs have also been found to exhibit other stable or metastable polymorphs at room conditions, as well as at HP. Next, we focus on the rich and varied HP behavior of the different phases of TM-SOs, with special emphasis on Fe_2O_3, Mn_2O_3 and V_2O_3.

Iron oxides are highly important materials in a large variety of disciplines, including materials science, geology, mineralogy, corrosion science, planetology and biomedicine. In the particular case of iron SO (iron (III) oxide or ferric oxide), there exist several polymorphs that are stable or metastable at room conditions: α, β, γ, ε and ζ. First, we will briefly describe these phases at room conditions and, afterwards, we will discuss their HP behavior.

The most common phase of Fe_2O_3 is the α phase, which has the corundum structure and is ubiquitously found in nature as the mineral form hematite. α-Fe_2O_3 is a wide-gap antiferromagnetic insulator, with a Morin transition around 250 K. It is weakly ferromagnetic from this temperature up to a Néel temperature around 950 K, above which it becomes paramagnetic. Besides temperature and doping, pressure is a highly convenient tool to modify the insulating/metallic state of this type of material. In particular, hematite can be considered as an archetypal Mott insulator [159] and is interesting from the point of view of metal-insulator PTs in highly correlated electron systems.

The mineral hematite is the main ore of iron and a very common phase in sedimentary rocks, where it occurs as a weathering or alteration product. It is, however, unlikely that Fe_2O_3 or any of its HP phases have much relevance in the Earth's lower mantle, where Fe^{3+} is mainly incorporated into other minerals, like bridgmanite or garnet, with the possible coexistence of reduced iron-rich metal (Fe^0) [160]. In contrast, it cannot be ruled out that Fe_2O_3 may be responsible for the presence of magnetic anomalies in the upper mantle arising from deep-subduction processes [161]. Thus, the HP behavior of hematite may still be highly relevant from a geophysical point of view, as it might play a key role in relation to the magnetic properties of recycled crust materials in subduction zones.

The γ phase of Fe_2O_3 is also found in the Earth's crust as the relatively common mineral maghemite, which is formed by weathering or low-temperature oxidation of magnetite and related spinel minerals. Maghemite, which is also a very common corrosion product, is metastable and transforms upon heating into the α phase. γ-Fe_2O_3 has the same cubic spinel ferrite structure (s.g. 227, $Fd\bar{3}m$) as magnetite ($Fe^{2+}Fe^{3+}_2O_4$), with 1/3 of Fe atoms tetrahedrally coordinated with oxygen (A sites), and 2/3 octahedrally coordinated with oxygen (B sites). Thus, maghemite can be considered as a Fe^{2+}-deficient magnetite, with all A sites filled with Fe^{3+} and only 5/6 of the total available B sites filled with Fe^{3+}; i.e., with 1/6 of vacant (V) B sites, so it can be noted as $(Fe^{3+})^A(Fe^{3+}_{5/3}V_{1/3})^BO_4$. γ-Fe_2O_3 exhibits soft ferrimagnetism at RT, with a Curie temperature around 950 K [162]. Most remarkably, ultrafine particles of γ-Fe_2O_3 exhibit super-paramagnetism, which is related to the thermally induced random flip of the magnetization in ferromagnetic nanoparticles. This phase is being intensively investigated in order to construct an appropriate theoretical framework for this phenomenon, which might set a limit on the storage density of hard disk drives.

A rare orthorhombic polymorph of Fe_2O_3, named ε-Fe_2O_3 (s.g. 33, $Pna2_1$), has also been found to be stable at room conditions. This phase seems to exist only in nanosized form and was first reported back in 1934 [163]. A few years later, it was synthesized and characterized by Schrader et al. [164]. The crystal arrangement of ε-Fe_2O_3 is considered as intermediate between the structures of the α and γ phases, with three octahedral FeO_6 units (two of which are distorted) and a FeO_4 tetrahedral unit. It is worth noting that this phase has recently been found in nature as the Al-bearing nanomineral luogufengite, which is an oxidation product of Fe-rich basaltic glass at HT [165]. It has also been found in ancient black-glazed wares from Jianyang county, China, produced during the Song dynasty (960–1279 AD) [166]. Recent studies have also shown that this polymorph is a non-collinear ferrimagnet [167]. Owing to its remarkable magnetic properties, such as giant coercive field, ferromagnetic resonance and magnetoelectric coupling [168], this phase is highly promising to develop a broad range of novel applications. Besides, the giant coercive field of lougufengite might explain the observed high-remanent magnezitation in some igneous and metamorphic rocks [169].

In contrast to ε-Fe_2O_3, β-Fe_2O_3 and the recently discovered "ζ-Fe_2O_3" phase [170] have not been found to occur in nature so far. These two rare polymorphs are metastable phases that have been obtained by synthetic methods and which only exist in nanocrystalline form. The beta phase exhibits the bixbyite structure, with trivalent Fe ions occupying two non-equivalent octahedral sites. It exhibits paramagnetic behavior at room conditions, with a Néel temperature just above 100 K. Antiferromagnetic ordering is observed below this temperature. Upon heating, β-Fe_2O_3 undergoes a PT to either the α or γ phases. In turn, "ζ-Fe_2O_3" displays a monoclinic structure (s.g. 15, $I2/a$) and was obtained by applying HP (>30 GPa) to β-Fe_2O_3 [169]. Surprisingly, this new phase (with a Néel temperature around 70 K) is metastable after releasing pressure, being able to withstand pressures up to ~70 GPa.

In spite of much effort to elucidate the structural, magnetic and electronic properties of α-Fe_2O_3 (hematite) at HT and HP [159,170–176], the HP behavior of the most important phase of Fe_2O_3 is still under debate. Only recently was a complete T-P phase diagram of α-Fe_2O_3 established [176]. According to that work, α-Fe_2O_3 transforms at RT to a HP phase above ~50 GPa, with an ~8.4% volume discontinuity. It is worth noting that in this work the HP phase was called "ζ-phase" and was indexed as triclinic ($P\bar{1}$) [175]. This phase is not equivalent to the monoclinic s.g. $I2/a$ phase stable at room conditions, also named "ζ-Fe_2O_3", which was produced by Tucek et al. [169]. It must be noted that the $P\bar{1}$ phase was previously indexed as monoclinic ($P2_1/n$) in reference [175]. The fact that this HP phase is relabeled as "double perovskite" in reference [170] and treated as s.g. $P2_1/n$ only adds confusion to the nomenclature of the different phases of such an important compound as Fe_2O_3.

As already mentioned, the single-crystal diffraction patterns of the $P\bar{1}$ HP phase were initially indexed in a monoclinic unit cell [173], but were subsequently refined into a triclinic s.g. $P\bar{1}$ structure [175]. However, previous works had assigned the HP phase (~50 GPa) of hematite to either a distorted corundum phase (Rh_2O_3-II-type structure, s.g. $Pbcn$) [159,170,172,177] or to a Pv structure (s.g. $Pbnm$) [129,178]. According to reference [175], the $P\bar{1}$ structure of Fe_2O_3 is closer to a distorted (and not "double", as in reference [176]) Pv structure. However, a more recent study up to ~79 GPa has put into question the very existence of the distorted Pv phase [174], in agreement with previous results [179]. The data and analysis of reference [159] indicate that the 1st-order PT around 50 GPa is accompanied by an insulator-to-metal transition with collapse of the magnetic properties. The structural PT would thus be very progressive [160,170], starting from the corundum-type structure and leading to a distorted Rh_2O_3-II-type structure. In contrast, Bykova et al. suggested that the Rh_2O_3-II-type structure only occurs below 50 GPa at HT and would be reconstructive, with a volume change of ~1.3% [173]. According to this work, the distorted Pv phase shows a PT at RT around ~67 GPa to a possibly metastable orthorhombic structure (s.g. 41, *Aba2*) or to a $CaIrO_3$-type PPv phase (s.g. *Cmcm*) [175]. In fact, the *Aba2* to PPv PT was observed at HT (above 1600 K). After quenching the samples to RT, the PPv polymorph was found to be stable up to 100 GPa and down to 26 GPa; however, hematite was recovered below this pressure. In turn, the metastable *Aba2* phase seems to be preserved at 78 GPa, up to 1550 K. This polymorph, also observed by Greenberg et al. [176] at RT, can be considered as a metastable HP phase at low temperature. It adopts a similar packing to the Rh_2O_3-II-type structure observed at low pressures, but, instead of octahedra, it consists of FeO_6 prisms. Recent results obtained by Sanson et al. are compatible with the existence of the *Aba2* polymorph [174]. However, these authors seem to rule out not only the $P2_1/n$ HP phase, but also the *Pbnm* one.

Regarding the PPv structure, Ono et al. had previously reported a PT from hematite to the PPv structure above 30 GPa at HT [131]. Moreover, a new phase was found above 50 GPa to an unknown phase that was tentatively assigned to an orthorhombic or a monoclinic lattice. In relation to the magnetic and electronic properties of this phase, Shim et al. showed that the magnetic ordering of PPv-type Fe_2O_3 is recovered after laser heating at 73 GPa [179]. According to these authors, the appearance of the PPv phase gives rise to the transformation of Fe^{3+} ions from a low-spin to a high-spin state, with Fe_2O_3 undergoing a semiconducting-to-metal PT. This effect might be relevant to understand the electromagnetic coupling between the Earth's mantle and the core. Previous research had already investigated with Mössbauer spectroscopy the high-spin to low-spin PT in α-Fe_2O_3, at HP associated to the structural PT [180,181].

Indeed, one of the main interests of studying hematite at HP, besides its crystal structure and phase diagram, has to do with its electronic properties and the nature of the HP-PT [171]. What drives the HP-PT in hematite? Is it the crystal structure, or the electronic structure? Note that one could expect that electronic PTs, such as the insulator-metal Mott PT observed at 50 GPa [159] or the high-spin to low-spin PT, are isostructural. According to this argument, the PT observed in hematite at ~50 GPa, accompanied by a large volume change, should drive the change in electronic properties. Badro et al. reached this conclusion and indicated that the electronic PT is isostructural and occurs after the crystal structure PT [171]. However, Sanson et al. have recently reached the opposite conclusion, i.e., that the structural PT is driven by the electronic transition [174]. These authors analyzed the local structure and observed no FeO_6 octahedra distortion at HP, in disagreement with previous works [170]. A careful analysis recently performed by Greenberg et al. suggests that the pressure-induced insulator-to-metal PT in hematite arises from site-selective delocalization of electrons [176]. According to this study, such site-selective Mott transition ought to be characterized by delocalization (i.e., metallization) of Fe 3*d* electrons on only half of the sites in the unit cell. This work suggests that the interplay between crystal structure and electronic correlations may yield a complex behavior that could be also relevant to understand the HP behavior of other TM-SOs.

With regard to the rest of Fe_2O_3 phases stable at room conditions (γ, β and ε), the amount of published works regarding their HP behavior is sizably lower in comparison to hematite. The phase diagram of γ-Fe_2O_3 (maghemite) is relatively simple, since this phase readily transforms to α-Fe_2O_3 around 35 GPa [182]. Therefore, much of the effort regarding this phase has been in relation to the comparison of the compression behavior between bulk and nanocrystalline samples (see reference [183] and references therein). Further details about polymorphic PTs of nanosized Fe_2O_3, including doping, temperature and HP, can be found in reference [184]. In the case of maghemite, Jiang et al. found that nanocrystalline γ-Fe_2O_3 transforms to hematite at ~27 GPa, below the PT pressure for bulk material [182]. As shown by Zhu et al., the γ-to-α PT might be initiated at the same pressure, around

16 GPa, both in bulk and nanocrystalline forms [183]. According to these authors, vacancies might play an important role in the structural PT. More recently, a combination of XRD and Mössbauer spectroscopy up to 30 GPa on vacancy-ordered maghemite has shown that the γ-to-α PT would be initiated at 13–16 GPa, giving rise to a particular texture in the transformed α-Fe_2O_3 material [185].

Finally, in relation to the Fe_2O_3 polymorphs, we must mention the recent HP study of ε-Fe_2O_3 by Sans et al. [186]. These authors found that this rare nanocrystalline phase, which is a promising magnetic material for a range of technological applications, is stable up to 27 GPa. Above this pressure, evidence for a PT, in which the tetrahedrally coordinated iron ions change towards quasi-octahedral coordination, was observed. Given that this phase has very high magnetic coercivity, it remains to be investigated whether its mineral form may play any relevant role in the magnetic remanence of mantle xenoliths, and in the magnetism of subducting slabs at depths corresponding to the mantle transition zone [161].

Manganese SO (Mn_2O_3) is unique among 3*d*-TM-SOs because its most stable phase at room conditions (α-Mn_2O_3) does not adopt the corundum structure. Instead, it crystallizes in the so-called cubic bixbyite structure that takes the name from the mineral form of this compound. In bixbyite, Mn^{3+} occupies five different crystallographic sites, each of them surrounded by a highly distorted octahedron of O atoms. Such distortion originates from the Jahn-Teller effect, due to the lowering of overall energy as a consequence of a distortion-induced loss of degeneracy in the 3*d*-orbitals of the sixfold-coordinated Mn^{3+} ions. As discussed by several authors, the Jahn-Teller distortion might explain the fact that Mn_2O_3 does not adopt the corundum structure at room conditions [158]. Also, it has been predicted that HP might suppress the Jahn-Teller effect, which would make Mn_2O_3 transform to a corundum-type polymorph and, afterwards, follow the usual sequence of pressure-induced PTs observed in the rest of TM-SOs [140]. α-Mn_2O_3 has been shown to be antiferromagnetic, with a Néel temperature of around 79 K [187] and has recently attracted considerable interest due to its potential use in chemical catalysis and magnetic devices, and also for energy conversion and storage applications.

Interestingly, the bixbyite structure actually occurs in doped samples and/or at temperatures slightly above RT. Indeed, it has been shown that Mn_2O_3 undergoes a subtle cubic-to-orthorhombic (*Pcab*) PT around RT and that the bixbyite structure is stabilized by the presence of impurities [188]. The low-temperature phase *Pcab* is just an orthorhombically distorted bixbyite structure. There exists another polymorph of Mn_2O_3 stable at room conditions, namely γ-Mn_2O_3. It is a tetragonal spinel-like polymorph (s.g. $I4_1/amd$), related to Mn_3O_4. This phase, which is ferrimagnetic below 39 K, can be viewed as the equivalent of maghemite in the case of the Fe_2O_3 system.

Several studies have investigated the HP phases of Mn_2O_3 [140,189–194]. It is now clear that α-Mn_2O_3 transforms at RT to a CaIrO₃-type PPv-like structure above 16 GPa [190–194] with a large volume collapse of ~12%. Thus, no Pv polymorph is observed between the bixbyite and PPv structures at RT. As pointed out by Santillán et al. [190], this might be due to the similarity between the two structures (cubic bixbyite and PPv). Yamanaka et al. also reported a monoclinic HP phase of Mn_2O_3 that has not been confirmed by any other study [189]. More recently, synthesis performed with a multianvil apparatus and subsequent study of the recovered material has shown that, at HTs (between ~850 and 1150 K) the Pv structure, also called ζ-Mn_2O_3, can be synthesized and recovered at room conditions [193]. This polymorph is formed by a Pv octahedral framework with strongly tilted MnO_6 octahedra. Its crystal structure was determined within the $a = 4a_p$ supercell, s.g. $F\overline{1}$ (s.g. $P\overline{1}$ in the standard setting, see reference [193]), where Mn cations are preferentially found in different sites depending on their valence state (Mn^{2+}, Mn^{3+}, Mn^{4+}). This phase has been shown to be a narrow and direct bandgap semiconductor (0.45 eV) with remarkable hardness [195], and to exhibit a 3:1 charge ordering between two different sites, and a unique orbital texture involving a remarkable alternation of orbitals states [196].

A corundum-type polymorph of Mn_2O_3 (ε-Mn_2O_3) was also recovered at room conditions after applying intermediate pressures (15 GPa) and HT (> 1200 K) [193]. This phase, which is about 1% denser than the bixbyite polymorph, seems to confirm the existence of the undistorted, corundum-type structure in the exotic Mn_2O_3. It should be noted that, in contrast to this phase and the ζ phase, the more usual HP PPv polymorph is not quenchable at room conditions. The multianvil apparatus experiments also show that reduction of Mn_2O_3 to Mn_3O_4 is observed below 12 GPa at HT (>1200 K) [193]. No HP studies of ε and γ phases are available to our knowledge.

Among TM-SOs, vanadium oxide (V_2O_3) is one of the most interesting materials both from fundamental and applied points of view, due to its remarkable electronic and magnetic properties. This compound has a very rich phase diagram at RP and is often considered a paradigmatic example of metal to Mott-insulator transition in strongly correlated materials. Depending on temperature and doping with small amounts of Cr or Ti, V_2O_3 exhibits different phases including a paramagnetic metallic state, a paramagnetic insulating state or a low-temperature antiferromagnetic insulating state [197,198]. V_2O_3 adopts a corundum-type structure at atmospheric conditions and is found in nature as the rare mineral karelianite, first found in glacial boulders from the North Karelia region, in Finland. A metastable polymorph of V_2O_3 with bixbyite structure was obtained by synthetic methods [199]. This phase transforms into the stable corundum phase above 800 K, and is found to exhibit a paramagnetic to canted antiferromagnetic transitions at about 50 K.

The application of HP strongly modifies the phase diagram of corundum-type V_2O_3. Several studies have investigated the Mott-insulator transition in V_2O_3 and Cr-doped V_2O_3 at HP with a range of different experimental techniques [200–205]. Here, we would like to first focus on the link between the structural and electronic properties of V_2O_3. In particular, McWhan and Remeika [200] studied V_2O_3 and $(V_{0.96}Cr_{0.04})_2O_3$ samples and found that the metallic state typically found in undoped V_2O_3 is recovered by compressing the Cr-doped sample. These authors also showed that the c/a ratio of the trigonal lattice is related to the insulating/metallic state as a function of Cr-doping level and as a function of pressure. In particular, they found that the a parameter of their Cr-doped sample, which was larger than that of undoped V_2O_3 at RP, dropped to the value of V_2O_3 at 4 GPa. In contrast, the c parameter was found to barely increase at HP in both samples, implying a remarkable increase in the c/a ratio in the Cr-doped sample as the metallic state was approached. In fact, they found a sharp increase in c/a ratio just at the insulator–metal transition. These authors also showed that undoped V_2O_3 becomes even more metallic at HP, with a suppression of the antiferromagnetic-insulating phase above 0.3 GPa.

More recently, Lupi et al. [198] studied with submicron resolution the Mott-insulator transition in Cr-doped V_2O_3. These authors found that, with decreasing temperature (which yields the antiferromagnetic insulator phase), microscopic domains become metallic and coexist with an insulating matrix. This seems to explain why the paramagnetic metallic phase is a poor metal. They attributed the observed phase separation to a thermodynamic instability around the Mott-insulator transition, showing that such instability is reduced at HP. Therefore, they concluded that the Mott-insulator transition is a more genuine Mott transition in compressed samples, which also exhibited an abrupt increase in the c/a ratio around the insulator-to-metal transition (i.e., around 0.3 GPa). This result confirms the close relationship between the structural properties of V_2O_3, the nature of the Mott-insulator transition and the role of HP in the modification of the electronic properties of this interesting compound.

In contrast to other TM-SOs, the compression of V_2O_3 is highly anisotropic (the a axis is nearly three times more compressible than the c axis) [206] and the unusual behavior of the c/a ratio is closely related to the Mott-insulator transition. It was also shown that the atomic positions in V_2O_3 tend to those of the ideal hcp lattice at HP, with large changes in interionic distances and bond angles upon compression [206]. Zhang et al. observed a pressure-induced PT to an unknown phase around 30 GPa [207]. Ovsyannikov et al. found two different PTs above 21 GPa and 50 GPa, respectively, and attributed the first HP phase to an Rh_2O_3-II-type orthorhombic structure [157]. With regard to the second HP phase, they conclude that it seems not to be related to the Th_2S_3- and α-Gd_2S_3-type structures predicted by Zhang et al. in V_2O_3, of around 30–50 GPa [208]. In conclusion, more work is needed to unveil the HP behavior of both the corundum and bixbyite polymorphs of V_2O_3.

Besides the compounds discussed above, other TM-SOs have been synthesized and/or exist in nature that crystallize in the corundum-structure at room conditions: the $3d$-SOs Ti_2O_3, Cr_2O_3, and Co_2O_3, and the $4d$-SO Rh_2O_3. There are also a few reports about the possible synthesis and characterization at room conditions of hexagonal Ni_2O_3 [209,210] and $5d$-SO Au_2O_3 [211]. However, more work would be required to unambiguously demonstrate the stability of these two latter compounds at room conditions. Note that no natural occurrence of SOs with Ni or Co ions, both of which are more stable as divalent cations, has been found so far. On the other hand, no HP study is known for Au_2O_3.

The amount of HP investigations on the above mentioned materials is variable. Minomura and Drickammer [212] reported HP resistance measurements on several compounds at RT. They found a resistance maximum around 11 GPa in Ni_2O_3 that could be indicative of a PT. Unfortunately, no additional HP investigations can be found in the literature on this material. In the case of Co_2O_3, Chenavas et al. reported a low spin to high spin transition in quenched material, subject to 8 GPa and 850 °C, from a synthesis process involving different chemical reactions [213]. Therefore, the fundamental properties of both Ni_2O_3 and Co_2O_3 at HP have yet to be fully investigated.

Chromia (Cr_2O_3) is another important TM-SO, both from technological and fundamental points of view. This material with corundum structure occurs in nature as the uncommon mineral eskolaite. However, it is probably better known for its high stability and hardness, which makes it a commonly used green pigment and abrasive material. Cr_2O_3 is a typical antiferromagnetic insulator with a Néel temperature T_N ~307 K, but it is remarkable as it shows an intriguing linear magnetoelectric effect that has attracted some research attention in relation to spintronic applications. Although the existence of HP polymorphs of Cr_2O_3 have not yet been well-established [214,215], it is clear that HP affects the magnetic properties of this compound. For instance, as recently shown by Kota et al., hydrostatic pressure may increase the T_N of corundum-type Cr_2O_3, which could provide an enhancement of the magnetoelectric operating temperature in compressed samples [215]. However, there is an ongoing controversy in the literature in relation to the sign of dT_N/dP (see reference [215] and references therein). With regard to its crystal structure at HP, no PT was found either by Finger and Hazen up to 5 GPa [206], or by Kantor et al. up to ~70 GPa [216]. In contrast, Shim et al. observed several changes around 15–30 GPa on pure synthetic Cr_2O_3 under cold compression that were compatible with a PT to a monoclinic V_2O_3-type ($I2/a$) structure [217]. According to these authors, a second PT could occur above 30 GPa at HT. The new HP-HT phase might be explained by either orthorhombic, Pv, or Rh_2O_3-II-type structures. In fact, the orthorhombic Rh_2O_3-II-type phase had been previously predicted to exist at ~15 GPa by first-principles calculations [218]. However, the recent results by Golosova et al. up to 35 GPa, also on synthetic Cr_2O_3, seem to rule out both PTs [219]. While these authors report a somewhat lower $(1/T_N)(dT_N/dP)$ value in relation to reference [215] (0.0091 GPa^{-1} vs. 0.016 GPa^{-1}), their structural data point towards a slight anisotropy in the pressure behavior of the c/a axis at around 20 GPa, which could be of magnetic origin. Additional work is thus required to fully understand the HP behavior of chromia.

In the case of Ti_2O_3, which can be found in nature as the rare tistarite mineral, HP-XRD studies were first conducted by McWhan and Remeika [200] with the aim of comparing the results obtained for this compound with those of V_2O_3. Later on, Nishio-Hamane et al. reported a PT from corundum-type Ti_2O_3 to an orthorhombic Th_2S_3-type (s.g. *Pnma*) structure above 19 GPa and 1850 K, being the new polymorph metastably recovered at room conditions [220]. The same structure had been previously predicted at ultra HP for the case of Al_2O_3 [221]; hence, the synthesis of the Th_2S_3-type polymorph in Ti_2O_3 has been a remarkable finding, shedding new light on the fundamental properties of the TM-SOs family. In Ti_2O_3, the Th_2S_3-type structure was found to be ~10% denser than the corundum-type phase. Subsequently, Ovsyannikov et al. [222] synthesized polycrystalline Ti_2O_3 samples with Th_2S_3-type structure with the aim of studying its structural stability up to 73 GPa and 2200 K. The HP behavior of the structural, optical and electronic properties of this unusual, golden-colored polymorph of Ti_2O_3 were later thoroughly investigated [223,224], together with the T-P phase diagram of Ti_2O_3 [224].

With regard to 4*d*-SOs, there are a few reports on the existence of such compounds at room conditions, the most relevant of which is Rh_2O_3. Two low-pressure polymorphs of this material are known, namely Rh_2O_3-I and Rh_2O_3-III. While the stable phase at room conditions, Rh_2O_3-I, adopts a corundum-type structure, the Rh_2O_3-III polymorph crystallizes in an orthorhombic *Pbca* structure and is only stable at RP and HT (>750 °C) [225]. In addition, an HP phase, known as Rh_2O_3-II, was synthesized by Shannon and Prewitt at 6.5 GPa and 1200 °C [226]. This phase was found to adopt an orthorhombic *Pbna* structure, similar to that of corundum as it contains RhO_6 octahedra that share their faces. In Rh_2O_3-II, however, only two edges of each octahedra are shared with other octahedra. In spite of the use of HP to synthesize Rh_2O_3-II, only a limited number of HP studies on this phase can be found in the literature. Zhuo and Sohlberg investigated theoretically the stability of this phase in comparison to Rh_2O_3-I and Rh_2O_3-III, which allowed them to predict the phase diagram of Rh_2O_3 [227]. According to their results, Rh_2O_3-II is stable below ~5 GPa and HT (>700 K), while the Rh_2O_3-III form is the main phase above 5 GPa, regardless of temperature. Unfortunately, no experimental HP studies dealing with the structural properties of any of the Rh_2O_3 phases have been published so far. Experimental work on this material, and, in particular, on the Rh_2O_3-III polymorph, due to its connection to the HP phases of other SOs, as described above, is still necessary.

Possible stable forms of other 4*d*-SOs, like Mo_2O_3 and W_2O_3, have been reported in relation to the growth of thin films or coatings (see bibliography in reference [228]). Recent DFT calculations predict that corundum-type Mo_2O_3 might be stabilized at 15 GPa, while a much higher pressure (60 GPa) is obtained by these authors for W_2O_3 [228]. Similar conclusions have been reached by other authors in relation to Hf_2O_3 and Zr_2O_3 using crystal prediction methods [157,229]. These two SOs might be metastable at HP. So far, however, there is no experimental evidence for the existence and stability of these SOs at HP. Much experimental and theoretical effort is still required to understand the stability and behavior of these two SOs upon compression, which could lead to the discovery of new HP phases with remarkable properties.

To finish this section, we summarize in Table 3 the phase transition pressures and HP phases observed in the best studied TM-SOs.

Table 3. Summary of the high pressure (HP) phases and phase transition (PT) pressures, P_T (in GPa) reported for the best studied transition-metal sesquioxides (TM-SO) materials: Fe_2O_3, Mn_2O_3, V_2O_3 and Ti_2O_3. The fourth column shows the resulting phase obtained at HP starting from the phase displayed in the first column (DPv: distorted perovskite; PPv: post-perovskite; LH: laser heating).

Compound	P_T	Temperature	HP Phase	Reference
α-Fe_2O_3 (hematite, $R\bar{3}c$)	54	RT	DPv, $P\bar{1}$	[175]
α-Fe_2O_3 (hematite, $R\bar{3}c$)	67	RT	*Aba*2	
α-Fe_2O_3 (hematite, $R\bar{3}c$)	40	LH	Rh_2O_3-II, *Pbcn*	[173]
α-Fe_2O_3 (hematite, $R\bar{3}c$)	68	LH	PPv, *Cmcm*	
β-Fe_2O_3 ($Ia\bar{3}$)	30	RT	"ζ-Fe_2O_3"	[169]
γ-Fe_2O_3 (maghemite, $Fd\bar{3}m$)	35	RT	α-Fe_2O_3	[182]
ε-Fe_2O_3 (luogufengite, $Pna2_1$)	27	RT	ε'-Fe_2O_3	[186]
α-Mn_2O_3 (bixbyite, $Ia\bar{3}$)	16–25	RT	PPv, *Cmcm*	[190]
α-Mn_2O_3 (bixbyite, $Ia\bar{3}$)	18	800 K	Pv, $P\bar{1}$	[193]
α-Mn_2O_3 (bixbyite, $Ia\bar{3}$)	13	1100 K	ε-Mn_2O_3, $R\bar{3}c$	
V_2O_3 (karelianite, $R\bar{3}c$)	21–27	LH	Rh_2O_3-II, *Pbcn*	[157]
V_2O_3 (karelianite, $R\bar{3}c$)	50	LH	PPv, *Cmcm* (?)	
Ti_2O_3 (tistarite, $R\bar{3}c$)	19	1850 K	Th_2S_3-type, *Pnma*	[220]

2.4. Group-15 Sesquioxides

Group-15-SOs are relevant in a multitude of technological [230–232] and medical [233–235] applications, but the main interest in the study of the properties of these compounds resides in the physical–chemical interactions of their chemical bonds. The progressive increase in the strength of the stereoactive cationic lone electron pair (LEP) going up along the group 15 column, i.e., as the cation becomes lighter, dominates the formation and compressibility of the resulting crystallographic structures. In other words, the influence of the cation LEP determines the different existing polymorphs of group-15-SOs and it is possible to observe several trendlines of their behavior at HP. According to the literature, the group-15-SO polymorphs can be obtained from a defective fluorite structure where oxygen vacancy arrays are applied along different crystallographic directions [236,237]. These arrays define most of the polymorphs found at room conditions in these SOs. Nevertheless, the crystalline phases formed are quite different from the rest of SOs, mainly due to the above-mentioned influence of the LEP.

Structural phases belonging to the cubic family are observed in group-15-SOs since the most usual structures of arsenic and antimony SO crystallizes in the s.g. $Fd\bar{3}m$. They are obtained by applying an array model along the (111) direction to a defective fluorite structure. They have been experimentally found in c-As_2O_3 (whose mineral form is called arsenolite) and α-Sb_2O_3 (whose mineral form is called senarmontite). These SOs are also noted as As_4O_6 and Sb_4O_6, respectively, because this notation reflects the molecular arrangement of these two molecular or zero-dimensional solids. The molecular units X_4O_6 (X= As, Sb) are disposed in a closed-compact adamantane-type molecular cages, joined together by van der Waals (vdW), and chalcogen bonding among the closest arsenics belonging to different molecular units. In this case, the cations belonging to the same molecular unit form pseudo-tetrahedral units following the same disposition as their counterpart, the white phosphorus (P_4). The molecular cages are stabilized by the cationic LEP oriented towards the outside of the cage, which confers great stability to the molecular unit. The HP behavior of both Sb_4O_6 and As_4O_6 dimorphs is quite different. They undergo different PTs [238–242], and both compounds show a pressure-induced amorphization driven by mechanical instability [238–240]. The origin of this instability is associated with the steric repulsion caused by the increase in the interaction between different molecular units when pressure increases [238]. It has been shown in a HP study using different pressure-transmitting media that the instability is enhanced by the loss of the hydrostatic conditions [241]. However, up to 15 GPa, both compounds behave differently. Whereas two 2nd-order isostructural PTs were observed in Sb_4O_6 at 3.5 and 12 GPa, respectively [242], no pressure-induced PT was reported in As_4O_6 [239,240]. The PTs exhibited by Sb_4O_6 at 3.5 and 12 GPa were suggested to be due to the loss of its molecular character and related to changes in the hybridization of Sb *5s* and O *2p* electrons, i.e., they are related to the decrease in LEP stereoactivity at HP [242].

At 15 GPa, Sb_4O_6 tends to form a new HP structure that can be defined by a distortion of the structural lattice due to the loss of the molecular character [238]. This HP phase crystallizes in a tetragonal structure (s.g. $P\bar{4}2_1c$) similar to that of β-Bi_2O_3 [243] that will be described later. At HP (9-11 GPa) and HT (573–773 K), cubic Sb_4O_6 undergoes a 1st-order PT towards a new polymorph called γ-Sb_2O_3 [244]. This new phase crystallizes in an orthorhombic structure s.g. $P2_12_12_1$, where the trigonal pyramid form given by the coordination of the Sb atom surrounded by O atoms in the cubic phase is changed by a tetrahedral unit in the orthorhombic phase, where the LEP acts as a pseudo-ligand. The tetrahedral units are arranged, forming chains or rods along the *a-axis* that show strong similarities with the β-Sb_2O_3 polymorph (s.g. *Pccn*), whose $(Sb_2O_3)_\infty$ rods are aligned along the *c-axis*. Due to its rod-like structure, orthorhombic β-Sb_2O_3 is a quasi-molecular or acicular solid with two independent rods in the unit cell. In this structure, each Sb atom is bonded to three oxygen atoms, with their LEP oriented towards the void formed by the rod arrangement. Thus, this SO can be described by rods and linear voids oriented along the *c-axis*, being isostructural to the ε-Bi_2O_3 phase [245]. No HP study of ε-Bi_2O_3 is known so far.

HP studies of β-Sb_2O_3 revealed a 1st-order PT around 15 GPa [246,247]. Nevertheless, the nature of its HP phase is still under debate. Zou et al. tentatively assigned it to a monoclinic structure with s.g. $P2_1/c$. β-Sb_2O_3 also shows strong similarities with the tetragonal phase of bismuth SO (named β-Bi_2O_3), which crystallizes in s.g. $P\bar{4}2_1c$ [243]. In fact, β-Bi_2O_3 is a metastable polymorph obtained by heating the most stable phase of the bismuth SO (α-Bi_2O_3) to 650 °C [248]. This material can be considered a defective fluorite structure, where an (100) array of vacants is applied [236]. This polymorph contains a linear channel, as well as β-Sb_2O_3, with the cation LEP oriented towards the center of this empty channel. β-Bi_2O_3 is very interesting from a physical–chemical point of view because it is a clear example of an isostructural 2nd-order PT. This structure is characterized by a Bi atom located at an 8e Wyckoff site and two O atoms located at 8e and 4d Wyckoff sites, with the Bi atom coordinated to six O atoms, forming a distorted pyramid. Above 2 GPa, several Wyckoff coordinates of β-Bi_2O_3 symmetrize, remaining fixed up to 12 GPa [243]. Thus, the six different Bi-O interatomic distances observed at RP in the polyhedral unit only become three different Bi-O bond lengths, thus reducing the eccentricity of the Bi atom drastically, up to 2 GPa. This PT is even more explicit when analyzing the compressibility of the unit cell volume below and above 2 GPa. Finally, a progressive broadening of the peaks leads to a pressure-induced amorphization above 12 GPa that is completed at 20 GPa [243].

α-Bi_2O_3 crystallizes in a monoclinic s.g. $P2_1/c$ structure, where Bi atoms are located in two inequivalent crystallographic 4e sites and O atoms are located in three inequivalent 4e Wyckoff positions. One of the Bi atoms is penta-coordinated and the other is hexa-coordinated. This polymorph remains in the same structure up to 20 GPa, where a pressure-induced amorphization occurs [249]. The amorphization was explained by a frustrated PT towards the HPC-Bi_2O_3 polymorph, since the HPC-phase is thermodynamically more stable than the α-phase above 6 GPa, according to DFT calculations. This ongoing PT is likely inhibited by the kinetic barriers between both phases. The correlation between the amorphous phase and the HPC polymorph was revealed, thanks to the analysis of the diffraction pattern as a function of the interionic distances, using the process described in reference [250]. Results indicated average interatomic Bi-O distances closer to those of the HPC-Bi_2O_3 than to any other polymorph. HPC-Bi_2O_3 crystallizes in the hexagonal s.g. $P6_3mc$ structure, and it is the HP phase of the HP-Bi_2O_3 phase, above 2 GPa [251]. In turn, HP-Bi_2O_3 with s.g. $P31c$ was obtained from α-Bi_2O_3 at HP and HT (6 GPa and 880 °C) [252]. The strong polymorphism exhibited by Bi_2O_3 does not require the application of HP. It must be noted that α-Bi_2O_3 at HT (above 400 °C) shows a PT to R-Bi_2O_3 (s.g. $P2_1/c$) [252], which, despite belonging to the same s.g. as α-Bi_2O_3, has a completely different structure. No HP study of R-Bi_2O_3 is known.

In the same s.g. $P2_1/n$, related to the α- and R-Bi_2O_3 phases, one can find two additional As_2O_3 polymorphs, whose names are claudetite-I [253] and claudetite-II [254]. The naturally stable phase, claudetite-I, has been only characterized by HP-RS measurements to our knowledge, revealing a possible PT between 7–13 GPa and a lack of pressure-induced amorphization up to 40 GPa [255]. Claudetite-II is a synthesized quasi-layered material, similar to As_4O_6, where the molecular unit is an open shell forming AsO_3E pseudo-tetrahedra (where E denotes the cation LEP) [256]. In this case, the LEPs of neighboring cations are oriented towards opposite neighboring layers. This particular structure shows several pressure-induced 2nd-order PTs around 2 and 6 GPa [257], given by two abrupt breakdowns of inversion centers that lead to doubling and hexaplicating the original unit cell, respectively. Additionally, a PT around 10.5 GPa leads to the non-centrosymmetric HP phase β-As_2O_3 (s.g. $P2_1$).

Another group-15-SO crystallizing in the monoclinic s.g. $P2_1/n$ structure is the phosphorous SO (P_2O_3), which keeps the molecular character of cubic arsenic and antimony SOs, i.e., P_4O_6 [258]. Nevertheless, the symmetric distribution of the different P_4O_6 molecular cages is distorted in the case of monoclinic P_2O_3. The HP study of P_2O_3 exhibits strong difficulties, since it is only stable at low-temperatures, below RT. Thus, the HP behavior of this compound is still unknown.

To finish, we summarize in Table 4 the 1st- and 2nd-order PTs observed in group-15-SOs.

Table 4. Experimental and theoretical (DFT) data of the PT pressures (in GPa) for group-15-SOs. PIA stands for pressure-induced amorphization and MEW for methanol-ethanol mixture.

s.g.	Compound	2nd-Order PT	1st-Order PT	PIA	Technique	Reference
$Fd\bar{3}m$	As_4O_6	-	-	18.7	DFT	[240]
		-	-	15	XRD (MEW)	[239]
		-	-	-	XRD (He)	[241]
	Sb_4O_6	3.5	10		XRD, RS, DFT (MEW)	[242]
		-	25	30	XRD (Ne)	[238]
$P2_1/c$	mII-As_2O_3 (claudetite II)	2	6	11	-	XRD (He)	[256]
	α-Bi_2O_3	-	-	20	XRD, RS, DFT (MEW, Ar)	[249]
	mI-As_2O_3 (claudetite I)	-	-	-	RS	[253]
$Pccn$	β-Sb_2O_3	7	15	33	XRD, RS (Ne)	[246]
		-	13.5	-	XRD, RS (MEW)	[247]
$P\bar{4}2_1c$	β-Bi_2O_3	2	-	12	XRD, RS, DFT (MEW)	[243]

2.5. Mixed-Valence Sesquioxides

As mentioned in the introduction, there are A_2O_3 compounds where the A has two different valences, which constitute the subfamily MV-SOs. MV-SOs are highly interesting because they feature cations with different valences and consequently can be differently coordinated, thus resulting in a distortion in the structures of normal SOs. Examples of this kind of SOs are Sn_2O_3 and Pb_2O_3, where Sn and Pb cations show both 2+ and 4+ valences. While Pb_2O_3 has been known for almost a century [259] and crystallizes in the monoclinic s.g. $P2_1/a$ [260], Sn_2O_3 has just recently been predicted [261] and experimentally found to crystallize in the triclinic s.g. $P\bar{1}$ [262,263]. No HP studies on these two interesting but complex compounds have been reported to our knowledge.

3. Conclusions and Future Prospects

A great deal of work has been done on sesquioxides under compression and many pressure-induced phase transitions have been elucidated. In particular, the pressure-induced phase transitions of rare-earth sesquioxides and group-13 sesquioxides are quite well known, except for Tl_2O_3. The knowledge of pressure-induced phase transitions in transition-metal sesquioxides are not so well known, and those of group-15 sesquioxides have been recently studied, but still require more work. Finally, pressure-induced phase transitions in mixed-valence sesquioxides are completely unknown. It must be stressed that most of the high-pressure work done on sesquioxides corresponds to cubic, tetragonal, hexagonal, rhombohedral, and orthorhombic phases, while most of the high-pressure work still to be done on sesquioxides corresponds to complex monoclinic and triclinic phases. The study of these complex phases was rather difficult in the past, and is a challenge that must be accomplished in the 21st century with current and future experimental and theoretical techniques.

It must be noticed that the crystalline structures of the different subfamilies of sesquioxides and their pressure-induced phase transitions have some features in common. The archetypic structure of rare-earth sesquioxides is the C-type structure, which is isostructural to the bixbyite structure of several transition metal sesquioxides (Mn_2O_3) and group-13 sesquioxides (In_2O_3 and Tl_2O_3). On the other hand, the archetypic structure of group-13 and transition-metal sesquioxides is the corundum structure (Al_2O_3). In this context, it is remarkable that the pressure-induced transitions of C-type rare-earth sesquioxides, including Sc_2O_3 and Y_2O_3, are markedly different to those of C-type group-13 and transition-metal sesquioxides. In fact, these last compounds seem to have pressure-induced transitions that are closer to those of corundum-type structures of group-13 and transition-metal sesquioxides. On the other hand, the crystalline structures and pressure-induced phase transitions of group-15 sesquioxides are completely different to those of other sesquioxides, due to the stereochemically active lone electron pair of group-15 cations, which lead to more open-framework structures than those present in other subfamilies. Consequently, we can speculate that a similar behavior is expected to be observed in future experiments of mixed-valence sesquioxides under compression, since this subfamily is also characterized by having some cations with lone electron pairs.

Finally, it should be mentioned that we are not aware of studies of negative pressures in sesquioxides. In principle, it is possible to reach absolute negative pressures (of the order of a few GPa) [264] where novel structures might appear [265]. In fact, negative pressures have been found in several solids, such as $CaWO_4$ nanocrystals [266] and ice polymorphs [267]. Therefore, the application of negative pressures to materials is a route that can also be explored in sesquioxides, since it would lead to the expansion of the crystalline lattice, leading to the possible instability of the stable phase at room conditions in certain compounds, thus opening a new avenue to obtain novel and exotic metastable phases.

Author Contributions: F.J.M. contributed to the conceptualization and writing of the introduction, Sections 2.1 and 2.5 and conclusions; A.L.J.P. contributed to the writing of Section 2.2; J.I. contributed to the writing of Section 2.3 and J.A.S.T. contributed to the writing of Section 2.4 and the revision of the manuscript.

Funding: This research was funded by Spanish Ministerio de Ciencia, Innovación y Universidades under grants MAT2016-75586-C4-1/2/3-P, FIS2017-83295-P, PGC2018-094417-B-100, and RED2018-102612-T (MALTA-Consolider-Team network) and by Generalitat Valenciana under grant PROMETEO/2018/123 (EFIMAT). J. A. S. also acknowledges Ramón y Cajal Fellowship for financial support (RYC-2015-17482).

References

1. Liu, L.-G.; Bassett, W.A. *Elements, Oxides, Silicates, High-Pressure Phases with Implications for the Earth's Interior*; Oxford Monographs on Geology and Geophysics; Oxford University Press: Oxford, UK, 1986; Volume 4.
2. Adachi, G.-Y.; Imanaka, N. The Binary Rare Earth Oxides. *Chem. Rev.* **1998**, *98*, 1479–1514. [CrossRef] [PubMed]
3. Smyth, J.R.; Jacobsen, S.D.; Hazen, R.M. Comparative Crystal Chemistry of Dense Oxide Minerals. In *Reviews in Mineralogy and Geochemistry*; Mineralogical Society of America: Chantilly, VA, USA, 2000; Volume 41.
4. Zinkevich, M. Thermodynamics of rare earth sesquioxides. *Prog. Mat. Sci.* **2007**, *52*, 597–647. [CrossRef]
5. Manjón, F.J.; Errandonea, D. Pressure-induced structural phase transitions in materials science. *Phys. Stat. Sol. B* **2009**, *246*, 9–31. [CrossRef]
6. Foex, M.; Traverse, J.P. Remarques sur les transformations cristallines presentees a haute temperature par les sesquioxydes de terres rares. *Rev. Int. Hautes Temp. Refract.* **1966**, *3*, 429–453.
7. Hoekstra, H.R.; Gingerich, K.A. High-Pressure B-Type Polymorphs of Some Rare-Earth Sesquioxides. *Science* **1964**, *14*, 1163–1164. [CrossRef]
8. Sawyer, J.O.; Hyde, B.G.; Eyring, L. Pressure and Polymorphism in the Rare Earth Sesquioxides. *Inorg. Chem.* **1965**, *4*, 426–427. [CrossRef]
9. Vegas, A.; Isea, R. Distribution of the M-M Distances in the Rare Earth Oxides. *Acta Cryst. B* **1998**, *54*, 732–740. [CrossRef]
10. Vegas, A. *Structural Models of Inorganic Crystals*; Editorial Universitat Politècnica de Valencia: Valencia, Spain, 2018.
11. McClure, J.P. High Pressure Phase Transitions in the Lanthanide Sesquioxides. Ph.D. Thesis, University of Nevada, Las Vegas, NV, USA, 2009.
12. Jiang, S.; Liu, J.; Lin, C.L.; Bai, L.G.; Xiao, W.S.; Zhang, Y.F.; Zhang, D.C.; Li, X.D.; Li, Y.C.; Tang, L.Y. Pressure-induced phase transition in cubic Lu_2O_3. *J. Appl. Phys.* **2010**, *108*, 083541. [CrossRef]
13. Lin, C.-M.; Wu, K.-T.; Hung, T.-L.; Sheu, H.-S.; Tsai, M.-H.; Lee, J.-F.; Lee, J.-J. Phase transitions in Lu_2O_3 under high pressure. *Solid State Commun.* **2010**, *150*, 1564–1569. [CrossRef]

14. Yusa, H.; Kikegawa, T. *Photon Factory Activity Report #28 Part B 223*; Kishimoto, S., Ed.; High Energy Accelerator Research Organization (KEK): Ibaraki, Japan, 2010.

15. Meyer, C.; Sanchez, J.P.; Thomasson, J.; Itié, J.P. Mossbauer and energy-dispersive x-ray-diffraction studies of the pressure-induced crystallographic phase transition in C-type Yb_2O_3. *Phys. Rev. B* **1995**, *51*, 12187–12193. [CrossRef]

16. Yusa, H.; Kikegawa, T.; Tsuchiya, T. *Photon Factory Activity Report #27 Part B 195*; Iwano, K., Ed.; High Energy Accelerator Research Organization (KEK): Ibaraki, Japan, 2009.

17. Lonappan, D. High Pressure Phase Transformation Studies on Rare Earth Sesquioxides. Ph.D. Thesis, Indira Gandhi Centre for Atomic Research, Tamil Nadu, India, 2012.

18. Pandey, S.D.; Samanta, K.; Singh, J.; Sharma, N.D.; Bandyopadhyay, A.K. Anharmonic behavior and structural phase transition in Yb_2O_3. *AIP Adv.* **2013**, *3*, 122123. [CrossRef]

19. Sahu, P.C.; Lonappan, D.; Chandra Shekar, N.V. High Pressure Structural Studies on Rare-Earth Sesquioxides. *J. Phys. Conf. Ser.* **2012**, *377*, 012015. [CrossRef]

20. Irshad, K.A.; Anees, P.; Sahoo, S.; Sanjay Kumar, N.R.; Srihari, V.; Kalavathi, S.; Chandra Shekar, N.V. Pressure induced structural phase transition in rare earth sesquioxide Tm_2O_3: Experiment and *ab initio* calculations. *J. Appl. Phys.* **2018**, *124*, 155901. [CrossRef]

21. Yan, D.; Wu, P.; Zhang, S.P.; Liang, L.; Yang, F.; Pei, Y.L.; Chen, S. Assignments of the Raman modes of monoclinic erbium oxide. *J. Appl. Phys.* **2013**, *114*, 193502. [CrossRef]

22. Ren, X.T.; Yan, X.Z.; Yu, Z.H.; Li, W.T.; Wang, L. Photoluminescence and phase transition in Er_2O_3 under high pressure. *J. Alloy. Compd.* **2017**, *725*, 941–945. [CrossRef]

23. Lonappan, D.; Chandra Shekar, N.V.; Ravindran, T.R.; Sahu, P.C. High-pressure phase transition in Ho_2O_3. *Mater. Chem. Phys.* **2010**, *120*, 65–67. [CrossRef]

24. Jiang, S.; Liu, J.; Li, X.D.; Bai, L.G.; Xiao, W.S.; Zhang, Y.F.; Lin, C.L.; Li, Y.C.; Tang, L.Y. Phase transformation of Ho_2O_3 at high pressure. *J. Appl. Phys.* **2011**, *110*, 013526. [CrossRef]

25. Pandey, S.D.; Samanta, K.; Singh, J.; Sharma, N.D.; Bandyopadhyay, A.K. Raman scattering of rare earth sesquioxide Ho_2O_3: A pressure and temperature dependent study. *J. Appl. Phys.* **2014**, *116*, 133504. [CrossRef]

26. Yan, X.Z.; Ren, X.T.; He, D.W.; Chen, B.; Yang, W.G. Mechanical behaviors and phase transition of Ho_2O_3 nanocrystals under high pressure. *J. Appl. Phys.* **2014**, *116*, 033507. [CrossRef]

27. Sharma, N.D.; Singh, J.; Dogra, S.; Varandani, D.; Poswal, H.K.; Sharma, S.M.; Bandyopadhyay, A.K. Pressure-induced anomalous phase transformation in nano-crystalline dysprosium sesquioxide. *J. Raman Spectrosc.* **2011**, *42*, 438–444. [CrossRef]

28. Jiang, S.; Liu, J.; Lin, C.L.; Bai, L.G.; Zhang, Y.F.; Li, X.D.; Li, Y.C.; Tang, L.Y.; Wang, H. Structural transformations in cubic Dy_2O_3 at high pressures. *Solid State Commun.* **2013**, *169*, 37–41. [CrossRef]

29. Chen, H.Y.; He, C.Y.; Gao, C.X.; Ma, Y.M.; Zhang, J.H.; Wang, X.J.; Gao, S.Y.; Li, D.M.; Kan, S.H.; Zou, G.T. The structural transition of Gd_2O_3 nanoparticles induced by high pressure. *J. Phys. Condens. Matter* **2007**, *19*, 425229. [CrossRef]

30. Chen, H.Y.; He, C.Y.; Gao, C.X.; Ma, Y.M.; Zhang, J.H.; Gao, S.Y.; Lu, H.L.; Nie, Y.G.; Li, D.M.; Kan, S.H.; et al. Structural Transition of Gd_2O_3: Eu Induced by High Pressure. *Chin. Phys. Lett.* **2007**, *24*, 158–160. [CrossRef]

31. Zhang, F.X.; Lang, M.; Wang, J.W.; Becker, U.; Ewing, R.C. Structural phase transitions of cubic Gd_2O_3 at high pressures. *Phys. Rev. B* **2008**, *78*, 064114. [CrossRef]

32. Dilawar, N.; Varandani, D.; Mehrotra, S.; Poswal, H.K.; Sharma, S.M.; Bandyopadhyay, A.K. Anomalous high pressure behaviour in nanosized rare earth sesquioxides. *Nanotechnology* **2008**, *19*, 115703. [CrossRef] [PubMed]

33. Bai, L.G.; Liu, J.; Li, X.D.; Jiang, S.; Xiao, W.S.; Li, Y.C.; Tang, L.Y.; Zhang, Y.F.; Zhang, D.C. Pressure-induced phase transformations in cubic Gd_2O_3. *J. Appl. Phys.* **2009**, *106*, 073507. [CrossRef]

34. Zou, X.; Gong, C.; Liu, B.B.; Li, Q.J.; Li, Z.P.; Liu, B.; Liu, R.; Chen, Z.Q.; Zou, B.; et al. X-ray diffraction of cubic Gd_2O_3/Er under high pressure. *Phys. Stat. Sol. B* **2011**, *248*, 1123–1127. [CrossRef]

35. Zhang, C.C.; Zhang, Z.M.; Dai, R.C.; Wang, Z.P.; Ding, Z.J. High Pressure Luminescence and Raman Studies on the Phase Transition of Gd_2O_3:Eu^{3+} Nanorods. *J. Nanosci. Nanotechnol.* **2011**, *11*, 9887–9891. [CrossRef] [PubMed]

36. Yang, X.; Li, Q.J.; Liu, Z.D.; Bai, X.; Song, H.W.; Yao, M.G.; Liu, B.; Liu, R.; Gong, C.; Lu, S.C.; et al. Pressure-Induced Amorphization in Gd_2O_3/Er^{3+} Nanorods. *J. Phys. Chem. C* **2013**, *117*, 8503–8508. [CrossRef]
37. Chen, G.; Haire, R.G.; Peterson, J.R. Effect of pressure on cubic (C-type) Eu_2O_3 studied via Eu^{3+} luminescence. *High Press. Res.* **1991**, *6*, 371–377. [CrossRef]
38. Chen, G.; Stump, N.A.; Haire, R.G.; Peterson, J.R. Study of the phase behavior of Eu_2O_3 under pressure via luminescence of Eu^{3+}. *J. Alloy. Compd.* **1992**, *181*, 503–509. [CrossRef]
39. Dilawar, N.; Varandani, D.; Pandey, V.P.; Kumar, M.; Shivaprasad, S.M.; Sharma, P.K.; Bandyopadhyay, A.K. Structural Transition in Nanostructured Eu_2O_3 Under High Pressures. *J. Nanosci. Nanotechnol.* **2006**, *6*, 105–113. [PubMed]
40. Jiang, S.; Bai, L.G.; Liu, J.; Xiao, W.S.; Li, X.D.; Li, Y.C.; Tang, L.Y.; Zhang, Y.F.; Zhang, D.C.; Zheng, L.R. The Phase Transition of Eu_2O_3 under High Pressures. *Chin. Phys. Lett.* **2006**, *26*, 076101.
41. Irshad, K.A.; Chandra Shekar, N.V.; Srihari, V.; Pandey, K.K.; Kalavathi, S. High pressure structural phase transitions in Ho: Eu_2O_3. *J. Alloy. Compd.* **2017**, *725*, 911–915. [CrossRef]
42. Irshad, K.A.; Chandra Shekar, N.V. Anomalous lattice compressibility of hexagonal Eu_2O_3. *Mat. Chem. Phys.* **2017**, *195*, 88–93. [CrossRef]
43. Yu, Z.H.; Wang, Q.L.; Ma, Y.Z.; Wang, L. X-ray diffraction and spectroscopy study of nano-Eu_2O_3 structural transformation under high pressure. *J. Alloy. Compd.* **2017**, *701*, 542–548. [CrossRef]
44. Guo, Q.X.; Zhao, Y.S.; Jiang, C.; Mao, W.L.; Wang, Z.W. Phase transformation in Sm_2O_3 at high pressure: In situ synchrotron X-ray diffraction study and ab initio DFT calculation. *Solid State Commun.* **2008**, *145*, 250–254. [CrossRef]
45. Jiang, S.; Liu, J.; Lin, C.L.; Li, X.D.; Li, Y.C. High-pressure x-ray diffraction and Raman spectroscopy of phase transitions in Sm_2O_3. *J. Appl. Phys.* **2013**, *113*, 113502. [CrossRef]
46. Liu, D.; Lei, W.W.; Li, Y.W.; Ma, Y.M.; Hao, J.; Chen, X.H.; Jin, Y.X.; Liu, D.; Yu, S.D.; Cui, Q.L.; et al. High-Pressure Structural Transitions of Sc_2O_3 by X-ray Diffraction, Raman Spectra, and Ab Initio Calculations. *Inorg. Chem.* **2009**, *48*, 8251–8256. [CrossRef]
47. Yusa, H.; Tsuchiya, T.; Sata, N.; Ohishi, Y. High-Pressure Phase Transition to the Gd_2S_3 Structure in Sc_2O_3: A New Trend in Dense Structures in Sesquioxides. *Inorg. Chem.* **2009**, *48*, 7537–7543. [CrossRef]
48. Ovsyannikov, S.V.; Bykova, E.; Bykov, M.; Wenz, M.D.; Pakhomova, A.S.; Glazyrin, K.; Liermann, H.-P.; Dubrovinsky, L. Structural and vibrational properties of single crystals of Scandia, Sc_2O_3 under high pressure. *J. Appl. Phys.* **2015**, *118*, 165901. [CrossRef]
49. Husson, E.; Proust, C.; Gillet, P.; Itié, J.P. Phase transitions in yttrium oxide at high pressure studied by Raman spectroscopy. *Mater. Res. Bull.* **1999**, *34*, 2085–2092. [CrossRef]
50. Bai, X.; Song, H.W.; Liu, B.B.; Hou, Y.Y.; Pan, G.H.; Ren, X.G. Effects of High Pressure on the Luminescence Properties of Nanocrystalline and Bulk Y_2O_3:Eu^{3+}. *J. Nanosci. Nanotechnol.* **2008**, *8*, 1404–1409. [CrossRef] [PubMed]
51. Jovanic, B.R.; Dramicanin, M.; Viana, B.; Panic, B.; Radenkovic, B. High-pressure optical studies of Y_2O_3:Eu^{3+} nanoparticles. *Radiat. Eff. Defects Solids* **2008**, *163*, 925–931. [CrossRef]
52. Wang, L.; Pan, Y.X.; Ding, Y.; Yang, W.G.; Mao, W.L.; Sinogeikin, S.V.; Meng, Y.; Shen, G.Y.; Mao, H.-K. High-pressure induced phase transitions of Y_2O_3 and Y_2O_3:Eu^{3+}. *Appl. Phys. Lett.* **2009**, *94*, 061921. [CrossRef]
53. Wang, L.; Yang, W.; Ding, Y.; Ren, Y.; Xiao, S.G.; Liu, B.B.; Sinogeikin, S.V.; Meng, Y.; Gosztola, D.J.; Shen, G.Y.; et al. Size-Dependent Amorphization of Nanoscale Y_2O_3 at High Pressure. *Phys. Rev. Lett.* **2010**, *105*, 095701. [CrossRef]
54. Halevy, I.; Carmon, R.; Winterrose, M.L.; Yeheskel, O.; Tiferet, E.; Ghose, S. Pressure-induced structural phase transitions in Y_2O_3 sesquioxide. *J. Phys. Conf. Ser.* **2010**, *215*, 012003. [CrossRef]
55. Dai, R.C.; Zhang, Z.M.; Zhang, C.C.; Ding, Z.J. Photoluminescence and Raman Studies of Y_2O_3:Eu^{3+} Nanotubes Under High Pressure. *J. Nanosci. Nanotechnol.* **2010**, *10*, 7629–7633. [CrossRef]
56. Dai, R.C.; Wang, Z.P.; Zhang, Z.M.; Ding, Z.J. Photoluminescence study of SiO_2 coated Eu^{3+}:Y_2O_3 core-shells under high pressure. *J. Rare Earth* **2010**, *28*, 241–245. [CrossRef]
57. Yusa, H.; Tsuchiya, T.; Sata, N.; Ohishi, Y. Dense Yttria Phase Eclipsing the A-Type Sesquioxide Structure: High-Pressure Experiments and ab initio Calculations. *Inorg. Chem.* **2010**, *49*, 4478–4485. [CrossRef]

58. Bose, P.P.; Gupta, M.K.; Mittal, R.; Rols, S.; Achary, S.N.; Tyagi, A.K.; Chaplot, S.L. High Pressure Phase Transitions in Yttria, Y_2O_3. *J. Phys. Conf. Ser.* **2012**, *377*, 012036. [CrossRef]

59. Srivastava, A.M.; Renero-Lecuna, C.; Santamaría-Pérez, D.; Rodríguez, F.; Valiente, R. Pressure-induced Pr^{3+} $^3P_0$ luminescence in cubic Y_2O_3. *J. Lumin.* **2014**, *146*, 27–32. [CrossRef]

60. Jiang, S.; Liu, J.; Li, X.-D.; Li, Y.-C.; He, S.-M.; Zhang, J.-C. High-Pressure Phase Transitions of Cubic Y_2O_3 under High Pressures by In-situ Synchrotron X-Ray Diffraction. *Chin. Phys. Lett.* **2019**, *36*, 046103. [CrossRef]

61. Ibáñez, J.; Sans, J.A.; Cuenca-Gotor, V.; Oliva, R.; Blázquez, O.; Gomis, O.; Rodríguez-Hernández, P.; Muñoz, A.; Rodríguez-Mendoza, U.R.; Velázquez, M.; et al. Experimental and theoretical study of Tb_2O_3 under compression. Manuscript in preparation for 2020.

62. Zhang, Q.; Wu, X.; Qin, S. Pressure-induced phase transition of B-type Y_2O_3. *Chin. Phys. B* **2017**, *26*, 090703. [CrossRef]

63. Chen, G.; Peterson, J.R.; Brister, K.E. An Energy-Dispersive X-Ray Diffraction Study of Monoclinic Eu_2O_3 under Pressure. *J. Solid State Chem.* **1994**, *111*, 437–439. [CrossRef]

64. Atou, T.; Kusaba, K.; Tsuchida, Y.; Utsumi, W.; Yagi, T.; Syono, Y. Reversible B-type- A-type transition of Sm_2O_3 under high pressure. *Mater. Res. Bull.* **1989**, *24*, 1171–1176. [CrossRef]

65. Hongo, T.; Kondo, K.; Nakamura, K.G.; Atou, T. High pressure Raman spectroscopic study of structural phase transition in samarium oxide. *J. Mater. Sci.* **2007**, *42*, 2582–2585. [CrossRef]

66. Guo, Q.X.; Zhao, Y.S.; Jiang, C.; Mao, W.L.; Wang, Z.W.; Zhang, J.Z.; Wang, Y.J. Pressure-Induced Cubic to Monoclinic Phase Transformation in Erbium Sesquioxide Er_2O_3. *Inorg. Chem.* **2007**, *46*, 6164–6169. [CrossRef]

67. Pandey, K.K.; Garg, N.; Mishra, A.K.; Sharma, S.M. High pressure phase transition in Nd_2O_3. *J. Phys. Conf. Ser.* **2012**, *377*, 012006. [CrossRef]

68. Jiang, S.; Liu, J.; Bai, L.G.; Li, X.D.; Li, Y.C.; He, S.M.; Yan, S.; Liang, D.X. Anomalous compression behaviour in Nd_2O_3 studied by x-ray diffraction and Raman spectroscopy. *AIP Adv.* **2018**, *8*, 025019. [CrossRef]

69. Lipp, M.J.; Jeffries, J.R.; Cynn, H.; Park Klepeis, J.H.; Evans, W.J.; Mortensen, D.R.; Seidler, G.T.; Xiao, Y.; Chow, P. Comparison of the high-pressure behavior of the cerium oxides Ce_2O_3 and CeO_2. *Phys. Rev. B* **2016**, *93*, 064106. [CrossRef]

70. Hirosaki, N.; Ogata, S.; Kocer, C. Ab initio calculation of the crystal structure of the lanthanide Ln_2O_3 sesquioxides. *J. Alloy. Compd.* **2003**, *351*, 31–34. [CrossRef]

71. Marsella, L.; Fiorentini, V. Structure and stability of rare-earth and transition-metal oxides. *Phys. Rev. B* **2004**, *69*, 172103. [CrossRef]

72. Petit, L.; Svane, A.; Szotec, Z.; Temmerman, W.M. First-principles study of rare-earth oxides. *Phys. Rev. B* **2005**, *72*, 205118. [CrossRef]

73. Wu, B.; Zinkevich, M.; Chong, W.A.N.G.; Aldinger, F. Ab initio energetic study of oxide ceramics with rare-earth elements. *Rare Met.* **2006**, *25*, 549–555. [CrossRef]

74. Singh, N.; Saini, S.M.; Nautiyal, T.; Auluck, S. Electronic structure and optical properties of rare earth sesquioxides (R_2O_3, R = La, Pr, and Nd). *J. Appl. Phys.* **2006**, *100*, 083525. [CrossRef]

75. Mikami, M.; Nakamura, S. Electronic structure of rare-earth sesquioxides and oxysulfides. *J. Alloy. Compd.* **2006**, *408–412*, 687–692. [CrossRef]

76. Wu, B.; Zinkevich, M.; Aldinger, F.; Wen, D.Z.; Chen, L. Ab initio study on structure and phase transition of A- and B-type rare-earth sesquioxides Ln_2O_3 (Ln = La–Lu, Y, and Sc) based on density function theory. *J. Solid State Chem.* **2007**, *180*, 3280–3287. [CrossRef]

77. Rahm, M.; Skorodumova, N.V. Phase stability of the rare-earth sesquioxides under pressure. *Phys. Rev. B* **2009**, *80*, 104105. [CrossRef]

78. Jiang, H.; Gomez-Abal, R.I.; Rinke, P.; Scheffler, M. Localized and Itinerant States in Lanthanide Oxides United by GW@LDA+U. *Phys. Rev. Lett.* **2009**, *102*, 126403. [CrossRef] [PubMed]

79. Gillen, R.; Clark, S.J.; Robertson, J. Nature of the electronic band gap in lanthanide oxides. *Phys. Rev. B* **2013**, *87*, 125116. [CrossRef]

80. Richard, D.; Muñoz, E.L.; Rentería, M.; Errico, L.A.; Svane, A.; Christensen, N.E. Ab initio LSDA and LSDA+U study of pure and Cd-doped cubic lanthanide sesquioxides. *Phys. Rev. B* **2013**, *88*, 165206. [CrossRef]

81. Richard, D.; Errico, L.A.; Rentería, M. Structural properties and the pressure-induced C → A phase transition of lanthanide sesquioxides from DFT and DFT+U calculations. *J. Alloy. Compd.* **2016**, *664*, 580–589. [CrossRef]

82. Ogawa, T.; Otani, N.; Yokoi, T.; Fisher, C.A.J.; Kuwabara, A.; Moriwake, H.; Yoshiya, M.; Kitaoka, S.; Takata, M. Density functional study of phase stabilities and Raman spectra of Yb_2O_3, Yb_2SiO_5 and $Yb_2Si_2O_7$ under pressure. *Phys. Chem. Chem. Phys.* **2018**, *20*, 16518–16527. [CrossRef]

83. Pathak, A.K.; Vazhappily, T. Ab Initio Study on Structure, Elastic, and Mechanical Properties of Lanthanide Sesquioxides. *Phys. Stat. Sol. B* **2018**, *255*, 1700668. [CrossRef]

84. Catlow, C.R.A.; Guo, Z.X.; Miskufova, M.; Shevlin, S.A.; Smith, A.G.H.; Sokol, A.A.; Walsh, A.; Wilson, D.J.; Woodley, S.M. Advances in computational studies of energy materials. *Philos. Trans. R. Soc. A Math. Phys. Eng. Sci.* **2010**, *368*, 3379–3456. [CrossRef]

85. Caracas, R.; Cohen, R.E. Prediction of a new phase transition in Al_2O_3 at high pressures. *Geophys. Res. Lett.* **2005**, *32*, 1–4. [CrossRef]

86. Funamori, N. High-Pressure Transformation of Al_2O_3. *Science* **1997**, *278*, 1109–1111. [CrossRef]

87. Cynn, H.; Isaak, D.G.; Cohen, R.E.; Nicol, M.F.; Anderson, O.L. A high-pressure phase transition of corundum predicted by the potential induced breathing model. *Am. Mineral.* **1990**, *75*, 439–442.

88. Jephcoat, A.P.; Hemley, R.J.; Mao, H.K. X-ray diffraction of ruby ($Al_2O_3:Cr^{3+}$) to 175 GPa. *Phys. B C* **1988**, *150*, 115–121. [CrossRef]

89. Dewaele, A.; Torrent, M. Equation of state of α-Al_2O_3. *Phys. Rev. B* **2013**, *88*, 064107. [CrossRef]

90. Costa, T.M.H.; Gallas, M.R.; Benvenutti, E.V.; da Jornada, J.A.H. Study of Nanocrystalline γ-Al_2O_3 Produced by High-Pressure Compaction. *J. Phys. Chem. B* **1999**, *103*, 4278–4284. [CrossRef]

91. Hart, H.V.; Drickamer, H.G. Effect of high pressure on the lattice parameters of Al_2O_3. *J. Chem. Phys.* **1965**, *43*, 2265–2266. [CrossRef]

92. Marton, F.C.; Cohen, R.E. Prediction of a high-pressure phase transition in Al_2O_3. *Am. Miner.* **1994**, *79*, 789–792.

93. Mashimo, T.; Tsumoto, K.; Nakamura, K.; Noguchi, Y.; Fukuoka, K.; Syono, Y. High-pressure phase transformation of corundum (α-Al_2O_3) observed under shock compression. *Geophys. Res. Lett.* **2000**, *27*, 2021–2024. [CrossRef]

94. Ono, S.; Oganov, A.R.; Koyama, T.; Shimizu, H. Stability and compressibility of the high-pressure phases of Al_2O_3 up to 200 GPa: Implications for the electrical conductivity of the base of the lower mantle. *Earth Planet. Sci. Lett.* **2006**, *246*, 326–335. [CrossRef]

95. Zhao, J.; Hearne, G.R.; Maaza, M.; Laher-Lacour, F.; Witcomb, M.J.; Le Bihan, T.; Mezouar, M. Compressibility of nanostructured alumina phases determined from synchrotron x-ray diffraction studies at high pressure. *J. Appl. Phys.* **2001**, *90*, 3280–3285. [CrossRef]

96. Thomson, K.T.; Wentzcovitch, R.M.; Bukowinski, M.S.T. Polymorphs of alumina predicted by first principles: Putting pressure on the ruby pressure scale. *Science* **1996**, *274*, 1880–1882. [CrossRef]

97. Lin, J.F.; Degtyareva, O.; Prewitt, C.T.; Dera, P.; Sata, N.; Gregoryanz, E.; Mao, H.K.; Hemley, R.J. Crystal structure of a high-pressure/high-temperature phase of alumina by in situ X-ray diffraction. *Nat. Mater.* **2004**, *3*, 389–393. [CrossRef]

98. Jahn, S.; Madden, P.A.; Wilson, M. Dynamic simulation of pressure-driven phase transformations in crystalline Al_2O_3. *Phys. Rev. B* **2004**, *69*, 020106. [CrossRef]

99. Tsuchiya, J.; Tsuchiya, T.; Wentzcovitch, R.M. Transition from the Rh_2O_3(II)-to-$CaIrO_3$ structure and the high-pressure-temperature phase diagram of alumina. *Phys. Rev. B* **2005**, *72*, 020103. [CrossRef]

100. Garcia-Domene, B.; Ortiz, H.M.; Gomis, O.; Sans, J.A.; Manjón, F.J.; Muñoz, A.; Rodríguez-Hernández, P.; Achary, S.N.; Errandonea, D.; Martínez-García, D.; et al. High-pressure lattice dynamical study of bulk and nanocrystalline In_2O_3. *J. Appl. Phys.* **2012**, *112*, 123511. [CrossRef]

101. García-Domene, B.; Sans, J.A.; Gomis, O.; Manjón, F.J.; Ortiz, H.M.; Errandonea, D.; Santamaría-Pérez, D.; Martínez-García, D.; Vilaplana, R.; Pereira, A.L.J.; et al. Pbca-type In_2O_3: The high-pressure post-corundum phase at room temperature. *J. Phys. Chem. C* **2014**, *118*, 20545–20552. [CrossRef]

102. Gurlo, A. Structural stability of high-pressure polymorphs in In_2O_3 nanocrystals: Evidence of stress-induced transition? *Angew. Chem. Int. Ed.* **2010**, *49*, 5610–5612. [CrossRef]

103. Liu, D.; Lei, W.W.; Zou, B.; Yu, S.D.; Hao, J.; Wang, K.; Liu, B.B.; Cui, Q.L.; Zou, G.T. High-pressure x-ray diffraction and Raman spectra study of indium oxide. *J. Appl. Phys.* **2008**, *104*, 083506. [CrossRef]

104. Qi, J.; Liu, J.F.; He, Y.; Chen, W.; Wang, C. Compression behavior and phase transition of cubic In_2O_3 nanocrystals. *J. Appl. Phys.* **2011**, *109*, 063520. [CrossRef]

105. Tang, S.; Li, Y.; Zhang, J.; Zhu, H.; Dong, Y.; Zhu, P.; Cui, Q. Effects of microstructures on the compression behavior and phase transition routine of In_2O_3 nanocubes under high pressures. *RSC Adv.* **2015**, *5*, 85105–85110. [CrossRef]

106. Yusa, H.; Tsuchiya, T.; Sata, N.; Ohishi, Y. Rh_2O_3(II)-type structures in Ga_2O_3 and In_2O_3 under high pressure: Experiment and theory. *Phys. Rev. B* **2008**, *77*, 064107. [CrossRef]

107. Yusa, H.; Tsuchiya, T.; Tsuchiya, J.; Sata, N.; Ohishi, Y. α-Gd_2S_3-type structure in In_2O_3: Experiments and theoretical confirmation of a high-pressure polymorph in sesquioxide. *Phys. Rev. B* **2008**, *78*, 092107. [CrossRef]

108. García-Domene, B.; Sans, J.A.; Manjón, F.J.; Ovsyannikov, S.V.; Dubrovinsky, L.S.; Martinez-Garcia, D.; Gomis, O.; Errandonea, D.; Moutaabbid, H.; Le Godec, Y.; et al. Synthesis and High-Pressure Study of Corundum-Type In_2O_3. *J. Phys. Chem. C* **2015**, *119*, 29076–29087. [CrossRef]

109. Sans, J.A.; Vilaplana, R.; Errandonea, D.; Cuenca-Gotor, V.P.; García-Domene, B.; Popescu, C.; Manjón, F.J.; Singhal, A.; Achary, S.N.; Martinez-Garcia, D.; et al. Structural and vibrational properties of corundum-type In_2O_3 nanocrystals under compression. *Nanotechnology* **2017**, *28*, 205701. [CrossRef] [PubMed]

110. Lipinska-Kalita, K.E.; Kalita, P.E.; Hemmers, O.A.; Hartmann, T. Equation of state of gallium oxide to 70 GPa: Comparison of quasihydrostatic and nonhydrostatic compression. *Phys. Rev. B* **2008**, *77*, 094123. [CrossRef]

111. Lipinska-Kalita, K.E.; Chen, B.; Kruger, M.B.; Ohki, Y.; Murowchick, J.; Gogol, E.P. High-pressure x-ray diffraction studies of the nanostructured transparent vitroceramic medium K_2O-SiO_2-Ga_2O_3. *Phys. Rev. B* **2003**, *68*, 035209. [CrossRef]

112. Luan, S.; Dong, L.; Jia, R. Analysis of the structural, anisotropic elastic and electronic properties of β-Ga_2O_3 with various pressures. *J. Cryst. Growth* **2019**, *505*, 74–81. [CrossRef]

113. Machon, D.; McMillan, P.F.; Xu, B.; Dong, J. High-pressure study of the β-to-α transition in Ga_2O_3. *Phys. Rev. B* **2006**, *73*, 094125. [CrossRef]

114. Wang, H.; He, Y.; Chen, W.; Zeng, Y.W.; Stahl, K.; Kikegawa, T.; Jiang, J.Z. High-pressure behavior of β-Ga_2O_3 nanocrystals. *J. Appl. Phys.* **2010**, *107*, 033520. [CrossRef]

115. Claussen, W.F.; Mackenzie, J.D. Crystallization of B_2O_3 at High Pressures. *J. Am. Chem. Soc.* **1959**, *81*, 1007. [CrossRef]

116. Prewitt, C.T.; Shannon, R.D. Crystal structure of a high-pressure form of B_2O_3. *Acta Crystallogr. Sect. B Struct. Crystallogr. Cryst. Chem.* **1968**, *24*, 869–874. [CrossRef]

117. Brazhkin, V.V.; Katayama, Y.; Inamura, Y.; Kondrin, M.V.; Lyapin, A.G.; Popova, S.V.; Voloshin, R.N. Structural transformations in liquid, crystalline, and glassy B_2O_3 under high pressure. *JETP Lett.* **2003**, *78*, 393–397. [CrossRef]

118. Nicholas, J.; Sinogeikin, S.; Kieffer, J.; Bass, J. Spectroscopic evidence of polymorphism in vitreous B_2O_3. *Phys. Rev. Lett.* **2004**, *92*, 3–6. [CrossRef] [PubMed]

119. Lee, S.K.; Mibe, K.; Fei, Y.; Cody, G.D.; Mysen, B.O. Structure of B_2O_3 glass at high pressure: A 11B solid-state NMR study. *Phys. Rev. Lett.* **2005**, *94*, 27–30. [CrossRef] [PubMed]

120. Kulikova, L.F.; Dyuzheva, T.I.; Nikolaev, N.A.; Brazhkin, V.V. Single-crystal growth of the high-pressure phase B_2O_3 II. *Crystallogr. Rep.* **2012**, *57*, 332–335. [CrossRef]

121. Burianek, M.; Birkenstock, J.; Mair, P.; Kahlenberg, V.; Medenbach, O.; Shannon, R.D.; Fischer, R.X. High-pressure synthesis, long-term stability of single crystals of diboron trioxide, B_2O_3, and an empirical electronic polarizability of $[3]B^{3+}$. *Phys. Chem. Miner.* **2016**, *43*, 527–534. [CrossRef]

122. Gomis, O.; Santamaría-Pérez, D.; Ruiz-Fuertes, J.; Sans, J.A.; Vilaplana, R.; Ortiz, H.M.; García-Domene, B.; Manjón, F.J.; Errandonea, D.; Rodríguez-Hernández, P.; et al. High-pressure structural and elastic properties of Tl_2O_3. *J. Appl. Phys.* **2014**, *116*, 133521. [CrossRef]

123. Weir, S.T.; Mitchell, A.C.; Nellis, W.J. Electrical resistivity of single-crystal Al_2O_3 shock-compressed in the pressure range 91-220 GPa (0.91-2.20 Mbar). *J. Appl. Phys.* **1996**, *80*, 1522–1525. [CrossRef]

124. Syassen, K. Ruby under pressure. *High Press. Res.* **2008**, *28*, 75–126. [CrossRef]

125. Smyth, J.R.; Jacobsen, S.D.; Hazen, R.M. Comparative Crystal Chemistry of Dense Oxide Minerals. *Rev. Miner. Geochem.* **2000**, *41*, 157–186. [CrossRef]

126. Song, H.I.; Kim, E.S.; Yoon, K.H. Phase transformation and characteristics of beta-alumina. *Phys. B C* **1988**, *150*, 148–159. [CrossRef]

127. Engürlü, S.; Taslicukur Öztürk, Z.; Kuskonmaz, N. Investigation of the Production of β-Al_2O_3 Solid Electrolyte from Seydişehir α-Al_2O_3. *Süleyman Demirel Üniversitesi Fen Bilimleri Enstitüsü Derg.* **2017**, *21*, 816. [CrossRef]

128. Duan, W.; Wentzcovitch, R.M.; Thomson, K.T. First-principles study of high-pressure alumina polymorphs. *Phys. Rev. B* **1998**, *57*, 10363–10369. [CrossRef]

129. Oganov, A.R.; Ono, S. The high-pressure phase of alumina and implications for Earth's D" layer. *Proc. Natl. Acad. Sci. USA* **2005**, *102*, 10828–10831. [CrossRef]

130. Hama, J.; Suito, K. The evidence for the occurrence of two successive transitions in Al_2O_3 from the analysis of Hugoniot data. *High Temp. High Press.* **2002**, *34*, 323–334. [CrossRef]

131. Ono, S.; Kikegawa, T.; Ohishi, Y. High-pressure phase transition of hematite, Fe_2O_3. *J. Phys. Chem. Solids* **2004**, *65*, 1527–1530. [CrossRef]

132. Oganov, A.R.; Ono, S. Theoretical and experimental evidence for a post-perovskite phase of $MgSiO_3$ in Earth's D" layer. *Nature* **2004**, *430*, 445–448. [CrossRef]

133. Vaidya, S.N. High-pressure high-temperature transitions in nanocrystalline γ Al_2O_3, γ Fe_2O_3 and TiO_2. *Bull. Mater. Sci.* **1999**, *22*, 287–293. [CrossRef]

134. Mishra, R.S.; Lesher, C.E.; Mukherjee, A.K. High-Pressure Sintering of Nanocrystalline gamma Al_2O_3. *J. Am. Ceram. Soc.* **1996**, *79*, 2989–2992. [CrossRef]

135. Vaidya, S.N.; Karunakaran, C.; Achary, S.N.; Tyagi, A.K. New polymorphs of alumina. *High Press. Res.* **1999**, *16*, 147–160. [CrossRef]

136. Vaidya, S.N.; Karunakaran, C.; Achary, S.N.; Tyagi, A.K. New polymorphs of alumina: Part II mu and lambda alumina. *High Press. Res.* **1999**, *16*, 265–278. [CrossRef]

137. Bekheet, M.F.; Schwarz, M.R.; Lauterbach, S.; Kleebe, H.J.; Kroll, P.; Riedel, R.; Gurlo, A. Orthorhombic In_2O_3: A metastable polymorph of indium sesquioxide. *Angew. Chem. Int. Ed.* **2013**, *52*, 6531–6535. [CrossRef]

138. Karazhanov, S.Z.; Ravindran, P.; Vajeeston, P.; Ulyashin, A.; Finstad, T.G.; Fjellvåg, H. Phase stability, electronic structure, and optical properties of indium oxide polytypes. *Phys. Rev. B* **2007**, *76*, 075129. [CrossRef]

139. Shannon, R.D. New high pressure phases having the corundum structure. *Solid State Commun.* **1966**, *4*, 629–630. [CrossRef]

140. Prewitt, C.T.; Shannon, R.D.; Rogers, D.B.; Sleight, A.W. The C rare earth oxide-corundum transition and crystal chemistry of oxides having the corundum structure. *Inorg. Chem.* **1969**, *8*, 1985–1993. [CrossRef]

141. Atou, T.; Kusaba, K.; Fukuoka, K.; Kikuchi, M.; Syono, Y. Shock-induced phase transition of M_2O_3 (M = Sc, Y, Sm, Gd, and In)-type compounds. *J. Solid State Chem.* **1990**, *89*, 378–384. [CrossRef]

142. Epifani, M.; Siciliano, P.; Gurlo, A.; Barsan, N.; Weimar, U. Ambient Pressure Synthesis of Corundum-Type In_2O_3. *J. Am. Chem. Soc.* **2004**, *126*, 4078–4079. [CrossRef]

143. Yu, D.; Wang, D.; Qian, Y. Synthesis of metastable hexagonal In_2O_3 nanocrystals by a precursor-dehydration route under ambient pressure. *J. Solid State Chem.* **2004**, *177*, 1230–1234. [CrossRef]

144. Sorescu, M.; Diamandescu, L.; Tarabasanu-Mihaila, D.; Teodorescu, V.S. Nanocrystalline rhombohedral In_2O_3 synthesized by hydrothermal and postannealing pathways. *J. Mater. Sci.* **2004**, *39*, 675–677. [CrossRef]

145. Åhman, J.; Svensson, G.; Albertsson, J. A reinvestigation of β-gallium oxide. *Acta Crystallogr. Sect. C Cryst. Struct. Commun.* **1996**, *52*, 1336–1338. [CrossRef]

146. Geller, S. Crystal structure of β-Ga_2O_3. *J. Chem. Phys.* **1960**, *33*, 676–684. [CrossRef]

147. Remeika, J.P.; Marezio, M. Growth of α-Ga_2O_3 single crystals at 44 kbars. *Appl. Phys. Lett.* **1966**, *8*, 87–88. [CrossRef]

148. Tsuchiya, T.; Yusa, H.; Tsuchiya, J. Post-Rh_2O_3(II) transition and the high pressure-temperature phase diagram of gallia: A first-principles and x-ray diffraction study. *Phys. Rev. B* **2007**, *76*, 174108. [CrossRef]

149. Kishimura, H.; Matsumoto, H. Evaluation of the shock-induced phase transition in β-Ga_2O_3. *Jpn. J. Appl. Phys.* **2018**, *57*, 125503. [CrossRef]

150. Gurr, G.E.; Montgomery, P.W.; Knutson, C.D.; Gorres, B.T. The crystal structure of trigonal diboron trioxide. *Acta Crystallogr. Sect. B Struct. Crystallogr. Cryst. Chem.* **1970**, *26*, 906–915. [CrossRef]

151. Switzer, J.A. The n-Silicon/Thallium(III) Oxide Heterojunction Photoelectrochemical Solar Cell. *J. Electrochem. Soc.* **1986**, *133*, 722–728. [CrossRef]

152. Phillips, R.J.; Shane, M.J.; Switzer, J.A. Electrochemical and photoelectrochemical deposition of thallium(III) oxide thin films. *J. Mater. Res.* **1989**, *4*, 923–929. [CrossRef]

153. Van Leeuwen, R.A.; Hung, C.J.; Kammler, D.R.; Switzer, J.A. Optical and electronic transport properties of electrodeposited thallium(III) oxide films. *J. Phys. Chem.* **1995**, *99*, 15247–15252. [CrossRef]

154. Bhattacharya, R.N.; Yan, S.L.; Xing, Z.; Xie, Y.; Wu, J.Z.; Feldmann, M.; Chen, J.; Xiong, Q.; Ren, Z.F.; Blaugher, R.D. Superconducting Thallium Oxide and Mercury Oxide Films. *MRS Online Proc. Libr. Arch.* **2000**, *659*. [CrossRef]

155. Ma, C.; Rossman, G.R. Tistarite, Ti_2O_3, a new refractory mineral from the Allende meteorite. *Am. Mineral.* **2009**, *94*, 841–844. [CrossRef]

156. Xue, K.H.; Blaise, P.; Fonseca, L.R.C.; Nishi, Y. Prediction of semimetallic tetragonal Hf_2O_3 and Zr_2O_3 from first principles. *Phys. Rev. Lett.* **2013**, *110*, 065502. [CrossRef]

157. Ovsyannikov, S.V.; Trots, D.M.; Kurnosov, A.V.; Morgenroth, W.; Liermann, H.P.; Dubrovinsky, L. Anomalous compression and new high-pressure phases of vanadium sesquioxide, V_2O_3. *J. Phys. Condens. Matter* **2013**, *25*, 385401. [CrossRef]

158. Goodenough, J.B.; Hamnett, A.; Huber, G.; Hullinger, F.; Leiß, M.; Ramasesha, S.K.; Werheit, H. *Physics of Non-Tetrahedrally Bonded Binary Compounds III/Physik der Nicht-Tetraedrisch Gebundenen Binären Verbindungen III*; Madelung, O., Ed.; Springer: Berlin, Germany, 1984.

159. Pasternak, M.P.; Rozenberg, G.K.; Machavariani, G.Y.; Naaman, O.; Taylor, R.D.; Jeanloz, R. Breakdown of the mott-hubbard state in Fe_2O_3: A first-order insulator-metal transition with collapse of magnetism at 50 GPa. *Phys. Rev. Lett.* **1999**, *82*, 4663–4666. [CrossRef]

160. Frost, D.J.; Liebske, C.; Langenhorst, F.; McCammon, C.A.; Trønnes, R.G.; Rubie, D.C. Experimental evidence for the existence of iron-rich metal in the Earth's lower mantle. *Nature* **2004**, *428*, 409–412. [CrossRef]

161. Kupenko, I.; Aprilis, G.; Vasiukov, D.M.; McCammon, C.; Chariton, S.; Cerantola, V.; Kantor, I.; Chumakov, A.I.; Rüffer, R.; Dubrovinsky, L.; et al. Magnetism in cold subducting slabs at mantle transition zone depths. *Nature* **2019**, *570*, 102–106. [CrossRef]

162. Shokrollahi, H. A review of the magnetic properties, synthesis methods and applications of maghemite. *J. Magn. Magn. Mater.* **2017**, *426*, 74–81. [CrossRef]

163. Forestier, H.; Guiot-Guillain, G. Une nouvelle variété ferromagnétique de sesquioxyde de fer. *CR Acad. Sci. (Paris)* **1934**, *199*, 720–722.

164. Schrader, R.; Büttner, G. Eine neue Eisen(III)-oxidphase: ϵ-Fe_2O_3. *Z. Anorg. Allg. Chem.* **1963**, *320*, 220–234. [CrossRef]

165. Xu, H.; Lee, S.; Xu, H. Luogufengite: A new nano-mineral of Fe_2O_3 polymorph with giant coercive field. *Am. Mineral.* **2017**, *102*, 711–719. [CrossRef]

166. Dejoie, C.; Sciau, P.; Li, W.; Noé, L.; Mehta, A.; Chen, K.; Luo, H.; Kunz, M.; Tamura, N.; Liu, Z. Learning from the past: Rare ϵ-Fe_2O_3 in the ancient black-glazed Jian (Tenmoku) wares. *Sci. Rep.* **2015**, *4*, 4941. [CrossRef]

167. Tronc, E.; Chanéac, C.; Jolivet, J.P. Structural and Magnetic Characterization of ϵ-Fe_2O_3. *J. Solid State Chem.* **1998**, *139*, 93–104. [CrossRef]

168. Tuček, J.; Zbořil, R.; Namai, A.; Ohkoshi, S. ϵ-Fe_2O_3: An Advanced Nanomaterial Exhibiting Giant Coercive Field, Millimeter-Wave Ferromagnetic Resonance, and Magnetoelectric Coupling. *Chem. Mater.* **2010**, *22*, 6483–6505. [CrossRef]

169. Tuček, J.; Machala, L.; Ono, S.; Namai, A.; Yoshikiyo, M.; Imoto, K.; Tokoro, H.; Ohkoshi, S.; Zbořil, R. Zeta-Fe_2O_3—A new stable polymorph in iron(III) oxide family. *Sci. Rep.* **2015**, *5*, 15091. [CrossRef]

170. Rozenberg, G.K.; Dubrovinsky, L.S.; Pasternak, M.P.; Naaman, O.; Le Bihan, T.; Ahuja, R. High-pressure structural studies of hematite (Fe_2O_3). *Phys. Rev. B* **2002**, *65*, 064112. [CrossRef]

171. Badro, J.; Fiquet, G.; Struzhkin, V.V.; Somayazulu, M.; Mao, H.K.; Shen, G.; Le Bihan, T. Nature of the high-pressure transition in Fe_2O_3 hematite. *Phys. Rev. Lett.* **2002**, *89*, 205504. [CrossRef]

172. Ito, E.; Fukui, H.; Katsura, T.; Yamazaki, D.; Yoshino, T.; Aizawa, Y.; Kubo, A.; Yokoshi, S.; Kawabe, K.; Zhai, S.; et al. Determination of high-pressure phase equilibria of Fe_2O_3 using the Kawai-type apparatus equipped with sintered diamond anvils. *Am. Mineral.* **2009**, *94*, 205–209. [CrossRef]

173. Bykova, E.; Bykov, M.; Prakapenka, V.; Konôpková, Z.; Liermann, H.-P.; Dubrovinskaia, N.; Dubrovinsky, L. Novel high pressure monoclinic Fe_2O_3 polymorph revealed by single-crystal synchrotron X-ray diffraction studies. *High Press. Res.* **2013**, *33*, 534–545. [CrossRef]

174. Sanson, A.; Kantor, I.; Cerantola, V.; Irifune, T.; Carnera, A.; Pascarelli, S. Local structure and spin transition in Fe_2O_3 hematite at high pressure. *Phys. Rev. B* **2016**, *94*, 014112. [CrossRef]

175. Bykova, E.; Dubrovinsky, L.; Dubrovinskaia, N.; Bykov, M.; McCammon, C.; Ovsyannikov, S.V.; Liermann, H.P.; Kupenko, I.; Chumakov, A.I.; Rüffer, R.; et al. Structural complexity of simple Fe_2O_3 at high pressures and temperatures. *Nat. Commun.* **2016**, *7*, 10661. [CrossRef]

176. Greenberg, E.; Leonov, I.; Layek, S.; Konopkova, Z.; Pasternak, M.P.; Dubrovinsky, L.; Jeanloz, R.; Abrikosov, I.A.; Rozenberg, G.K. Pressure-Induced Site-Selective Mott Insulator-Metal Transition in Fe_2O_3. *Phys. Rev. X* **2018**, *8*, 31059. [CrossRef]

177. Liu, H.; Caldwell, A.; Benedetti, L.R.; Panero, W.; Jeanloz, R. Static compression of α-Fe_2O_3: Linear incompressibility of lattice parameters and high-pressure transformations. *Phys. Chem. Miner.* **2003**, *30*, 582–588. [CrossRef]

178. Olsen, J.S.; Cousins, C.S.G.; Gerward, L.; Jhans, H.; Sheldon, B.J. A study of the crystal structure of Fe_2O_3 in the pressure range up to 65 GPa using synchrotron radiation. *Phys. Scr.* **1991**, *43*, 327–330. [CrossRef]

179. Shim, S.H.; Bengtson, A.; Morgan, D.; Sturhahn, W.; Catalli, K.; Zhao, J.; Lerche, M.; Prakapenka, V. Electronic and magnetic structures of the postperovskite-type Fe_2O_3 and implications for planetary magnetic records and deep interiors. *Proc. Natl. Acad. Sci. USA* **2009**, *106*, 5508–5512. [CrossRef]

180. Syono, Y.; Ito, A.; Morimoto, S.; Suzuki, T.; Yagi, T.; Akimoto, S. Mössbauer study on the high pressure phase of Fe_2O_3. *Solid State Commun.* **1984**, *50*, 97–100. [CrossRef]

181. Nasu, S.; Kurimoto, K.; Nagatomo, S.; Endo, S.; Fujita, F.E. 57Fe Mössbauer study under high pressure; ε-Fe and Fe_2O_3. *Hyperfine Interact.* **1986**, *29*, 1583–1586. [CrossRef]

182. Jiang, J.Z.; Olsen, J.S.; Gerward, L.; Mørup, S. Enhanced bulk modulus and reduced transition pressure in γ-Fe_2O_3 nanocrystals. *Europhys. Lett.* **1998**, *44*, 620–626. [CrossRef]

183. Zhu, H.; Ma, Y.; Yang, H.; Ji, C.; Hou, D.; Guo, L. Pressure induced phase transition of nanocrystalline and bulk maghemite (γ-Fe_2O_3) to hematite (α-Fe_2O_3). *J. Phys. Chem. Solids* **2010**, *71*, 1183–1186. [CrossRef]

184. MacHala, L.; Tuček, J.; Zbořil, R. Polymorphous transformations of nanometric iron(III) oxide: A review. *Chem. Mater.* **2011**, *23*, 3255–3272. [CrossRef]

185. Hearne, G.; Pischedda, V. Pressure response of vacancy ordered maghemite (γ-Fe_2O_3) and high pressure transformed hematite (α-Fe_2O_3). *J. Solid State Chem.* **2012**, *187*, 134–142. [CrossRef]

186. Sans, J.A.; Monteseguro, V.; Garbarino, G.; Gich, M.; Cerantola, V.; Cuartero, V.; Monte, M.; Irifune, T.; Muñoz, A.; Popescu, C. Stability and nature of the volume collapse of ε-Fe_2O_3 under extreme conditions. *Nat. Commun.* **2018**, *9*, 4554. [CrossRef]

187. Grant, R.W.; Geller, S.; Cape, J.A.; Espinosa, G.P. Magnetic and crystallographic transitions in the α-Mn_2O_3-Fe_2O_3 system. *Phys. Rev.* **1968**, *175*, 686–695. [CrossRef]

188. Geller, S. Structure of α-Mn_2O_3, $(Mn_{0.983}Fe_{0.017})_2O_3$ and $(Mn_{0.37}Fe_{0.63})_2O_3$ and relation to magnetic ordering. *Acta Crystallogr. Sect. B Struct. Crystallogr. Cryst. Chem.* **1971**, *27*, 821–828. [CrossRef]

189. Yamanaka, T.; Nagai, T.; Okada, T.; Fukuda, T. Structure change of Mn_2O_3 under high pressure and pressure-induced transition. *Z. Kristallog.* **2005**, *220*, 938–945. [CrossRef]

190. Santillán, J.; Shim, S.H.; Shen, G.; Prakapenka, V.B. High-pressure phase transition in Mn_2O_3: Application for the crystal structure and preferred orientation of the $CaIrO_3$ type. *Geophys. Res. Lett.* **2006**, *33*, L15307. [CrossRef]

191. Shim, S.H.; LaBounty, D.; Duffy, T.S. Raman spectra of bixbyite, Mn_2O_3, up to 40 GPa. *Phys. Chem. Miner.* **2011**, *38*, 685–691. [CrossRef]

192. Mukherjee, G.D.; Vaidya, S.N.; Karunakaran, C. High Pressure and High Temperature Studies on Manganese Oxides. *Phase Transit.* **2002**, *75*, 557–566. [CrossRef]

193. Ovsyannikov, S.V.; Abakumov, A.M.; Tsirlin, A.A.; Schnelle, W.; Egoavil, R.; Verbeeck, J.; Van Tendeloo, G.; Glazyrin, K.V.; Hanfland, M.; Dubrovinsky, L. Perovskite-like Mn_2O_3: A path to new manganites. *Angew. Chem. Int. Ed.* **2013**, *52*, 1494–1498. [CrossRef]

194. Hong, F.; Yue, B.; Hirao, N.; Liu, Z.; Chen, B. Significant improvement in Mn_2O_3 transition metal oxide electrical conductivity via high pressure. *Sci. Rep.* **2017**, *7*, 44078. [CrossRef]

195. Ovsyannikov, S.V.; Karkin, A.E.; Morozova, N.V.; Shchennikov, V.V.; Bykova, E.; Abakumov, A.M.; Tsirlin, A.A.; Glazyrin, K.V.; Dubrovinsky, L. A hard oxide semiconductor with a direct and narrow bandgap and switchable p-n electrical conduction. *Adv. Mater.* **2014**, *26*, 8185–8191. [CrossRef]

196. Khalyavin, D.D.; Johnson, R.D.; Manuel, P.; Tsirlin, A.A.; Abakumov, A.M.; Kozlenko, D.P.; Sun, Y.; Dubrovinsky, L.; Ovsyannikov, S.V. Magneto-orbital texture in the perovskite modification of Mn_2O_3. *Phys. Rev. B* **2018**, *98*, 014426. [CrossRef]

197. McWhan, D.B.; Rice, T.M.; Remeika, J.P. Mott Transition in Cr-Doped V_2O_3. *Phys. Rev. Lett.* **1969**, *23*, 1384–1387. [CrossRef]

198. Lupi, S.; Baldassarre, L.; Mansart, B.; Perucchi, A.; Barinov, A.; Dudin, P.; Papalazarou, E.; Rodolakis, F.; Rueff, J.P.; Itié, J.P.; et al. A microscopic view on the Mott transition in chromium-doped V_2O_3. *Nat. Commun.* **2010**, *1*, 105. [CrossRef]

199. Weber, D.; Stork, A.; Nakhal, S.; Wessel, C.; Reimann, C.; Hermes, W.; Müller, A.; Ressler, T.; Pöttgen, R.; Bredow, T.; et al. Bixbyite-Type V_2O_3—A Metastable Polymorph of Vanadium Sesquioxide. *Inorg. Chem.* **2011**, *50*, 6762–6766. [CrossRef] [PubMed]

200. McWhan, D.B.; Remeika, J.P. Metal-Insulator Transition in $(V_{1-x}Cr_x)O_3$. *Phys. Rev. B* **1970**, *2*, 3734–3750. [CrossRef]

201. Jayaraman, A.; McWhan, D.B.; Remeika, J.P.; Dernier, P.D. Critical Behavior of the Mott Transition in Cr-Doped V_2O_3. *Phys. Rev. B* **1970**, *2*, 3751–3756. [CrossRef]

202. Carter, S.A.; Rosenbaum, T.F.; Lu, M.; Jaeger, H.M.; Metcalf, P.; Honig, J.M.; Spalek, J. Magnetic and transport studies of pure V_2O_3 under pressure. *Phys. Rev. B* **1994**, *49*, 7898–7903. [CrossRef] [PubMed]

203. Limelette, P.; Georges, A.; Jérome, D.; Wzietek, P.; Metcalf, P.; Honig, J.M. Universality and critical behavior at the Mott transition. *Science* **2003**, *302*, 89–92. [CrossRef]

204. Rodolakis, F.; Hansmann, P.; Rueff, J.-P.; Toschi, A.; Haverkort, M.W.; Sangiovanni, G.; Tanaka, A.; Saha-Dasgupta, T.; Andersen, O.K.; Held, K.; et al. Inequivalent Routes across the Mott Transition in V_2O_3 Explored by X-Ray Absorption. *Phys. Rev. Lett.* **2010**, *104*, 047401. [CrossRef]

205. Alyabyeva, N.; Sakai, J.; Bavencoffe, M.; Wolfman, J.; Limelette, P.; Funakubo, H.; Ruyter, A. Metal-insulator transition in V_2O_3 thin film caused by tip-induced strain. *Appl. Phys. Lett.* **2018**, *113*, 241603. [CrossRef]

206. Finger, L.W.; Hazen, R.M. Crystal structure and isothermal compression of Fe_2O_3, Cr_2O_3, and V_2O_3 to 50 kbars. *J. Appl. Phys.* **1980**, *51*, 5362–5367. [CrossRef]

207. Zhang, Q.; Wu, X.; Qin, S. Pressure-induced phase transition of V_2O_3. *Chin. Phys. Lett.* **2012**, *29*, 106101. [CrossRef]

208. Zhang, Q.; Wu, X.; Qin, S. A nine-fold coordinated vanadium by oxygen in V_2O_3 from first-principles calculations. *Eur. Phys. J. B* **2012**, *85*, 267. [CrossRef]

209. Aggarwal, P.S.; Goswami, A. An oxide of tervalent nickel. *J. Phys. Chem.* **1961**, *65*, 2105. [CrossRef]

210. Conell, R.S.; Corrigan, D.A.; Powell, B.R. The electrochromic properties of sputtered nickel oxide films. *Sol. Energy Mater. Sol. Cells* **1992**, *25*, 301–313. [CrossRef]

211. Jones, P.G.; Rumpel, H.; Schwarzmann, E.; Sheldrick, G.M. Gold (III) oxide. *Acta Cryst. B* **1979**, *35*, 1435–1437. [CrossRef]

212. Minomura, S.; Drickamer, H.G. Effect of Pressure on the Electrical Resistance of some Transition-Metal Oxides and Sulfides. *J. Appl. Phys.* **1963**, *34*, 3043–3048. [CrossRef]

213. Chenavas, J.; Joubert, J.C.; Marezio, M. Low-spin → high-spin state transition in high pressure cobalt sesquioxide. *Solid State Commun.* **1971**, *9*, 1057–1060. [CrossRef]

214. Rekhi, S.; Dubrovinsky, L.S.; Ahuja, R.; Saxena, S.K.; Johansson, B. Experimental and theoretical investigations on eskolaite (Cr_2O_3) at high pressures. *J. Alloy. Compd.* **2000**, *302*, 16–20. [CrossRef]

215. Kota, Y.; Yoshimori, Y.; Imamura, H.; Kimura, T. Enhancement of magnetoelectric operating temperature in compressed Cr_2O_3 under hydrostatic pressure. *Appl. Phys. Lett.* **2017**, *110*, 042902. [CrossRef]

216. Kantor, A.; Kantor, I.; Merlini, M.; Glazyrin, K.; Prescher, C.; Hanfland, M.; Dubrovinsky, L. High-pressure structural studies of eskolaite by means of single-crystal X-ray diffraction. *Am. Mineral.* **2012**, *97*, 1764–1770. [CrossRef]

217. Shim, S.H.; Duffy, T.S.; Jeanloz, R.; Yoo, C.S.; Iota, V. Raman spectroscopy and x-ray diffraction of phase transitions in Cr_2O_3 to 61 GPa. *Phys. Rev. B* **2004**, *69*, 144107. [CrossRef]

218. Dobin, A.Y.; Duan, W.; Wentzcovitch, R.M. Magnetostructural effects and phase transition in Cr_2O_3 under pressure. *Phys. Rev. B* **2000**, *62*, 11997–12000. [CrossRef]

219. Golosova, N.O.; Kozlenko, D.P.; Kichanov, S.E.; Lukin, E.V.; Liermann, H.-P.; Glazyrin, K.V.; Savenko, B.N. Structural and magnetic properties of Cr_2O_3 at high pressure. *J. Alloy. Compd.* **2017**, *722*, 593–598. [CrossRef]

220. Nishio-Hamane, D.; Katagiri, M.; Niwa, K.; Sano-Furukawa, A.; Okada, T.; Yagi, T. A new high-pressure polymorph of Ti_2O_3: Implication for high-pressure phase transition in sesquioxides. *High Press. Res.* **2009**, *29*, 379–388. [CrossRef]

221. Umemoto, K.; Wentzcovitch, R.M. Prediction of an U_2S_3-type polymorph of Al_2O_3 at 3.7 Mbar. *Proc. Natl. Acad. Sci. USA* **2008**, *105*, 6526–6530. [CrossRef] [PubMed]

222. Ovsyannikov, S.V.; Wu, X.; Shchennikov, V.V.; Karkin, A.E.; Dubrovinskaia, N.; Garbarino, G.; Dubrovinsky, L. Structural stability of a golden semiconducting orthorhombic polymorph of Ti_2O_3 under high pressures and high temperatures. *J. Phys. Condens. Matter* **2010**, *22*, 375402. [CrossRef] [PubMed]

223. Ovsyannikov, S.V.; Wu, X.; Karkin, A.E.; Shchennikov, V.V.; Manthilake, G.M. Pressure-temperature phase diagram of Ti_2O_3 and physical properties in the golden Th_2S_3-type phase. *Phys. Rev. B* **2012**, *86*, 024106. [CrossRef]

224. Ovsyannikov, S.V.; Wu, X.; Garbarino, G.; Núñez-Regueiro, M.; Shchennikov, V.V.; Khmeleva, J.A.; Karkin, A.E.; Dubrovinskaia, N.; Dubrovinsky, L. High-pressure behavior of structural, optical, and electronic transport properties of the golden Th_2S_3-type Ti_2O_3. *Phys. Rev. B* **2013**, *88*, 184106. [CrossRef]

225. Biesterbos, J.W.M.; Hornstra, J. The Crystal Structure of the high-temperature low-pressure form of Rh_2O_3. *J. Less Common Met.* **1973**, *30*, 121–125. [CrossRef]

226. Shannon, R.D.; Prewitt, C.T. Synthesis and structure of a new high-pressure form of Rh_2O_3. *J. Solid State Chem.* **1970**, *2*, 134–136. [CrossRef]

227. Zhuo, S.; Sohlberg, K. Origin of stability of the high-temperature, low-pressure Rh_2O_3 III form of rhodium sesquioxide. *J. Solid State Chem.* **2006**, *179*, 2126–2132. [CrossRef]

228. Becker, N.; Reimann, C.; Weber, D.; Lüdtke, T.; Lerch, M.; Bredow, T.; Dronskowski, R. A density-functional theory approach to the existence and stability of molybdenum and tungsten sesquioxide polymorphs. *Z. Kristallogr.* **2017**, *232*, 69–75. [CrossRef]

229. Zhang, J.; Oganov, A.R.; Li, X.; Xue, K.H.; Wang, Z.; Dong, H. Pressure-induced novel compounds in the Hf-O system from first-principles calculations. *Phys. Rev. B* **2015**, *92*, 184104. [CrossRef]

230. Ai, Z.; Huang, Y.; Lee, S.; Zhang, L. Monoclinic α-Bi_2O_3 photocatalyst for efficient removal of gaseous NO and HCHO under visible light irradiation. *J. Alloy. Compd.* **2011**, *509*, 2044–2049. [CrossRef]

231. Zheng, F.L.; Li, G.R.; Ou, Y.N.; Wang, Z.L.; Su, C.Y.; Tong, Y.X. Synthesis of hierarchical rippled Bi_2O_3 nanobelts for supercapacitor applications. *Chem. Commun.* **2010**, *46*, 5021–5023. [CrossRef] [PubMed]

232. Hu, M.; Jiang, Y.; Sun, W.; Wang, H.; Jin, C.; Yan, M. Reversible Conversion-Alloying of Sb_2O_3 as a High-Capacity, High Rate, and Durable Anode for Sodium Ion Batteries. *ACS Appl. Mater. Interfaces* **2014**, *6*, 19449–19455. [CrossRef] [PubMed]

233. Datta, A.; Giri, A.K.; Chakravorty, D. AC conductivity of Sb_2O_3-P_2O_5 glasses. *Phys. Rev. B* **1993**, *47*, 16242. [CrossRef]

234. Shen, Z.-X.; Chen, G.-Q.; Ni, J.-H.; Li, X.-S.; Xiong, S.-M.; Qiu, Q.-Y.; Zhu, J.; Tang, W.; Sun, G.-L.; Yang, K.-Q.; et al. Use of Arsenic Trioxide (As_2O_3) in the Treatment of Acute Promyelocytic Leukemia (APL): II. Clinical Efficacy and Pharmacokineticsin Relapsed Patients. *Blood* **1997**, *89*, 3354–3360. [CrossRef]

235. Shen, Z.-X.; Shi, Z.-Z.; Fang, J.; Gu, B.-W.; Li, J.-M.; Zhu, Y.-M.; Shi, J.-Y.; Zheng, P.-Z.; Yan, H.; Liu, Y.F.; et al. All-trans retinoic acid/As_2O_3 combination yields a high quality remission and survival in newly diagnosed acute promyelocytic leukemia. *Proc. Natl. Acad. Sci. USA* **2004**, *101*, 5328–5335. [CrossRef]

236. Matsumoto, A.; Koyama, Y.; Togo, A.; Choi, M.; Tanaka, I. Electronic structures of dynamically stable As_2O_3, Sb_2O_3, and Bi_2O_3 crystal polymorphs. *Phys. Rev. B* **2011**, *83*, 214110. [CrossRef]

237. Matsumoto, A.; Koyama, Y.; Tanaka, I. Structures and energetics of Bi_2O_3 polymorphs in a defective fluorite family derived by systematic first-principles lattice dynamics calculations. *Phys. Rev. B* **2010**, *81*, 094117. [CrossRef]

238. Zhao, Z.; Zeng, Q.; Zhang, H.; Wang, S.; Hirai, S.; Zeng, Z.; Mao, W.L. Structural transition and amorphization in compressed α-Sb_2O_3. *Phys. Rev. B* **2015**, *91*, 184112. [CrossRef]

239. Sans, J.A.; Manjón, F.J.; Popescu, C.; Cuenca-Gotor, V.P.; Gomis, O.; Muñoz, A.; Rodríguez-Hernández, P.; Contreras-García, J.; Pellicer-Porres, J.; Pereira, A.L.J.; et al. Ordered helium trapping and bonding in compressed arsenolite: Synthesis of $As_4O_6\cdot2He$. *Phys. Rev. B* **2016**, *93*, 054102. [CrossRef]

240. Cuenca-Gotor, V.P.; Gomis, O.; Sans, J.A.; Manjon, F.J.; Rodrıguez-Hernandez, P.; Muñoz, A. Vibrational and elastic properties of As_4O_6 and $As_4O_6\cdot2He$ at high pressures: Study of dynamical and mechanical stability. *J. Appl. Phys.* **2016**, *120*, 155901. [CrossRef]

241. Guńka, P.A.; Dziubek, K.F.; Gładysiak, A.; Dranka, M.; Piechota, J.; Hanfland, M.; Katrusiak, A.; Zachara, J. Compressed Arsenolite As_4O_6 and Its Helium Clathrate $As_4O_6\cdot2He$. *Cryst. Growth Des.* **2015**, *15*, 3740–3745. [CrossRef]

242. Pereira, A.L.J.; Gracia, L.; Santamaría-Pérez, D.; Vilaplana, R.; Manjón, F.J.; Errandonea, D.; Nalin, M.; Beltrán, A. Structural and vibrational study of cubic Sb_2O_3 under high pressure. *Phys. Rev. B* **2012**, *85*, 174108. [CrossRef]

243. Pereira, A.L.J.; Sans, J.A.; Vilaplana, R.; Gomis, O.; Manjón, F.J.; Rodríguez-Hernández, P.; Muñoz, A.; Popescu, C.; Beltrán, A. Isostructural Second-Order Phase Transition of β-Bi_2O_3 at High Pressures: An Experimental and Theoretical Study. *J. Phys. Chem. C* **2014**, *118*, 23189–23201. [CrossRef]

244. Orosel, D.; Dinnebier, R.E.; Blatov, V.A.; Jansen, M. Structure of a new high-pressure–high-temperature modification of antimony(III) oxide, γ-Sb_2O_3, from high-resolution synchrotron powder diffraction data. *Acta Cryst. B* **2012**, *68*, 1–7. [CrossRef]

245. Cornei, N.; Tancret, N.; Abraham, F.; Mentré, O. New ε-Bi_2O_3 Metastable Polymorph. *Inorg. Chem.* **2006**, *45*, 4886–4888. [CrossRef]

246. Zou, Y.; Zhang, W.; Li, X.; Ma, M.; Li, X.; Wang, C.-H.; He, B.; Wang, S.; Chen, Z.; Zhao, Y.; et al. Pressure-induced anomalies and structural instability in compressed β-Sb_2O_3. *Phys. Chem. Chem. Phys.* **2018**, *20*, 11430–11436. [CrossRef]

247. Geng, A.-H.; Cao, L.-H.; Ma, Y.-M.; Cui, Q.-L.; Wan, C.-M. Experimental Observation of Phase Transition in Sb_2O_3 under High Pressure. *Chin. Phys. Lett.* **2016**, *33*, 097401. [CrossRef]

248. Harwig, H.A. On the Structure of Bismuthsesquioxide: The α, β, γ, and δ-Phase. *Z. Anorg. Allg. Chem.* **1978**, *444*, 151–166. [CrossRef]

249. Pereira, A.L.J.; Errandonea, D.; Beltrán, A.; Gracia, L.; Gomis, O.; Sans, J.A.; García-Domene, B.; Miquel-Veyrat, A.; Manjón, F.J.; Muñoz, A.; et al. Structural study of α-Bi_2O_3 under pressure. *J. Phys. Condens. Matter* **2013**, *25*, 475402. [CrossRef] [PubMed]

250. Gavriliuk, A.G.; Struzhkin, V.; Lyubutin, S.; Eremets, I.; Trojan, A.; Artemov, V. Equation of state and high-pressure irreversible amorphization in $Y_3Fe_5O_{12}$. *JETP Lett.* **2006**, *83*, 37–41. [CrossRef]

251. Locherer, T.; Dasari, L.; Prasad, V.K.; Dinnebier, R.; Wedig, U.; Jansen, M. High-pressure structural evolution of HP-Bi_2O_3. *Phys. Rev. B* **2011**, *83*, 214102. [CrossRef]

252. Ghedia, S.; Locherer, T.; Dinnebier, R.; Prasad, D.L.V.K.; Wedig, U.; Jansen, M.; Senyshyn, A. High-pressure and high-temperature multianvil synthesis of metastable polymorphs of Bi_2O_3: Crystal structure and electronic properties. *Phys. Rev. B* **2010**, *82*, 024106. [CrossRef]

253. Frueh, A.J., Jr. The crystal structure of claudetite (monoclinic As_2O_3). *Am. Miner.* **1951**, *36*, 833–850.

254. Pertlik, F. Die Kristallstruktur der monoklinen Form von As_2O_3 (Claudetit II). *Mon. Chem.* **1975**, *106*, 755–762. [CrossRef]

255. Soignard, E.; Amin, S.A.; Mei, Q.; Benmore, C.J.; Yarger, J.L. High-pressure behavior of As_2O_3: Amorphous-amorphous and crystalline-amorphous transitions. *Phys. Rev. B* **2008**, *77*, 144113. [CrossRef]

256. Guńka, P.A.; Dranka, M.; Piechota, J.; Żukowska, G.Z.; Zalewska, A.; Zachara, J. As_2O_3 Polymorphs: Theoretical Insight into Their Stability and Ammonia Templated Claudetite II Crystallization. *Cryst. Growth Des.* **2012**, *12*, 5663–5670. [CrossRef]

257. Guńka, P.A.; Dranka, M.; Hanfland, M.; Dziubek, K.F.; Katrusiak, A.; Zachara, J. Cascade of High-Pressure Transitions of Claudetite II and the First Polar Phase of Arsenic(III) Oxide. *Cryst. Growth Des.* **2015**, *15*, 3950–3954. [CrossRef]

258. Jansen, M.; Moebs, M. Structural Investigations on Solid Tetraphosphorus Hexaoxide. *Inorg. Chem.* **1984**, *23*, 4486–4488. [CrossRef]

259. Clark, G.L.; Schieltz, N.C.; Quirke, T.T. A New Study of the Preparation and Properties of the Higher Oxides of Lead. *J. Am. Chem. Soc.* **1937**, *59*, 2305–2308. [CrossRef]

260. Bouvaist, J.; Weigel, D. Sesquioxyde de plomb, Pb_2O_3. I. Determination de la structure. *Acta Cryst. A* **1970**, *26*, 501–510. [CrossRef]

261. Seko, A.; Togo, A.; Oba, F.; Tanaka, I. Structure and Stability of a Homologous Series of Tin Oxides. *Phys. Rev. Lett.* **2008**, *100*, 045702. [CrossRef] [PubMed]

262. Zhao, J.-H.; Tan, R.-Q.; Yang, Y.; Xu, W.; Li, J.; Shen, W.-F.; Wu, G.-Q.; Yang, X.-F.; Song, W.-J. Synthesis mechanism of heterovalent Sn_2O_3 nanosheets in oxidation annealing process. *Chin. Phys. B* **2015**, *24*, 070505. [CrossRef]

263. Kuang, X.L.; Liu, T.M.; Zeng, W.; Peng, X.H.; Wang, Z.C. Hydrothermal synthesis and characterization of novel Sn_2O_3 hierarchical nanostructures. *Mat. Lett.* **2016**, *165*, 235–238. [CrossRef]

264. Imre, A.R. On the existence of negative pressure states. *Phys. Stat. Sol. (b)* **2007**, *244*, 893–899. [CrossRef]

265. McMillan, P.F. New materials from high-pressure experiments. *Nat. Mater.* **2002**, *1*, 19–25. [CrossRef]

266. Manjón, F.J.; Errandonea, D.; López-Solano, J.; Rodríguez-Hernández, P.; Muñoz, A. Negative pressures in CaWO₄ nanocrystals. *J. Appl. Phys.* **2009**, *105*, 094321. [CrossRef]

267. Matsui, T.; Yagasaki, T.; Matsumoto, M.; Tanaka, H. Phase diagram of ice polymorphs under negative pressure considering the limits of mechanical stability. *J. Phys. Chem.* **2019**, *150*, 041102. [CrossRef]

MDPI

St. Alban-Anlage 66

4052 Basel

Switzerland

Tel. +41 61 683 77 34

Fax +41 61 302 89 18

www.mdpi.com

Crystals Editorial Office

E-mail: crystals@mdpi.com

www.mdpi.com/journal/crystals

CPSIA information can be obtained
at www.ICGtesting.com
Printed in the USA
LVHW072026220920
666819LV00034B/1102